Lighting for Industry and Security

A handbook for providers and users of lighting

LIGHTING FOR INDUSTRY AND SECURITY

A handbook for providers and users of lighting

Stanley Lyons

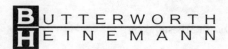

BUTTERWORTH
HEINEMANN

Butterworth-Heinemann Ltd
Linacre House, Jordan Hill, Oxford OX2 8DP

 PART OF REED INTERNATIONAL BOOKS

OXFORD   LONDON   BOSTON
MUNICH   NEW DELHI   SINGAPORE   SYDNEY
TOKYO   TORONTO   WELLINGTON

First published 1992

British Library Cataloguing in Publication Data
Lyons, Stanley L.
  Lighting for Industry and Security:
  Handbook for Providers and Users of
  Lighting
  I. Title
  621.32

  ISBN 0 7506 1084 0

Library of Congress Cataloging-in-Publication Data
Lyons, Stanley L. (Lewis)
  Lighting for industry and security: a handbook for providers and
  users of lighting/Stanley Lyons.
  p.   cm.
  Includes bibliographical references and index.
  ISBN 0 7506 1084 0
  1. Factories–Lighting. 2. Lighting. 3. Factories–Security
  measures. I. Title.
TK4399.F2L96   1992
  621.32′ 254–dc20   92–10329
                                             CIP

Composition by Genesis Typesetting, Laser Quay, Rochester, Kent
Printed and bound in Great Britain

Contents

Preface

This is a practical book to aid the reader who seeks to understand how to achieve good interior and exterior industrial lighting. It contains up-to-date information not readily available elsewhere, and incorporates much previously unpublished material. The text is divided into short sections, extensively cross-indexed for rapid reference.

This book is intended for lighting engineers, electrical engineers, building services engineers, specialist providers of security equipment and services, architects and designers, as well as for buyers and users of lighting.

Terms and principles are explained when first introduced, so that students and non-specialist readers should not be confused by jargon. I have described the effects of lighting without giving academic proofs of my statements, dealing more with the application of the science of lighting than with the science itself. It is hoped that this book will enable non-specialist users and specifiers to acquire sufficient understanding of the principles of good lighting to be able to brief the experts and lighting providers.

In this volume I have combined revised versions of the contents of two of my earlier books: *Exterior Lighting for Industry and Security* (Applied Science, 1980) and *Handbook of Industrial Lighting* (Butterworth, 1981). The former considered only exterior lighting, the latter was mainly concerned with interior lighting.

I have provided information about the new lightsources and lighting practices developed during the past decade. Reflecting today's needs, attention is given to the wise use of energy, and proper account is taken of the nature of modern industrial tasks, as well as developments in industrial architecture.

I have outlined the principles, the equipment and the design methods for many applications of interior and exterior lighting for industry, including lighting for security and emergency use, and portable and mobile lighting for temporary use. Good practice in electrical installations is reviewed, and some important aspects of the management of lighting are covered. The principles of exterior lighting relevant to security lighting are dealt with in detail. As explained in my book *Security of Premises – a Manual for Managers* (Butterworth, 1988), security lighting plays an important role in crime prevention, and its use is endorsed by the police and security consultants. Techniques and design methods are given to enable security lighting to be applied effectively and economically to protect industrial premises against night intrusion.

This book contains many references to good safety practices and the role of lighting in preventing industrial accidents. Emphasis is placed on quality assurance in work performance, and a survey of modern visual inspection techniques is provided.

It is recommended that this book should be read in conjunction with the *CIBSE Code for Interior Lighting*, the related CIBSE technical memoranda, and the publication *Interior Lighting Design*; for, while I explain and augment the data from those publications, I do not repeat their content. I take the reader through many aspects of the science and craft of lighting, sharing much practical know-how gained in a long career in industrial lighting.

The text refers to UK practice, and many British Standards and CIBSE publications are cited. However, because of the extensive harmonization of standards, in general the practices will be in accord with those of other EC countries. Safety practices and engineering procedures in the UK do not differ basically from those in the USA, Canada, New Zealand and Australia, but the reader should ensure that all practices and installations comply with local standards and legislation.

Stanley Lyons

NOTICE

In addition to a review of current practice in the fields of industrial lighting, security lighting and emergency lighting, this book contains proposals for the development of novel equipment and the introduction of innovative methods. The book is intended for use by persons with suitable qualifications in the relevant fields. In case of doubt, the user should obtain advice and assistance from a suitably qualified person before adopting any proposal made in this book. To the best of the author's and publishers' knowledge the information in this book is accurate and up to date at the time of publication. However neither the author nor the publishers can accept responsibility for any inaccuracy or error.

List of tables

Part 1

Fundamentals of light and vision

1 The eye and vision

Although it is possible to design lighting installations by rules of thumb and by repetition of standard layouts, an understanding of the basic scientific principles of illuminating engineering will enable the designer or specifier of lighting to develop judgement of the fitness for purpose and the economics of proposals under review, and to seek effective and practical solutions to the problems of achieving good lighting. Lighting, though science based, is not entirely a mathematical subject. Understanding the function of the human eye and how vision is affected by lighting will help specifiers and providers of lighting achieve better lighting results. Specifications and designs must be conceived to satisfy the eye as well as the lightmeter.

1.1 Physiology of the eye

1.1.1 Viewed in simplified cross-section (Figure 1.1), the human eye is seen to be roughly spherical, and is internally divided into two chambers. The smaller of these (the *anterior chamber*) is bounded at its front by the transparent window of the eye (the *cornea*) and behind by the crystalline lens. The lens is supported from the wall of the orb (the *sclera*) by a delicate muscle (the *ciliary muscle*) which exercises force on the lens to change its curvature as necessary to focus at different distances (a process called *accommodation*).

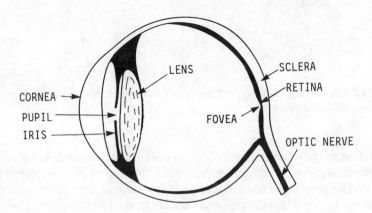

Figure 1.1 *Simplified cross-section of the human eye.*

Between the cornea and the lens lies a delicately thin diaphragm structure (the *iris*) at the centre of which is an aperture (the *pupil*). The muscles of the iris cause the pupil to contract or dilate, so that its size varies according to the brightness of what we see. The size of the pupil at any time may also be affected by drugs which cause it to dilate or contract, and by emotion – for example, fear may cause dilation. Although the variation in pupil size is important in adapting to the field of brightness, another important adaptive function also occurs in the retina (section 1.4).

1.1.2   The larger chamber of the eye (the *posterior chamber*) is bounded at its front by the lens, at its sides by the sclera and, at the rear, by a matrix of specialized light-sensitive nerve endings (the *retina*). The anterior chamber is filled with a clear fluid (the *aqueous humour*). The posterior chamber is filled with a jelly-like substance (the *vitreous humour*). These two substances keep the orb in shape by virtue of intraocular pressure.

1.1.3   The eye sits in a socket enfolded in a specialized tissue (the *conjunctiva*) which lubricates, irrigates and permits its free movement in the socket. External muscles direct the eyes to the object, the lines of sight converging according to the distance of the object from the eye. The angle between the converging axes is called the *angle of convergence* (Figure 1.2).

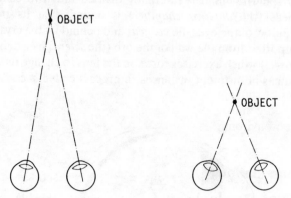

Figure 1.2 *The variable convergence angle of our eyes enables us to focus on objects at different distances.*

The action of the muscles controlling the eyelids is not entirely voluntary, but is partly controlled by reflexes which cause blinking or partial closing of the eyelids. The number of unconscious and involuntary blinks per minute (*blink rate*) tends to rise when the subject is under stress or is attempting to perform a task in insufficient light or while subjected to glare (section 1.5).

1.2 The process of vision

1.2.1 Light passing through the cornea reaches the lens after passing through the pupil. The lens projects an inverted real image on to the retina. Light energy in the image stimulates the nerve endings in the retina, triggering the generation of minute electric currents (*nervous stimuli*) which are conducted to the brain by the optic nerve.

1.2.2 The specialized nerve endings in the retina (the *light receptors*) are of two kinds: those sensitive to colour (the *cones*) and those sensitive only to degree of brightness (the *rods*). The rods and cones are not evenly distributed, but the cones are clustered in greater density at a near-central spot where deliberate vision occurs (the *fovea*). Rods are present in all parts of the retina, and extend to the edge of the visual disk to give the facility of peripheral vision.

1.2.3 The eyes of creatures have evolved to suit their needs. Predators generally have forward-facing eyes to pursue their prey efficiently, while herbivores generally have enhanced sideways vision for protection against the stalking predator when grazing head-down. Human eyes combine both these features, having the ability to focus on small objects close by or at a distance (*foveal vision*), while being able to orient the individual in space and to judge distances and speeds by virtue of a wide angle of vision (*peripheral vision*).

1.2.4 Lighting that illuminates only the 'object of special regard' (i.e. the critical part of a task) is insufficient for efficient vision; there must also be sufficient general illumination to provide for the subjects' spatial orientation and judgement of static and dynamic relationships of objects in their visual field, and for safe movement resulting from proper appreciation of the features of their environment (Figure 1.3).

1.3 Accommodation and stereoscopy

1.3.1 The range of human vision extends to infinity for, if we have normal vision, we can see the stars in sharp focus. Conversely, we can focus on a minute object at close range – for example, the dot on the top of this letter *i*.

When looking at distant objects, the curve of the lens of the eye is relatively flattened due to relaxation of the ciliary muscle; when observing close objects, the lens takes on greater curvature due to the action of that muscle. For this reason, close work is generally more tiring, particularly if performed in a poor light which may cause the subject to 'peer'. In this usage, the word

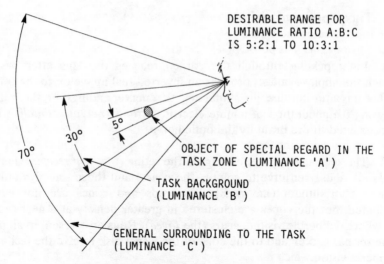

DESIRABLE RANGE FOR
LUMINANCE RATIO A:B:C
IS 5:2:1 TO 10:3:1

OBJECT OF SPECIAL REGARD IN THE
TASK ZONE (LUMINANCE 'A')

TASK BACKGROUND
(LUMINANCE 'B')

GENERAL SURROUNDING TO THE TASK
(LUMINANCE 'C')

IN THE TRANSVERSE PLANE THE ANGLES ARE APPROX.
45° AND 160°. THE OBLATE CONE OF VISION MAY BE
IN ANY ORIENTATION ACCORDING TO THE POSITION
OF THE TASK AND THE SUBJECT.

Figure 1.3 *Lighting for a task, its background and its surroundings.*

'peer' means to bring the eye closer to the task to gain the advantage of the apparent magnification due to proximity – a process which often results in excessive muscular tension throughout the body, with resulting increased fatigue (Figure 1.4). Peering also may bring the eyes close to danger of injury.

1.3.2   The process of accommodating the focus of the eyes to distances involves not only the adjustment of the curvature of the lens, but also the inward direction of each orb so that the centreline of vision of each is directed to the object of special regard.

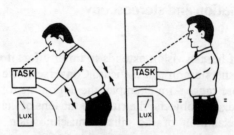

Figure 1.4 *There is an increase in muscular tension when the subject 'peers', i.e. brings his eyes closer to the task to gain the apparent magnification due to proximity, in an attempt to compensate for insufficient illuminance.*

The closer the object is to the eyes, the greater is the angle of convergence. The brain computes the visual inputs from the two eyes, co-ordinating them so that only one image is perceived. In certain diseases and following injury to the brain or optical system, double vision (*diploplia*) may occur. However, in normal vision the two images correspond, even though each eye sees a slightly different picture because it occupies a different position to its fellow (*parallax*). Although the angle of convergence is frequently adjusted, the act of perception includes the ability generally to avoid confusion due to parallax, and this is an important factor in our ability to judge speeds and distances.

1.3.3   We readily differentiate between our observation of a real scene or object and a photograph of the same scene or object. We recognize at once that the photograph is only a representation of depth, form and distance, and that the representation is flat. Similarly, it is impossible for us to confuse a real car 3 m long seen at a distance of 20 m with a small model car 10 cm long seen at a distance which makes its perceived image of identical size to that of the real car viewed as described. This is because normal vision gives us the experience of size, form and distance of objects which we term *stereoscopic vision*. Full stereoscopic vision cannot be experienced by a single eye, though a subject who has lost the sight of one eye after early childhood is able to make very good judgements of depth, form and distance by a sort of visual deductive reasoning based on his early experience of stereoscopic vision.

In industrial tasks such as driving, operating a crane etc., good stereoscopy is essential. To enhance stereoscopy, the skilful lighting engineer arranges that highlights, modelling due to flow of light, and the direction of shadows, are such as to amplify the information seen by the subject and help to prevent visual confusion.

1.3.4   Rapid accommodation to distance is important to our being able to see quickly and accurately. It is our common experience that the range and speed of our accommodation tend to worsen with age. It is usual for the near-point of accommodation to recede after about the age of 40 years. This is part of the reason for the adage: 'Older eyes need more light'. Older subjects need more light to develop sufficient acuity for efficient task performance (Figure 1.5).

Some functions of the eye can be explained by comparison with a camera: for example, when the lens on a camera is stopped-down (i.e. when the f-number is large), the depth of field is increased. The human vision parallel is that when the subject is presented with a brighter field of view, the pupil contracts ('stops down') and effectively increases the subject's depth of focus. This explains why a long-sighted person may see better at close distances if the lighting is adequate. If older subjects are provided with plentiful illuminance, the difference between their visual performance and that of younger persons is greatly diminished (Figure 1.6). However, a long-sighted person gains little from peering when the illuminance is inadequate (section

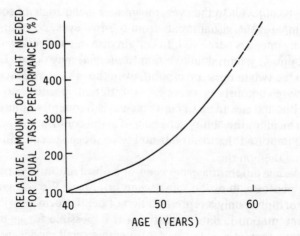

Figure 1.5 *Older subjects need more light to develop sufficient acuity for efficient task performance.*

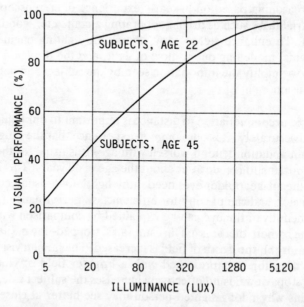

Figure 1.6 *Task performance improves with increased illuminance up to an optimum range, and the subjects suffer less fatigue. As a result, work quality may be better, or fewer errors made. At higher illuminances, the difference between the visual performance standards of older and younger subjects is reduced.*

1.3.1). This is because when looking at an object closer than the near-point of focus, the subject sees a larger but less distinct image.

1.3.5 By a simple experiment, we can demonstrate that the speed with which we accommodate our focus to different distances depends on the level of illumination. In a poorly lit place, face a dark wall, close one eye and hold one finger about 300 mm from the eye. At the same time, hold one finger of the other hand as far away as can be reached and roughly in line with the first finger. It will be found that it is difficult to focus rapidly in turn first on the near finger and then on the far, and the reverse. Next, repeat the experiment under good electric lighting or in daylight, and it will now be found possible to change focus rapidly between the two distances. This results from the speeding up of the accommodation with improved illuminance. Indeed, in bright sunshine, because of the increase in the depth of focus as a result of contraction of the iris, it is quite likely that some subjects will be able to focus on the two fingers simultaneously.

The experiment helps to show that when the illuminance is suitable for the subject, for the task and for the environment, speed of vision and visual performance improve, so that work can be done better, more accurately and with less fatigue.

1.4 Adaptation

1.4.1 We have seen how the diameter of the pupil varies in response to the brightness of the field of view (sections 1.1.1 and 1.3.4). This is only part of the adaptation process, providing a coarse degree of adjustment which deals mainly with excesses of brightness and large changes in brightness. In ordinary conditions the eye is commonly adapted to an intermediate state termed *mesopic vision*. If some part of the visual field is too bright for the iris to deal with, the subject suffers the sensation of *glare* (section 1.5).

The process of adaptation includes an adjustment that takes place within the retina itself, and the passage from the fully light-adapted state (*photopic vision*) to the fully dark-adapted state (*scotopic vision*) can take up to several hours, though most of the possible adaptation takes place in the first few seconds, and the process is substantially complete in a few minutes. Generally, adapting to dark conditions takes longer than adapting to bright conditions. The eye finds it difficult to cope with rapid increases in brightness when it is well advanced into dark adaptation, e.g. after a period in darkness, a small lightsource (e.g. a match being struck) seems very bright.

1.4.2 If there are frequent significant changes in field brightness, or if the eye has to cope with very diverse brightnesses in the field observed, the subject may soon become fatigued. Again, in pursuit of better lighting, we can deduce that better visual conditions will exist if the lighting is not only adequate in quantity, but if it is also reasonably evenly distributed so that the luminance pattern in the field of view is not excessively contrasty.

1.4.3   When the eye is adapted to a low field brightness (say, an illuminance of 5 lux), it will be found that the perception of colour becomes defective, and, as the illuminance is reduced further, vision becomes entirely monochromatic (*Purkinje Effect*) (section 1.8.5).

1.5   Glare

1.5.1   Light which comes to the eye, directly or reflected from objects, that embarrasses vision and handicaps performance of the visual task, is termed *glare*. It is convenient to discuss glare as *discomfort glare*, which does not – at least in the short term – affect the performance of the visual task, but which tends to bring about an earlier onset of fatigue, and *disability glare* (which handicaps the subject, reducing what can be seen, and – in an extreme case – so handicapping vision that all the subject can see is the glare source). Glare may be caused directly, e.g. by unshielded bright lamps, or indirectly, e.g. light reflected from a glossy surface or the surface of a liquid.

Some surfaces which appear to be matt, behave in a *specular* (mirror-like) manner when light strikes them at low incident angles. We are able to see printed characters on paper because of the contrast between the dark print and the light paper. However, semi-specular reflections from the surface of paper can reduce the *contrast rendering factor* of the print against the paper to the point where reading is difficult or impossible because of the *veiling glare*.

The degree of glare is subjective, and can be assessed or calculated (by reference to previous subjective assessments of glare experience) but cannot be measured. Glare sensation is not directly related to the light output (lumens) of the glare source, nor to its intrinsic brightness (candelas per square metre, cd/m^2). Commonly, when someone criticizes a lighting situation by saying 'there is too much light', almost invariably what is really meant is that there is too much *glare*, i.e. that the light is tending to produce a pattern of luminances which creates some degree of glare sensation – though this may be acceptable or imperceptible. It may be said that if the subject is seeing anything, he or she must be experiencing at least an imperceptible degree of glare, i.e. there is no concept of zero glare.

1.5.2   Glare may produce some degree of disability or discomfort. Disability glare is the effect we suffer when, for example, we are suddenly confronted with the undipped headlights of an oncoming vehicle on an unlit road at night.

Note that 12 hours later, in bright sunshine, a repetition of that situation would produce no disability glare, because the bright headlights would be seen against a background of far higher brightness. We say that the *luminance* of the headlights has not changed, but that their *luminosity* as perceived by the subject has changed, and very considerably. Discomfort glare, which produces discomfort without actually disabling the subject's vision, is tiring and reduces the efficiency of the worker.

1.5.3   When the eye is subjected to glare, it is unable to adjust itself instantly to the luminance conditions, though the iris contracts and the retina commences to adapt to the brightness presented to it (section 1.4.1). Thus, time is an important factor in glare sensation (section 3.4).

1.5.4   When glare is being experienced it is more difficult to distinguish small differences in brightness, i.e. *contrast sensitivity* is reduced. Thus, reduction in glare may improve the subject's visual performance, just as would an improvement in illuminance. It therefore should be an objective of lighting design to limit glare by good design (section 3.4), thereby enabling the desired visual performance to be achieved without wastefully high illuminances. This can reduce capital cost, running cost and energy consumption. However, the economic return for the extra cost of glare limitation cannot be of great magnitude. Nonetheless, glare limitation should be treated as an important factor to be considered in performing a lighting design.

1.6   Visual acuity and visual performance

1.6.1   Visual performance affected by lighting

When a person is working, at least 80% of the sensory data necessary for the performance of the task is obtained visually. *Visual performance* is the achievement by the worker in the carrying out of his or her *visual task* which may consist of observing small detail (e.g. reading from paper or a VDT screen, using instruments, gauges etc.), as well as monitoring the immediate environment (e.g. observing – usually unconsciously – for danger, movement of persons and objects, spatial relationships etc). In some tasks the worker will need to judge speeds and distances.

1.6.2   Visual acuity

Given the advantage of suitable lighting, the normal human eye can perform far better than is commonly assumed, readily resolving objects of good contrast down to 0.25–0.05 mm without optical aids. The ability of the eye to see small contrasts of brightness or small objects is termed its *visual acuity* or sharpness of vision, and is measured as the reciprocal of the angle subtended at the eye by the smallest detail that can be picked out.

The normal eye, under conditions of good contrast and adequate illuminance, can resolve small details down to about 1-minute of arc subtended at the eye. The subject's visual acuity is dependent from instant to instant on the contrast in the object, the contrast between the object and its immediate background, and the illuminance. It is also affected by the state of adaptation of the subject. Thus, the basic acuity of a subject can only be measured under controlled conditions of all these factors.

1.6.3   Measurement of a subject's acuity

Opticians measure acuity by asking the subject to read from a *Snellen Chart*. The chart's letters are of diminishing size, and are viewed under controlled conditions of contrast and illuminance. The limit of acuity is detected when the subject cannot distinguish between similar letters, e.g. a 'C' from an 'O' where the gap in the 'C' is just beyond the subject's power of resolution.

The Snellen Chart works because the conditions of the visual test are standardized. The standard test illuminance is 300 lx. If by mistake the illuminance on the chart were lower, then the test would indicate that the subject had a lower acuity than was actually the case; if the illuminance on the chart were higher than 300 lux (lx), the converse would apply. Similarly, the measurement of acuity is only possible because there is a known contrast ratio between the black letters and their white background; if the test card were darkened by dirt etc, then lower acuity assessments would result.

1.6.4   Determining the illuminance required for a task

Data obtained from Snellen Chart tests and similar tests form the basis of most recommendations as to the required illuminance for the efficient performance of tasks, for the reduction of fatigue in visual task performance, and for worker safety. An examination of the values (lux) recommended in the *CIBSE Code for Interior Lighting*[1] (see also appendix B) will show that tasks with small detail and low contrast generally require higher illuminances than tasks with large detail and high contrast. Note that our sensitivity to the brightness contrast ratio reduces in the presence of glare (section 1.5), and therefore the *Code* includes recommendations on the limitation of glare.

The Snellen Chart effect is reversible; by varying the contrast between the reflectance of the letters and background, or varying the illuminance on the chart, we can use a special form of the chart as a crude lightmeter. The effects are good enough to detect serious underlighting in a short-term test.

In practical engineering, when compiling lighting proposals, one refers to tables of recommended illuminances in the *Code*[1] (appendix B). For visual tasks of poor contrast, more illuminance must be provided to enable the subject to achieve the required standard of visual performance, and this is allowed for in the *Code*.

Standard illuminance recommendations have been compiled on a theoretical basis to determine the illuminance required to enable the normal eye to resolve a detail of known angular size, the lightest part of the critical detail having a reflectance factor between 0.1 and 0.8. The required illuminance can be determined by use of the nomogram (Figure 1.7), in which the example shows that an object subtending 5 minutes of arc to the eye at the normal viewing distance, and having a reflection factor R of 0.2, will require an illuminance of 700 lx for the subject to see it well. Adjusting such results empirically, a weighting factor of 1.5 should be applied to the resultant illuminance *E* if there are adverse circumstances, e.g. (a) if the consequences of error would be costly or dangerous, (b) if the subjects are over 40 years of age, (c) if the time available for seeing is fixed and brief, (d) if the object is moving, or (e) if the subject must wear eye protection. Calculations of this

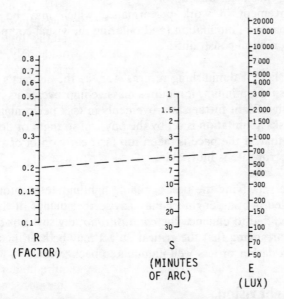

Figure 1.7 *Nomogram, after Weston, for determining the illuminance required for a specific visual task for known factors of 'S' and 'R', where 'R' is the reflection factor of the lightest part of the critical detail in the visual task, and 'S' is the apparent size of the critical detail in minutes of arc.*

kind, modified by extensive field practice, have been used to compile the recommendations given for the illumination of tasks generally (appendix B).

Consideration of Figure 1.7 will show that for a given factor R, the smallest detail that can be seen becomes smaller as the illuminance is increased. This is the so-called 'magnifying effect' of improved lighting, and, where the object is capable of being seen by this improvement of illuminance (or improvement of the reflection factor R), this will generally be preferred in the industrial situation to the use of magnifying lenses. However, note the limitation on the improvement in seeing obtained by 'peering' (section 1.3.1).

1.6.5   Difficult visual tasks

In organizations where really difficult visual tasks are performed routinely, the simple provision of enhanced general lighting and task lighting (section 9.5) can do much to improve the speed, ease and quality of visually-based decisions, e.g. in inspection (chapter 10). The adequacy of the operators' vision (corrected by spectacles if necessary) may be confirmed by eye tests or by vision screening (section 1.10). Only after the possibilities of improving visual performance by these reliable means have been explored should other methods be provided to enable difficult visual tasks to be performed, e.g. use of magnifying spectacles or loupes.

The improvement in work performance which may be achieved by providing enhanced illumination (and tailoring the visual components of the task) has two natural constraints:

- there is a 'law of diminishing returns' for, as the subject's performance approaches the optimum, it requires massive improvements in the lighting to obtain significant further improvements in task performance, and
- in many tasks a limitation is set by the physical strength or dexterity of the subject so that, if the pace is taken too fast, early onset of fatigue and/or increased error-rate is inevitable.

In addition to specifying the best available lighting method for a particular task, the skilled lighting engineer may advise that details of the visual task should be modified to enhance the *contrast ratio*, e.g. by introducing colour contrasts, or arranging that the critical object may be seen in silhouette, or seen against a darker or less well-illuminated background.

1.7   Low light vision

1.7.1   We have noted how the diameter of the pupil varies in response to the brightness presented to the eye (section 1.1), and the importance of the process of adaptation (section 1.4) and its relationship to the experience of glare (section 1.5). A major portion of the process of adaptation occurs in the retina, where, in the state of dark adaptation, the light receptors act in small groups rather than individually as in the light-adapted eye.

The process can be explained by an electrical analogy. Imagine that each receptor requires a minimum light-energy input in order to emit one quantum stimulus to the brain, just as a thyristor requires a minimum applied voltage before it will 'fire'. If the minimum light energy per receptor is not available, the receptors cannot 'see', but there is a mechanism whereby a group of receptors can pool the results of their individual energy inputs and create one stimulus as a group – just as a number of weak electrical cells connected in parallel can combine their outputs to produce a greater current. Thus, in low light vision, there are effectively fewer stimuli per unit area of the retina, and so the fine detail perceived by the brain is reduced, i.e. the subject's acuity is reduced.

1.7.2   The acuity reduction under low-light conditions (section 1.7.1) can be understood by a further analogy. Consider a picture printed by a fine lithographic screen (many dots per unit area) compared with one printed by a coarse screen (fewer dots per unit area). This reduction in acuity must be borne in mind when designing lighting for exterior work or for security, where the illuminances employed are far lower than those used in interior lighting applications.

For some outdoor tasks, it may be necessary to provide local or localized *task lighting* where visually difficult operations (such as reading, inspection, recognizing colours etc.) must be performed in an area lighted to a level only sufficient for safe movement and coarse tasks. Under low light conditions, the ability to perceive colour may be reduced or absent (section 1.8), and it may be necessary to modify the tasks or the environment to compensate for this (e.g. employ geometrical symbols associated with colours used to identify pipes, classes of steel stock etc.) (section 4.2).

1.7.3   The relaxed eye tends to focus as for distance vision when light-adapted but, when dark-adapted, the eye tends to focus at a distance roughly equal to the height of the subject. This appears to be nature's way of enabling us to move about safely in subdued light. Under low light conditions, particularly if presented with a *bland field* (a lack of any point of visual interest or special brightness in the field of vision) the eye lapses to a focus of around 1.5 to 2.0 m. This is termed *normal night myopia*.[2] The condition of night myopia is sometimes incorrectly termed 'night blindness'; the subject may easily fail to notice objects at a distance greater than about 2.0 m, even though they are of importance, i.e. they may affect his safety.

1.7.4   The bland field effect (section 1.7.3) appears also to operate at higher field illuminances. For example, it is believed that the effect is the cause of many motorway driving accidents in fog and mist, when the driver thinks he or she is 'peering into the fog ahead' but, in fact, is seeing no further than the front of the car.

1.7.5   The combination of bland field effect and night myopia (section 1.7.3) can be even more remarkable: a sentry or security guard patrolling at approximately 1.5 to 2.0 m from a wire fence may fail to see an attacker (or the inspecting officer!) standing just the other side of the fence – this being termed the *sleeping sentry syndrome*. In time of war, sentries have been court-martialled and shot for what appeared to be gross dereliction of duty in this respect. It is now known that under low-light conditions, the night-myopia/bland-field effect can occur, and this should be taken into account when locating patrol paths in relation to fencing in the design of security lighting systems (chapter 17).

1.7.6   Colour vision is affected by illuminance to a marked degree, and at low light levels the ability to distinguish colours is lost entirely (Purkinje Effect) (section 1.8).

1.8   Seeing colour

1.8.1   The system of light receptors in the retina has been referred to previously (sections 1.1.2, 1.2.2 and 1.7.1). The light receptors are of two kinds: the rods, which are sensitive to the spectral energy of light falling on them irrespective of its wavelength, and the cones, which are sensitive to the wavelengths of the spectral energy of the impinging light, and are thus able to send information about the colour of objects to the brain.

1.8.2   The cones are clustered densely around the fovea (section 1.2.2), but are less closely spaced towards the edges of the retina. Thus, seeing of colour is associated with deliberate (foveal) vision, while peripheral vision depends mainly on the edges of the retina which are most sensitive to movement and are relatively insensitive to colour. This may be demonstrated in a simple experiment. While looking straight forward, extend one arm sideways, when it will be found that if the hand is held steady it cannot be seen, but the fingers can be seen if they are moved. Next, repeat the experiment holding in the extended hand a small card having two different colours on its faces, say blue on one side and red on the other, and it will be found that while looking straight ahead one can see the card turn over in one's hand, but cannot determine whether it is the blue or red side that is presented to the eyes.

1.8.3   Objects which appear to us to be of a particular colour do so because they have the property of absorbing certain wavelengths of the visible spectrum and reflecting others. An object perceived as red reflects light from the long-wavelength (red) part of the visible spectrum while absorbing substantially all other wavelengths. It therefore follows that an object can only be seen to be a particular colour if energy at the wavelengths it is capable of reflecting is present in substantial amount in the light falling upon it.

Thus, a red object seen in light which contains a substantial amount of spectral energy in all parts of the spectrum, i.e. 'white' light (e.g. daylight or light from lamps of good colour rendering) can be perceived as a red object. But that same red object, seen by the light of a source which does not contain a substantial amount of spectral energy in all parts of the spectrum (e.g. near-monochromatic light from low-pressure sodium-vapour lamps (section 5.9)), will not appear to be of any particular colour, or will be perceived as a shade of yellow/black.

1.8.4   When a subject experiences vision under any 'non-white' illuminant, a process of *colour adaptation* takes place, the degree of adaptation varying with the period of exposure. A small degree of adaptation takes place almost at once; complete adaptation can take upwards of an hour, though most of it takes place within a few minutes.

This can be demonstrated by a subject donning a pair of spectacles with coloured lenses. If the lenses were pink, for example, on first putting the spectacles on everything would appear to the subject to be tinged with pink. After a few minutes, the pinkness would appear to fade away, and colours again seem to be normal. However, if, when the adaptation was well advanced, the pink-tinted lenses were removed, the illusion that everything was tinged with green would be experienced, i.e. the complementary colour to red/pink. The process of adaptation would again take place, and after a while the sensation of greenness would wear off. A similar process of adaptation occurs when leaving an area lit by daylight or a reasonably 'white' illuminant and entering one lit with a 'non-white' illuminant, e.g. high-pressure sodium-vapour lamps which give a yellow/golden appearance to a scene.

We therefore see that, provided the light from an illuminant contains sufficient spectral energy in those parts of the spectrum needed for the recognition of colour, some departure from a theoretical 'white' source may not be a serious disadvantage – at least for tasks which do not involve fine colour judgement (chapter 5).

1.8.5   As the illuminance is reduced, the ability of the eye to distinguish colour in photopic vision (which peaks at around 1500–3000 lx, perhaps up to 6000 lx) gradually deteriorates. Below about 150 lx colour judgement becomes markedly unreliable, and when the illuminance falls further the eye sees more by rods than by cones (vision becomes scotopic). One cannot distinguish any colours when the illuminance is below about 1 lx, all that we see being in monochrome (the Purkinje Effect).

1.8.6   The subject of lighting and colour is discussed in depth in chapter 4. The lighting requirements for the inspection of colour are discussed in section 10.5.

1.9   Vision and perception

1.9.1   We can distinguish between the process of *vision* (in which objects are seen and visual information about them is passed to the brain), and *perception* (a cerebral process in which visual information is processed and compared with remembered material). Optical illusions confuse us because we see something which we find difficult to perceive. Perception is a learned process; a primitive man could see a page of this book but would not perceive its information; someone without knowledge of fishing could see ripples in a pool but might not perceive the motion of the fish which caused them.

If an image is blurred, indistinct or incomplete, perception takes longer. Similarly, if the illuminance and contrast ratios in a visual scene are insufficient, perception takes longer. In many industrial situations this can lead to dangers not being perceived quickly enough for corrective or evasive action to be taken; in security lighting installations, such delayed perception might be perilous for the guard and result in lowered security.

1.9.2   We can consider a concept termed the *cybernetic loop* (Figure 1.8). Light from any source falls upon an object, and some is reflected towards the subject, enabling the object to be seen. Within the brain the act of perception takes place enabling the subject to decide upon a course of action and execute it. For example, a crane driver will cause the crane to perform a movement or modification of a movement already in the course of being performed. This changes the visual input, i.e. the crane driver observes the results of his or her actions. The cycle 'light from object to eye – vision – perception – decision – action – change in object – light from object to eye . . .' is termed the cybernetic loop. The loop breaks if any part fails, and only functions because of the visual input. Insufficient or unsuitable lighting will reduce the efficiency of work performance.

1.9.3   Although vision appears to be instantaneous (for example, we can see a flash of lightning lasting only for about 1 ms), the act of perception takes an appreciable time. In practical situations perception appears to be continuous (as in driving), but in unusual conditions the time taken to perceive a situation, evaluate it and decide upon action may be measured in seconds (for example, noticing something falling and then getting out of the way). It is observed that better vision is associated with faster perception. Thus, adequate lighting makes a positive contribution to safety in work situations.

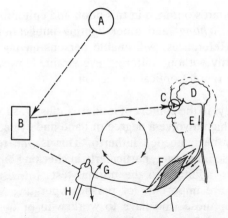

Figure 1.8 *The cybernetic loop. Light from a lightsource 'A' illuminates the workplace, and some falls on the object 'B'. Some light is reflected from the object 'B' towards the subject's eyes 'C', where vision takes place and data are transmitted to the subject's brain 'D'. Perception is followed by decisions which are transmitted by the nervous system 'E' to muscles 'F' which cause movements of limbs 'G' to move controls 'H'. Movements of controls 'H' cause changes in the object 'B', which are again monitored by the sequence C/D and operated by the sequence E/F/G/H continuously.*

1.9.4   The lengthening of perception-time described in section 1.9.1 is also observed in an unusual way if an image is blurred when first presented to the subject, and then becomes clear. For example, if when observing a scene on a closed-circuit television (cctv) monitor, the subject is presented with a slightly out-of-focus picture which has to be corrected, the perception time is considerably longer than if the same scene had been presented in sharp focus initially. This is one reason why fixed-focus cctv cameras having wide-angle lenses of great depth of focus are preferred for security surveillance, rather than cameras having adjustable zoom lenses. Perception tends to occur more rapidly if the picture is clear and steady, even if the image presented to the eye is very small.

1.10   Vision screening

Very few people have perfect vision – even those who wear properly-prescribed spectacles. Variance in visual performance from the theoretical norm is usually of little or no importance until the subject is required to perform visual tasks which demand visual abilities approaching the limit of visual performance. In most cases, the simple provision of good lighting to the recommended standards outlined in this book and epitomized by the *CIBSE Code for Interior Lighting*[1] and other lighting publications of the CIBSE (appendix B and References) will enable persons having normal vision to perform satisfactorily without suffering eye strain. However, there will be some subjects who require optical correction.

1.10.1   Increasingly, in the context of modern industry, subjects are required to perform tasks which involve a degree of hand and eye coordination which has not commonly been needed hitherto. This is one reason why vision screening tests are so important, particularly in checking binocular vision and eye coordination in addition to the normal tests for visual acuity. These faculties of vision are important for persons operating VDTs and similar equipment, and for those who have to work with or observe fast-moving machinery where a deficiency in binocular vision can cause considerable eyestrain. Visual deficiency may reduce operator efficiency, detract from job satisfaction, and can also be a factor in accident causation. These considerations lead to the recommendation that *vision screening* should be practised in industry.

Figure 1.9 *The 'Keystone' Vision Screener. (Photo: The Warwick-Evans Optical Co. Ltd).*

1.10.2   Vision screening is the process of testing a subject's eyes, using an apparatus such as the 'Keystone Vision Screener'(R) supplied by Warwick-Evans Optical Co Ltd (Appendix G) which enables basic tests to be carried out with great rapidity, and may be used by an operator with limited training (Figure 1.9). The tests include:

• *Usable vision*. Does the subject have normal acuity in each eye when they are tested separately?

- *Stereopsis*. Subjects who cannot judge distances are a danger to themselves and others in the presence of moving machinery, and the more so if they are required to drive a vehicle or a crane.
- *Acuity at required distances*. An operative who can count the legs of an ant at 3 m distance might not be able to focus accurately on a micrometer gauge at 300 mm. Tests will reveal if the subject can focus at specific distances, e.g. at the task distance needed for for VDT operators and typists. Task distances may typically be at 500 mm or 660 mm.
- *Colour vision*. Only specific testing will reveal if the subject has a defect in colour vision. While the possession of good colour vision is an obvious necessity for airline pilots, painters and designers, a bench-wirer or electrician working with coloured cable cores could make costly and potentially dangerous mistakes by having even very slightly defective colour vision, though he may have no difficulty in recognizing the red, amber and green of traffic lights.
- *Binocular vision*. This ascertains whether the two eyes work together properly. Any problems in this respect are common causes of eye strain, especially to VDT operators and others doing critical close work.
- *Tunnel vision*. Tests can be performed for detecting and evaluating the effect of other eye defects, such as tunnel vision (where the subject has defective peripheral vision).

1.10.3 Vision screening does not enable corrective lenses or orthoptics to be prescribed, but it does enable a trained operator to identify subjects whose vision may not be good enough for the demands of the visual tasks proposed for them, and thus enable them to be referred for professional advice.

A qualified optician may carry out tests for conditions such as *nystagmus*, in which the subject's eyes do not rest steadily at the point of vision, but hunt or vibrate about it, and *pathological night blindness* in which the mechanism of dark adaptation is defective, and the subject cannot see at all at low illuminances – a condition which should not be confused with normal night myopia (section 1.7.3). Eye examinations may reveal other pathological conditions of the eye, and also of the body generally.

2 Illumination principles

The creation of a good lighting installation is not simply a matter of arithmetic. There are some excellent computer programs for devising lighting layouts, but they do not automatically produce good lighting. Lighting design procedures must be performed against a background of an appreciation of the basic principles of good lighting.

The success of the designer's efforts cannot be judged entirely with a lightmeter, for what we are attempting is not merely to light the spaces, but to provide sufficient and suitable lighting for the people who will occupy those spaces. Without an awareness of the human needs to achieve good vision (chapter 1), we will be creating installations which might fall far short of desirable standards of effectiveness, comfort and safety, and which may lack a sense of rightness in their construction and performance.

2.1 Transmission of light

Light from any source travels in straight lines, at such a speed (300 000 km/s) that its time of departure and arrival may for practical purposes be regarded as simultaneous. Light is believed to be a form of electromagnetic radiation, in that it consists of energy being transmitted from the source in a manner that obeys to a limited extent the laws of wave motion (the *Wave Theory*). Light may also be considered to consist of photons of energy which behave in accordance with Planck's quantum theory[3] (the *Corpuscular Theory*). The Wave Theory enables plausible explanations to be given for the phenomenon of *polarization* (section 2.9). In practice, light is generally considered in terms of its wavelength, and Figure 2.1 shows the relationship of light to other forms of energy in the electromagnetic wave spectrum. Visible light has wavelengths in the range of about 380 to 780 nm, and the band of visible light merges with ultraviolet radiation below 400 nm and with infra-red radiation above 700 nm.

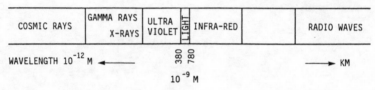

Figure 2.1 *The electromagnetic spectrum.*

2.2 Control of light

2.2.1 Reflection

If a ray of light were to impinge on the surface of a theoretical perfect absorber (a 'black body'), all the energy would be absorbed and none would be reflected; the body could therefore not be illuminated. In practice, when a ray of light arrives at a real opaque surface, some portion of the radiant light energy is reflected. The nature of the reflection will vary according to the nature of the surface. Thus, a specular (mirror-like) surface will reflect a high proportion of the incident light, the angle of reflection being equal to the angle of incidence. A non-planar specular reflector will distribute the reflected light according to its surface geometry, and this fact is exploited in the design of optical reflectors.

A light beam contains heat energy as well as visible light, and the heat may be reflected in a different pattern to that of the light, or may be absorbed at a surface to a greater or lesser degree. In the case of multi-coated dichroic reflectors, while the bulk of the visible light will be reflected according to the geometry of the reflector, a substantial proportion of the radiant heat (typically 60%) will be transmitted through the reflector and will be emitted from the back as radiant and convected heat – a feature that is exploited in low-voltage tungsten-halogen (LVTH) dichroic lamps (section 5.3).

When a ray of light strikes an opaque surface which is not perfectly specular, the light is reflected with some degree of diffusion, ranging from a small degree of scattering to diffuse reflection as would be obtained at a 'Lambertian' (non-specular totally diffusing) reflector. A luminaire reflector which has a highly diffusing non-glossy reflector does not redirect light in the same manner as, say, a white gloss reflector, but tends to behave simply as a large-area light source.

2.2.2 Absorption

No reflector is 100% efficient, but all reflectors, both specular and diffuse, absorb a proportion of the incident light and convert it to heat. Surfaces are selective in how they absorb incident light energy according to their colour and the spectral composition of the light. A red surface will be an efficient reflector of red light, but a green surface would reflect very little energy from a beam of red light.

2.2.3 Refraction

Transparent bodies (e.g. clear glass or water) transmit light rays, but attenuate them by absorption, the light loss depending on the transmission factor of the medium, and the absorbed light energy being converted to heat.

Light impinging on a transparent medium normal to its smooth planar surface enters with only a small loss due to reflection. Light entering a transparent medium at angles between the normal and the critical angle will be refracted (bent) towards the line of the ray path on entering the denser medium, and away from the line of the ray path on passing from the denser to the rarer medium, the angle of refraction being dependent on the refractive indices of the denser and rarer media. When light impinges on the surface of a transparent medium at angles beyond the critical angle no light will enter, but all will be reflected.

2.2.4   Diffusion

Diffusion occurs both at reflection (section 2.2.1) and on transmission through translucent media. Although there may be light loss due to absorption (section 2.2.3), diffusion does not necessarily involve loss of light. For example, translucent diffusers containing crystalline pigments have low light losses. However, diffusion can be effected in transparent media by the presence in suspension of opaque or optically dense particles, and such suspensions do absorb a significant proportion of the light energy while effecting diffusion.

There is no direct relationship between the diffusion factor of a medium and its absorption factor. Thus, it is possible to have a high degree of diffusion with low light losses, and media that do not diffuse efficiently but which absorb a significant proportion of the transient light. Diffusion can be effected by roughening the surface of a transparent medium, so that the surface microstructure comprises great numbers of minute refracting particles. A traditional light-diffusing medium is 'opal glass', which is a very thin membrane of glass, partially obscured with minute virtually-opaque particles, sandwiched between two layers of clear glass to give it mechanical strength. Some high quality diffusers have surfaces configured with very small prism sections which scatter and diffuse the light with low energy loss.

2.2.5   Dichroic reflection and transmission

Surfaces of reflectors may be treated with multiple microscopically thin coatings of transparent materials to form dichroic mirrors. The spectral energy in the various wavelengths comprising a ray (including infrared, visible and ultraviolet radiation) striking such a reflector will be selectively absorbed or reflected to yield a reflected ray which consists substantially of a selected wavelength or range of wavelengths.

A transparent medium similarly treated will selectively absorb or transmit energy of various wavelengths. Thus the dichroic reflector of a low-voltage tungsten-halogen (LVTH) lamp will allow a substantial proportion of radiant heat to pass through, while reflecting light of chosen wavelengths. This permits the creation of 'cool beam lamps' and coloured dichroic lamps.

2.3　The flow of light: modelling and shadow

2.3.1　In section 3.1 an analogue is used to describe the lumen, comparing it with a flow of 'a litre per minute'. In this respect, light can be described as flowing, in a similar manner to a fluid. But lighting designers use the term 'the flow of light' to mean also the way light falls from luminaires on to planes and surfaces, clothing what we see in gradations of luminance.

If we study any exterior or interior lighting scene, we will observe that, even in substantially uniform illuminance, objects tend to have some parts more brightly illuminated than others. There are also gradations of luminance, produced by diffuse and specular reflections from surfaces of objects, affected also by the shadows thrown by objects and formed upon them by their own contours, that aid our appreciation of any lighted scene. The formation of highlights (brilliant points of reflected light or mirror-images of luminaires etc) and modelling shadows are important to our recognition of the components of the visual scene, and enable us to perform rapid and correct perception of what we see (section 1.9).

2.3.2　A sound principle in designing exterior lighting is to ensure that every point on the ground receives light in significant proportions from at least two directions. The greater the number of directions from which light flows to the observed object, or the greater the area of the light source used, the softer will be the shadows. By use of an overall luminous ceiling, or by uplighting a light-coloured room, it is possible to create a virtually shadowless environment. But, lack of shadows does not necessarily aid our comprehension of the view. Indeed, by creating a major flow of light from one direction (to form highlights and shadows analogous to those experienced under daylight) we may achieve a pleasant and natural environment much more conducive to efficient work and safety.

2.3.3　If we created a lighted environment lacking strong visual features, with very uniform illuminance, very little glare and very little shadow, we might find that long occupancy of such an area produced strange effects upon the occupants. Such 'bland fields' do not stimulate and attract the eye, and can cause a loss of awareness and alertness. Indeed, under certain bland-field conditions, some subjects may drift into a state of light trance, i.e. become hypnotized. It is well known that this can occur under low-light conditions (section 1.7.3), but has also been known to occur under conditions of full light adaptation. The provision of some flow of light in the visual field, with highlights and modelling shadow, plus some clear visual interest, is essential in situations in which fast and accurate judgements of distance and speed are required (e.g. crane driving) or where a high state of alertness must be maintained over long periods (e.g. security guarding). The light-adapted bland-field effect is more marked if the eye is not stimulated by colour, which is one reason for avoiding the use of low-pressure sodium-vapour lamps (section 5.9) the light of which is virtually devoid of colour rendering.

2.3.4   We have seen how our comprehension of a lighted scene is affected, and how visual recognition and perception may be aided by the flow of light. The lighting engineer may arrange to throw light parallel to a plane which is being inspected, so that depressions are cast into shadow and raised portions preferentially illuminated, thus enabling the degree of flatness to be inspected. Floodlighting and area lighting may have a similar requirement. For example, a field which appears to be reasonably flat by the diffuse light of day, may be revealed as having considerable departures from flatness when illuminated by low-mounted floodlights.

The preferential direction of light flowing to an object reveals its form most clearly by producing 'modelling', an effect due as much to the formation of shadows as of highlighting. The effect can be put to good use in lighting design, and is an important feature of some lighting for inspection (chapter 10). Light which comes to the task from some unusual angle may throw shadows or create modelling and highlights which may be unusual, resulting in mistakes of perception. For example, light flowing upward from below at a glancing angle to a vertical surface may cause the illusion of depressed areas of the surface seeming to be raised, and vice versa.

2.3.5   An example of the use of modelling in area lighting is the lighting of railway tracks with transverse light flow so that ground obstructions can be more readily seen by workers walking in the area. In construction lighting, the use is well known of a type of task light termed a 'plasterer's lamp' (section 18.2.1) which has a large source (such as fluorescent tube) placed close to the surface being plastered, so that surface imperfections can be readily seen. The same idea may be employed for screeding and flatting-off concrete by night, by arranging for light to flow substantially parallel to the surface being worked upon.

2.3.6   If an area is lit by a single luminaire, objects will be preferentially illuminated on the side facing the luminaire, and portions not receiving light directly from the luminaire will be substantially shadowed. In interior installations, this effect is lessened by the multiple reflections from the room surfaces, and the more so if those surfaces have a high reflection factor. The shadowed surfaces of an object, and the shadows cast by the object, will usually be in considerable brightness contrast to the illuminated face of the object. The shadowed areas will only receive such ambient light as may be available from other sources or by reflections from other objects and surfaces, or which arrives because of scattering of light due to smoke, mist, fog or moisture in the atmosphere. The point to note here is that, in the absence of ambient light, the brightness (luminance) in shadowed areas will be fairly uniform.

The situation is different when the scene is illuminated by a large-area source (or a cluster of luminaires which act as a large-area source) or by two or more widely-spaced luminaires. Under these conditions, an object casts a shadow from each light source (Figure 2.2), and there is usually an area close

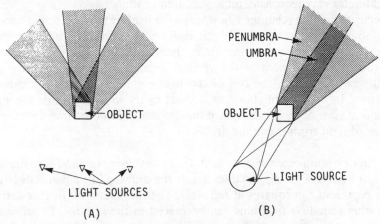

Figure 2.2 *(A) Umbra and penumbra cast by an object lit by several sources; (B) umbra and penumbra cast by object lit by a large-area source.*

to the object which is shadowed from all the luminaires. The dense shadow due to total shielding from all parts of a large-area source, or from all the sources concerned is termed the 'umbra'. The less dense shadow which is softened by light coming from part of a large-area source or coming from one or more of the other sources concerned is termed the 'penumbra'.

2.3.7   The umbration and penumbration associated with an object may be very important in seeing that object. For example, in a security lighting situation, an intruder wearing camouflaged clothing (or at least dark clothing that matches fairly well with the background) may not be easily visible, but his shadow may be the important feature by which he may be detected (chapter 17). An understanding of such matters may be important in the design of directional lighting systems (section 9.4).

2.4   Uniformity and diversity of illuminance

2.4.1   In exterior lighting installations, achieving a high degree of uniformity usually adds considerably to the complexity of the lighting installation and hence to its cost. However, there are few outdoor lighting applications which require substantially uniform illuminance; it is mainly for the lighting of sports fields (particularly to television standards) that a high standard of uniformity is specified. In outdoor industrial applications, quite a considerable variation in illuminance from one part of the lighted area to another may be tolerable for function, though there may be aesthetic considerations in some locations where at least an appearance of uniformity is desirable, even if the lightmeter reveals considerable variation of illuminance.

In security lighting (chapter 17), where area lighting must be relied upon to reveal an intruder (often at a great distance), uniformity of the vertical illuminance may be more important than uniformity of the horizontal illuminance. Measurement of the vertical illuminance for this purpose will normally be made with the cell of the lightmeter facing back towards the luminaires. In such installations, it is usual to try to arrange that the main direction of view of those who guard the site shall be in the same direction as the flow of light from the luminaires.

2.4.2   In designing exterior area lighting, a convenient method is to draw up the site at a scale of 1:500, and then to rule the drawing into squares of 10 mm (which represent 5 m squares at full size). Then, illuminance figures derived from isolux (equilux) diagrams can be placed in the squares. From such a diagram, a good estimate of the uniformity may be made. If the ratio of total horizontal illuminance (lux) in any two adjacent 5 m squares does not exceed 1.5:1, the illuminance will appear to be substantially uniform. At a ratio of 2:1, the uniformity will probably be good enough for most industrial purposes. If the ratio exceeds about 3:1 between adjacent 5 m squares, the lighting will appear to be patchy or noticeably uneven.

The eye tends to assume that luminance that appears to be reasonably uniform is actually uniform, and is easily deceived in this respect. However, if there is a contractual obligation to provide a stated uniformity ratio, the designer must adjust his design to achieve this. It is not so much the variation in luminance that is striking to the eye (taking the scene overall) so much as the gradient of change. To give an example of this, in the lighting of very large areas from towers, over the whole lighted area (which may extend more than 0.5 km from the towers), the variation may be as great as 100:1, but if the gradation of illuminance is gentle, a person walking across the area might hardly be conscious of the variation. If, however, there are local steep variations, these are likely to be noticed, and, if extreme, can be a handicap to vision.

2.4.3 As regards the uniformity of illuminance in interior lighting, the general considerations are similar to the foregoing notes on exterior lighting (sections 2.4.1 and 2.4.2), with the following observations. In order to meet the requirement for a level of illuminance suitable to the task, a uniformity ratio of 0.8 (minimum illuminance to average illuminance) over the task area should be ensured. The illuminance in areas adjacent to the task should not be less than one-third of that on the task itself. In practice, an illuminance uniformity of 0.8 is usually easily achieved by ensuring that the luminaire spacing does not exceed the luminaire manufacturer's recommendations as to spacing/mounting-height ratio.

2.4.4 A special situation regarding uniformity relates to interior uplighting, particularly interiors having light-coloured ceilings, walls and furnishings. In uplighting, all the flux from the luminaires is directed upward, so that the ceiling becomes the secondary source. Because of the extensive area of this secondary source, gradations of illuminance within the interior are very gentle. Paradoxically, in uplighter installations, by careful positioning of the luminaires it is possible to achieve a remarkable degree of uniformity. Yet, in areas lit with a single uplighter or with widely spaced uplighters, large ratios of minimum to average illuminance may seem hardly noticeable (section 9.8).

2.5 Measures of illuminance

2.5.1 Required illuminance

The illuminance needed by a subject to perform a visual task relates to the apparent size of the smallest detail to be seen (the apparent size taking into account the actual size and distance from the eye), the ratio of the reflection factors of the parts of the object of special regard, and between it and its immediate background (section 1.6). The reflection factor ratio between the immediate background and the general surroundings helps to determine the level of general illuminance required as compared with the task illuminance (chapter 9).

The requirements of the subject in resolving colour must be taken into account (chapter 4). Even with light sources of good colour rendering, an illuminance of the order of 600–1000 lx is required to bring the eye to full light adaptation so that colour vision can operate efficiently (section 1.4).

What ever the technical requirements for illuminance, the designer must provide a natural and comfortable lighted environment for the wellbeing of the occupants, and this may require the application of a far greater general illuminance than is needed for visual performance outside the immediate visual task zone. The designer should also take into account factors such as whether the task is to be performed intermittently or continuously – the latter usually requiring a higher illuminance to compensate for a tendency to progressive fatigue.

The illuminance recommendations given in this book referring to various applications of lighting are given as illuminances on the horizontal plane. These recommendations are summarized in appendices B and C. Where a task or activity not mentioned in this book must be catered for, it is possible to obtain a guiding recommendation from these references. In general, provided that the visual requirements are the same, an illuminance recommended for performing a given task indoors may be applied without modification for the lighting of a similar task out of doors.

Outdoor environments do not usually have convenient walls and other surfaces to reflect light and thereby provide the worker with a background brightness as high as that commonly achieved indoors. It may, therefore, sometimes be necessary to install a light-coloured screen to provide a suitable visual background to tasks performed out of doors such as cutting, threading and bending pipes, or cutting wood at a sawbench.

2.5.2   Horizontal illuminance

Illuminances, as commonly discussed, are those on the horizontal plane unless otherwise stated. The symbol for *horizontal illuminance* is E_h. Values of E_h are appropriate and convenient for dealing with interior general lighting, and also applicable to most outdoor situations. However, few outdoor tasks are concerned with seeing objects only on the horizontal plane, perception in the round being more commonly required. For many tasks, e.g. driving or security surveillance, it is the illuminance on planes facing the observer, i.e. the *vertical illuminance*, that determines the clarity and visibility.

2.5.3   Vertical illuminance

The *vertical illuminance* symbol E_v, must always relate to a specified direction. It is the value E_v with which we are most concerned when providing illumination for building floodlighting and for closed circuit television (cctv) cameras in security lighting (chapter 14). Measurement of the illuminance in the vertical plane may be made with the light-cell of the lightmeter vertical. Such a measurement is in one plane only.

2.5.4   Cylindrical illuminance

If we wish to measure the sum of the vertical illuminances due to light flowing to a point of measurement from various directions, we can introduce the concept of *mean cylindrical illuminance*, which involves the idea of a small cylinder with its axis vertical placed at the point of measurement. If it were possible to measure the illuminance on all parts of the curved surface of the cylinder, we would be able to compute all the vertical illuminance contributions from all directions in plan. In practice, four measurements of the vertical illuminance made at a point, with the light-cell rotated through 90° between measurements, and the four readings averaged, will give a very good approximation of the mean cylindrical illuminance. Special photocells are available to do the job conveniently and probably more accurately.

2.5.5   Scalar illuminance

To introduce a further concept, the reasoning that lead to the explanation of the mean cylindrical illuminance can be extended. If we imagine that a small sphere is located at the point of measurement, and if it were possible to summate and average all the lumens arriving on the surface of that sphere, we would be able to obtain the mean spherical illuminance or *scalar illuminance* symbol E_s at that point. This measure takes account of light reaching a point of measurement from all directions. It is mainly used for discussing lighting in very critical visual conditions, e.g. for fine inspection (chapter 10), and is not usually appropriate for exterior lighting.

A good approximation of the scalar illuminance measurement may be made by six photocell measurements, turning the cell so that it is directed to the six directions represented by the faces of an imaginary cube.

2.5.6   Illuminance vector

The concept of the *illuminance vector* is again based on a theoretical small sphere placed at the point of measurement. This would give a useful measure of the total flow of light to that point, but would tell nothing about the

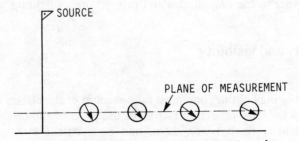

Figure 2.3 *Change in vector direction with change in position of measurement relative to a source.*

direction of light flow, i.e. which parts of the sphere were best lit. The vector is the axis through two diametrically opposed points where the maximum difference between the values of illuminance occurs on the surface of the small sphere placed at the measurement point, and the vector direction changes according to the position of measurement relative to a source (Figure 2.3).

2.5.7   Illuminance ratios

When discussing the flow of light on to vertical surfaces in any plane we could express the relative magnitude of the cylindrical illuminance (E_c) to horizontal (planar) illuminance (E_h) as the *cylindrical/planar ratio* (E_c/E_h), and this would provide a measure of how well vertical surfaces would be illuminated generally in comparison with the horizontal surfaces.

This could be a matter of importance in exterior lighting where the designer is concerned with mounting heights and angles of throw, but, in the author's opinion, this concept is unlikely to be of much value in the design of interior lighting systems. It is claimed by some workers that an accurate appreciation of the revealment and modelling effects of display lighting can be obtained by consideration of the scalar/vector ratio, but how this could assist in a practical study of industrial lighting is not clear. Perhaps the ratio of the mean cylindrical illuminance to the planar illuminance (E_c/E_h) would be more helpful for ensuring that light will come in sufficient quantities to illuminate all sides of a task, for example, to illuminate all faces of a large machine in the middle of a workshop.

In practice, the foregoing simplified descriptions of the various special measures of illuminance will only become of value to lighting engineers if they are able to connect their mathematical derivation with the appearance of lighted planes and objects under various conditions of light flow. It is strongly recommended that those who do not have extensive experience of lighting work should carry out some experiments, lightmeter in hand, making measurements of lighting conditions, both interior and exterior, and thus gain a practical sense of the magnitudes and proportions of lighting measures.

2.6   Clarity and visibility

2.6.1   In designing lighting, one must take account of the effects of contrast. These may include the reflection factors of the parts of the task to be seen, and the contrast ratio between them, and between the task or object of special regard and the immediate background. If the reflectances and the ratios of these reflectances are low, then a greater illuminance will be required to achieve the required level of task performance as compared with another task in which the apparent size to be picked out is the same, but in which there are higher reflectances and higher contrast ratios.

The clarity with which we see something is very much a function of the quantity and quality of illuminance provided. Even if the illuminance is sufficient to enable us to generate enough visual acuity for the visual task, vision may be difficult if the quality of the lighting is unhelpful. For example, while we would have no difficulty in reading a page of print in a flow of light approximately normal to the page, if the lighting came at a very flat glancing angle, we would find shadows forming in any slight depressions in the paper and in the surface texture of the paper. All this information about the page is distracting, and detracts from the clarity of the print. Similarly, if the surface of the page is glossy, reflections of the luminaires will produce a *veiling effect*, again reducing the clarity of the image on the page. These are common examples of how clarity of vision may be impaired by poor lighting quality. The first of these effects may be avoided by controlling the angle of incidence of the light flow, and the second by changing the relative positions of the page and the offending luminaire, or employing a matt paper, i.e. by improving the contrast rendering factor. Other methods used to improve clarity include using luminaires which give only limited illuminance in directions which are likely to give rise to troublesome reflections, using polarized light (section 2.9) wearing polarizing glasses, or interposing a polarizing screen (sections 2.9 and 10.5).

2.6.2   Considerations of visibility out of doors at night are similar to the foregoing. It has been found that the horizontal illuminance (i.e. the illuminance measured on a horizontal plane – usually the ground) may not be a reliable indicator of how easily a distant object may be seen. Probably the best theoretical measure would be the mean cylindrical illuminance, but in practice – and especially in security lighting installations – it is the vertical illuminance that is the best indicator of the visibility that will be produced. In security lighting, the vertical illuminance is usually measured with the photocell placed at 1 m above ground level and facing back towards the luminaires. The reason for this is that apparent brightness (luminosity) of distant horizontal surfaces is much attenuated by the cosine effect, and it is the vertical surfaces of objects (e.g. an intruder seen at a distance against a background) that we wish to make visible. By attention to these things, we can make a dark-clothed person visible at night with the naked eye against a background of low reflectance at distances of up to 0.5 km (chapter 17).

2.6.3   Visibility out of doors may be reduced by fog, rain etc, and we allow for this in design. Similarly, in designing interior installations, we may insert an absorption factor in our lumen method calculations to allow for atmospheric moisture, steam, smoke, dust etc. (section 15.4).

It is important to distinguish between an allowance made for reduction in illuminance due to absorption, and a reduction in visibility due to 'glare back' in outdoor installations. Even if the measured illuminance is adequate, visibility may be masked by atmospheric conditions which reflect light back to the observer and raise his adaptation level, thus making it more difficult for him to detect a low-contrast object at a distance. There is a given wisdom that security guards get a better view of a surveyed field by mounting a watchtower. In fact, at night, they will probably more easily see an intruder at a distance if their eye-level is as low as possible. The cosine effect then reduces the apparent brightness of the ground, and the horizontal flow of light will reveal the intruder better. This is also, of course, an argument in favour of placing security lighting floodlights at relatively low mounting heights. Note that visibility is affected by the colour-rendering properties of the lightsources employed (section 4.3).

2.7   General, task, and localized lighting

2.7.1   *General lighting* is that which is distributed over the whole lighted area. The usual objective is to distribute this light as evenly as possible (section 2.4). In many applications of exterior lighting, the lighting requirements are not similar over the whole area, but certain areas where tasks are performed, or where there are potential dangers, may need preferential treatment.

2.7.2   The lighting in any part of the area where it is necessary to provide an illuminance higher than the general lighting level may be augmented by *task lighting* (also called *local lighting*), for example by additional portable or fixed lighting under the control of the worker at that position. Local lighting may employ various kinds of task lights, including low-voltage luminaires and those which may be directed by the worker to a required angle (section 9.5). Some task lights may be focused or dimmed. Out of doors, the principles are similar, though the equipment may be different, i.e. it may consist of tripod lamps (powered from the reduced-voltage mains supply) or jenny-lights (powered by their own small internal combustion engine) (chapter 18). Local lighting luminaires may be part of a fixed installation, placed at points of need, for use all the time or as required. It is usual for the local lighting to be provided with local switching.

2.7.3   It is sometimes erroneously thought that all that is needed is to put plenty of light at the points where work is carried out, leaving the spaces in between lit only by the spill-light from the local lights. This can be a dangerous practice, for even if the spill-light is sufficient for safe movement, local lights may be switched off, leaving other areas without sufficient light for safety.

In interior lighting, unless the local lighting forms part of a well-thought-out scheme, it cannot be expected to make good deficiencies of a poor general lighting scheme. Trying to improve poor general lighting by additional local lighting usually proves uneconomical, for local lighting usually uses lamps that are of smaller power and hence of lower efficacy than those commonly used for overhead general lighting. Local lighting should be used only to raise the illuminance in the locality of tasks requiring it, and should not be relied upon to provide any part of the general lighting.

The level of general lighting should be sufficient to ensure safety of movement and to permit the performance of undemanding tasks irrespective of any contribution from local lighting. The general lighting should provide a field luminance sufficient to prevent there being excessive contrast between the task luminance and the surroundings.

2.7.4   There is a compromise which can be effective and economical: this is to have a general lighting scheme in which some of the general lighting luminaires are *localized* to the task areas. These luminaires therefore serve the purposes of both general lighting and a degree of task lighting (without the benefit of being under the control of workers at the task points, and without their being able to direct the light in chosen angles). In such schemes, all the luminaires should be switched on when lighting is needed, and no switches should be provided to control individual luminaires. If tasks needing more illuminance can be located adjacent to the general lighting luminaires, this can work quite well – for both indoor and outdoor situations.

Outdoors, lighting towers and masts may be located with the needs of local tasks in mind. A localized scheme will probably use higher efficacy lamps than those commonly used for task lighting, so some energy savings may result.

2.8   Flicker and stroboscopy

2.8.1   The term 'discontinuous light' is used to describe light that is not of a steady quantity, but which varies in amplitude or is intermittent in nature. All light sources which operate from alternating current (a.c.) supplies have some degree of rippling or continuous variation in light output at a frequency twice that of the supply (e.g. 100 and 120 Hz on 50 and 60 Hz supplies respectively) (Figure 2.4).

Some light sources intentionally generate discontinuous light, for instance stroboscopic flash lamps. Light which is substantially steady in nature may be

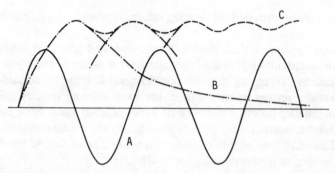

Figure 2.4 *The rippling quality of light from a lamp operating on alternating current. (A) curve of lamp current; (B) growth and decay of light from a fluorescent or HID lamp due to energy in one half-cycle; (C) net effect of a succession of half-cycles of lamp current, the frequency being at twice the supply frequency.*

broken into discontinuity by being viewed through the spokes of a rotating wheel, or by reflections from a vibrating surface.

2.8.2   Light pulsating in a regular manner may have an amplitude waveform similar to a succession of half-waves of sinusoidal shape. More commonly the half-waves are overlapped to produce a fairly steady quantity with a ripple of amplitude (Figure 2.4).

  2.8.3   Terms such as 'flickering' or 'pulsating' do not accurately describe light variations. A light which is *on* for substantially longer periods than it is *off* is described as an 'interrupting light', while one which is *off* for substantially longer periods than it is *on* is described as an 'occulting light'. Because of the phenomenon of *persistence of vision*, pulses separated by very short time intervals at a frequency of interruption or occulting that exceeds the 'flicker fusion frequency' of the subject are perceived as a steady light. It is this effect that enables us to perceive cinema and television pictures as continuous steady or moving images, and not as a rapid succession of still images or a travelling pulsating light-dot, respectively.

2.8.4   Discontinuous light of almost any frequency can produce a *stroboscopic effect*, in which a rotating or reciprocating object appears to be stationary, or moving slowly, or even to be moving in the opposite direction to actuality. This phenomenon is a potent cause of accidents in industry. Some items (e.g. routing spindles used in making wood mouldings) having speeds at, or close to, synchronous speed (i.e. at a multiple of the frequency of the a.c. supply) present a serious danger. Apart from advising the exercise of great care in the use of stroboscopic flash lamps (used to examine moving objects), no more need be said about them here. However, the avoidance of a fortuitous stroboscopic effect is of great importance (section 2.8.6).

2.8.5 Apart from the stroboscopic effect, there are other possible dangers associated with discontinuous light. Some subjects are sensitive to it, and can fall into a trance or hypnotic state in which they are accident prone.[15] Subjects who have a disposition to epilepsy may have a fit triggered by discontinuous light of any periodicity, and particularly at frequencies of around 3–8 Hz.

2.8.6 The degree of discontinuity of the light output from fluorescent lamps and high-intensity discharge (HID) lamps operating on high-frequency electronic ballasts is negligible, but HID lamps and fluorescent lamps operating on conventional inductive ballasts have a ripple content in their light output at twice mains frequency, and sometimes – because of a rectifying effect occurring at one electrode – at mains frequency also. Light from fluorescent lamps and phosphor-coated HID lamps generally has a fairly good form factor because of the relatively slow decay of luminance from the phosphors which has a 'smoothing effect' on the output. It is possible to take steps to reduce the stroboscopic effect from lighting installations (section 2.8.7), these being important in the case of installations of high-pressure sodium (SON) lamps, metal-halide and mercury-halide lamps and polyphosphor fluorescent lamps operating on inductive ballasts.

2.8.7 In any situation where stroboscopy might occur, any of the following measures will diminish the effect:

- Use high-frequency electronic ballasts instead of inductive ballasts.
- Light the moving object with lamps fed from two or three out-of-phase supplies, for example by taking supplies to adjacent lamps from two or three phases of a three-phase supply. In large installations of general lighting, the supplies to the luminaires may be taken from three phases to bring adjacent luminaires out of phase synchronism (Figure 2.5).

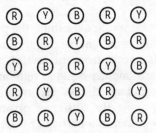

Figure 2.5 *Supplies to overhead luminaires spread over three phases, R, Y, B, to reduce stroboscopic effect.*

- Employ fluorescent luminaires having two lamps on a *lead–lag* ballast (section 2.8.8).
- Select lamps with low flicker characteristics such as fluorescent lamps with conventional phosphors (not polyphosphors), or fluorescent-coated HID lamps.
- Locally swamp the general lighting on the object with a task light containing one or more low-voltage GLS or PAR lamps, or low-voltage tungsten-halogen lamps.
- Use (exceptionally) general lighting service (GLS), pressed glass aluminized reflector (PAR) or tungsten–halogen (TH) lamps fed from a d.c. supply to swamp the light from a.c.-fed lamps.

2.8.8 Lead-lag luminaires are twin-tube fluorescent-lamp luminaires in which, instead of a normal power-factor correction (p.f.c.) capacitor for each lamp, one tube is uncorrected for power factor, while the other is over-corrected with a series p.f.c. capacitor. This results in the light output pulses from the two tubes being out of phase, one leading and one lagging, giving the effect of doubling the periodicity of ripple. The improvement in form factor reduces the tendency to produce stroboscopy. This equipment is declining in popularity (one reason being that the tube on the leading circuit tends to have a shorter life) and will probably soon be entirely supplanted by the use of high-frequency electronic ballasts.

2.9   Polarization

2.9.1   In the language of the Wave Theory (section 2.1), a ray of common (i.e. unpolarized) light may be considered as vibrating in all directions across the axis of the ray, but in polarized light these vibrations occur in only one plane normal to the axis of the ray. Common light may be considered as though composed of rays each of which is a cylinder; polarized light may be considered as though composed of rays each of which is a flat ribbon. This analogy permits a description of the phenomena of polarization and depolarization.

Certain crystals (e.g. Icelandic Spar) exhibit the property of internal double reflection, i.e. a ray of common light passing through the crystal divides into two polarized rays which emerge from the crystal at slightly different angles, the plane of polarization of one ray being at right angles to that of the other. If these two differently polarized rays are recombined, common light is produced again (Figure 2.6). By cutting a prism of Icelandic Spar, and re-cementing the parts together in a particular way, it can be arranged that it transmits only one polarized ray. Such a prism is termed a *Nicol Prism* or polarizer.

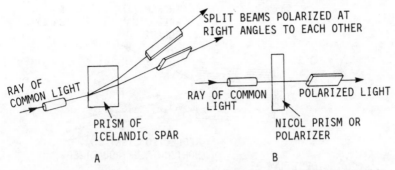

Figure 2.6 *Polarization of light by passage through certain crystals: (A) double refraction in a prism of Iceland spar; (B) polarization of light passing through a Nicol prism or polarizer.*

Other crystals (e.g. tourmaline, iodosulphate of quinine or herapatite) have the property of transmitting only one polarized ray. Such polarizers may be compared with a grating composed of narrow flat bars through which a flat ribbon (polarized light) will pass readily, but a cylinder (common light) will not pass. Similarly, the polarized light will only pass through the grating if its flat side is parallel to the grating (i.e. of the same plane of polarization). Thus, polarized light will pass unimpeded through a polarizer in the same plane of polarization, and will be completely stopped by a polarizer rotated through 90° about the axis of the ray.

2.9.2 Light can also be polarized by reflection, the degree of polarization depending on the angle that the incident ray is reflected, and on the number of times it is reflected. Maximum polarization in one reflection occurs at one particular angle. *Brewster's Law* states that the index of refraction is the tangent of the angle of polarization. The maximum polarizing angle for water is 53°11', for ordinary glass is 56°45' and for multiplate polarizers made of plastic materials is around 57°. At angles of 20° greater or smaller than the polarizing angle, some eight reflections are required to achieve complete polarization. Multiplate polarizers, also known as 'pile-of-plates' polarizers (Figure 2.7), are a means of polarizing common light, and may be incorporated into luminaires so that their light is substantially polarized.

2.9.3 Ruled gratings which produce a polarizing effect are widely employed for photographic and instrument purposes. Polarized light is used in certain inspection processes (chapter 10), and is applied in very critical colour-matching tasks to prevent the apparent change of colour due to light being reflected from the surface of particles in a self-coloured or pigmented article.

Of increasing importance is the use of polarized light to minimize reflections which otherwise impede vision. For example, when an operator is

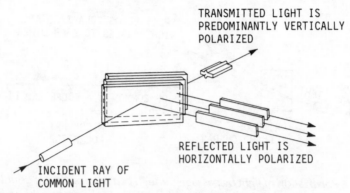

TRANSMITTED LIGHT IS
PREDOMINANTLY VERTICALLY
POLARIZED

REFLECTED LIGHT IS
HORIZONTALLY POLARIZED

INCIDENT RAY OF
COMMON LIGHT

Figure 2.7 *Polarization of light by reflection at a multiplate polarizer. Both the reflected and transmitted light effects may be employed in the design of polarizing luminaires.*

looking into a mass of clear liquid (e.g. water in a canister cooling pond at a nuclear power station), veiling reflections from the surface of the water can be greatly reduced by the operator wearing polarizing spectacles. Troublesome reflections from cover glasses on indicating instruments can be minimized by illuminating them with suitably polarized light to reduce brightness at selected angles of view.

There are some examples of polarized light being used in general lighting systems where reflections would otherwise be troublesome, for example in radar rooms and in railway signal boxes; but the idea of employing the polarizing principle in general lighting has not so far been greatly exploited in the UK.

2.10   Electric lighting and daylighting compared

2.10.1   Our eyes evolved over millions of years in light which came from the sun. Biologically speaking, light which flows from a direction other than generally downward is an experience for which our instincts and reflexes may not always be prepared.

Daylight varies continuously in its direction, intensity and colour properties throughout the day. Its hours of availability vary with the season and the lattitude (Figure 2.8), and the illuminance varies with changing weather. In comparison, electric lighting is constant in its properties, and its hours of availability are controllable. Because of its variability, daylight produces visual conditions ranging from ideal to positively unsuitable, from instant to instant, and from point to point – even in the same workplace. Because of its

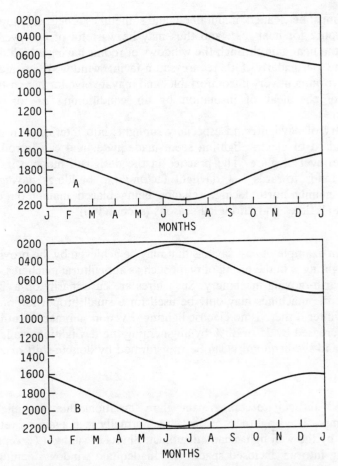

Figure 2.8 *Approximate number of daylight hours (A) in southern England (52°N), (B) in the Clyde Valley (56°N). (Based on civil twilight data, the Nautical Almanac.).*

controllability, electric lighting can be designed to produce excellent visual conditions continuously, in virtually any location.

2.10.2 Because there is no charge for daylight, it is easy to assume that daylight in buildings is free. If operations in commerce and industry were restricted to daylight hours, the cost per hour would be significantly greater than if the work were performed at the most convenient times by daylight or electric light as appropriate. No matter how well a building is fenestrated, in most situations it is impossible for natural light to satisfy the requirements for work except for relatively short periods in each week.

Windows are actually costly. The maintenance cost per unit area of window opening is considerably higher than for an equivalent area of blank wall.

Windows must be cleaned, maintained, and sometimes repaired. They can be a major route for heat loss, and thus may be wasteful of energy. During summer, the heat gain through the windows of rooms having a high daylight factor (and particularly if there are south-facing windows) can make the indoor environment very uncomfortable, and may involve further costs in the removal of the heat of insolation by air conditioning or forced-draft ventilation.

The entry of daylight, and especially sunlight, into interiors can make a good standard of electric lighting seem inadequate, and can introduce an intolerable level of glare. The practice in the design of modern industrial buildings tends towards low daylight factors, i.e. buildings having small windows, mainly north facing, reliance being placed mainly on electric lighting at all times and during all seasons (section 9.6).

2.10.3   An example of cost savings that may be achieved by the provision of exterior lighting is in the hiring of plant such as agricultural machines, boring rigs and earth-moving machinery. Such hires are customarily on a '24-hour basis', yet the machines may only be used for a small proportion of the 24 hours in winter if there is no electric lighting. Even in summer the utilization may be increased by 25 or 30% by augmenting the daylighting available for working, and the economics can be transformed by double-shift or 24-hour working.

2.10.4   It is difficult to compare the glare sensations due to sunlight and electric lamps. The brightness of the sun is normally seen against a relatively bright sky by the well light-adapted subject. Shafts of sunlight (except when penetrating into an enclosed space with inadequate window openings) are generally seen in an ambient brightness that makes them acceptable. No electric lighting installation is free from some degree of discomfort glare (section 1.5.1). With ordinary care and design, such installations should not produce disabling glare (section 1.5.2). Glare in interiors can be controlled by good design. Similarly, in exterior lighting installations, by carefully siting and correct aiming of the luminaires, it is possible to minimize the glare effect – at least in selected directions of view – so that good visual conditions can be produced at critical locations.

2.10.5   When security lighting is used to help defend premises (chapter 17), the premises may actually have better protection by night than by day; at night, it is possible not only to provide illumination to enable early detection of an intruder, but also to give concealment of the defenders by deliberately producing glare or luminance contrasts – a possibility that cannot be realized by natural light.

2.10.6 The term 'artificial lighting' is deprecated, and the term *electric lighting* is preferred. Light produced by electric lamps is not artificial though its spectral composition differs from that of daylight or sunlight.

2.10.7 Equal illuminances of daylight and electric light do not necessarily produce equal visibility in practical situations. In exterior installations, the low illuminance at dawn and dusk may need to be supplemented with electric lighting, so that for a short time the total illuminance available for tasks may be somewhat greater than daylight alone, and may be somewhat greater than would normally be provided by electric lighting in the absence of daylight. This is to allow for the lack of revealing power of the soft diffused light of morning and evening twilight, when colour distortion may be far greater than is experienced with electric lighting.

Assuming that the normal electric site lighting illuminance is in the range 5–50 lx, a suitable overlap will be achieved by operating the electric lighting from 30 min after sunset to 30 min before sunrise. If the electric lighting is switched by a photocell, it will usually prove satisfactory to set the switching level to twice the electric lighting illuminance. For example, if the installation produces 10 lx, set the controller to switch on the electric lamps when the daylight illuminance falls to 20 lx, and keep the lamps in operation until the daylight level reaches 20 lx again at dawn.

These remarks on the overlap between electric lighting and daylighting hold good in temperate latitudes where there is a significant period of twilight. However, near the equator, while twilights are brief, the more so under clear sky conditions, photocell control is not satisfactory, and a suitable timer controller should be used. On plains near the equator, the transition from daylight of some thousands of lux to night conditions of near total darkness can occur in about 20 min. Where the sun sets behind mountains, there is a crepuscular period which may last an hour or more, during which the site is in shadow but the sky is bright. Closer to the poles, there is greater seasonal variation of day length, and longer twilights, and indeed, in the summer months, the sky may never completely darken at night. In temperate regions, particularly under heavy cloud, daylight may need to be supplemented with electric lighting for outdoor working for a considerable part of each day (Figure 2.8).

2.10.8 On the darkest night out of doors there is always some light, though moonless nights in equatorial regions are considerably darker than those in temperate regions. Full moonlight is about 0.2 lx over most of the world, perhaps up to 0.25 lx near the equator. On a moonless night, clear starlight may be of the order of 0.01 lx. Near towns, especially under conditions of cloud or high humidity, or if there is much atmospheric pollution, the outdoor illuminance may be far higher than this due to light from roadlighting etc. reflected from clouds or the atmosphere, and readings of around 0.1 lx are found in unlit areas. Measurements on moonless nights with a sensitive lightmeter will enable the *district brightness* to be assessed. The district brightness must be taken into account when determining the required illuminance at outdoor installations, particularly in planning lighting at low illuminances as for security lighting (chapter 17).

2.11   Exterior and interior lighting compared

2.11.1 In most interior lighting installations (other than uplighting installations), the luminaires are disposed overhead in regularly-spaced patterns so that a substantially uniform horizontal illuminance is produced on the working plane (usually assumed to be 0.85 m above floor level), or on the floor. In well-designed uplighter installations, the uniformity may be generally similar to the results obtained with overhead luminaires, with the observation that even extreme departures from uniformity may have such gentle gradients of illuminance that they are hardly noticeable (sections 2.4.4 and 9.8). Overhead general lighting luminaires are usually designed to give restricted flow of light at those angles which could direct glare to a seated or standing subject. Typically, the horizontal illuminance at any point is somewhat greater than the vertical illuminance at that point (section 2.5.3).

2.11.2 In most exterior area lighting installations, the luminaires are disposed at one or more sides of the lighted area, and a high level of uniformity is generally not easily achieved or necessary (except in special applications such as the lighting of sports fields). Again, the eye is often deceived into believing that exterior area lighting is more-or-less uniform, even though the diversity ratio may be 2:1 or more – the more gentle the gradient, the more acceptable the diversity (section 2.4). In typical industrial and commercial exterior lighting installations, the direct glare effect of the luminaires varies according the position of the observer. Typically, the vertical illuminance in such installations tends to be of the same order as the horizontal illuminance at least in certain directions from the point of measurement. However, the E_h/E_v ratio increases at points further from the luminaires, and a ratio of as much as 1:20 at extreme range is not uncommon.

2.11.3 No reliable way of computing exterior glare indices has yet been devised. However, an approximate assessment method can be used for exterior lighting (section 3.4.3). It is not possible to make a quantified comparison between the glare effects of interior and exterior lighting installations.

2.11.4 There are curious differences between the perceived effects of natural light and electric light inside buildings. Although many interiors use a mixture of daylight and electric lighting, the environmental luminance patterns produced by the two are different – as is the human response – and the difference has never been fully explained. The classic example of this phenomenon is that 'bad light stops play' in cricket at around 1000 lx. Yet, indoors, the same cricketers will happily play at 750 lx. Another example is that workers in a workplace devoid of natural light but provided with adequate illuminance from electric lighting express satisfaction with their environment, but other workers, provided with similar electric lighting but having sight of some daylight (possibly that entering by a distant window) may complain of being deprived of sufficient light, and envy those working closer to the windows.

Anomalies abound where daylight and electric lighting both contribute to the illumination. In exterior lighting practice it has been found necessary to provide considerable overlap in the switching of electric lighting as daylight fades and again at dawn (section 2.10), yet it is common practice in the lighting of agricultural buildings to provide a slightly lower standard of general lighting in buildings having windows than in windowless ones.

3 Lighting terms and units

Newcomers to the field of lighting sometimes have difficulty in communicating their needs and intentions simply because they have not absorbed the language of lighting and come to understand the common units. The lighting art is a fairly specialized branch of applied physics, and though good lighting can be achieved without great knowledge of the finer points of calculation, it is essential that the practitioner grasps the fundamental concepts of the mathematical relationships in lighting.

3.1 Quantity of light: the lumen

3.1.1 The *lumen* (abbreviation lm) is the SI unit of luminous flux, and describes the quantity of light emitted by a light source or received on a surface. It is called the measure of luminous flux, and is analogous to the flow of 'a litre of water per minute'. In other words, the number of lumens emitted by a lamp is a *rate of flow* of luminous energy from the lamp. Power input to a lamp is measured in *watts* (symbol W). Generally, larger power lamps of a given type produce more lumens per watt than smaller sizes of the same type of lamp. The number of lumens per watt of a lamp is termed its *efficacy*.

3.1.2 A 100-W tungsten-filament lamp (GLS lamp) has an output of around 1260 lm at 240 V, that is, it has an efficacy of around 12.6 lm/W. A 20 W fluorescent tube (MCF) has an output of around 1300 lm, which is roughly the same, but its efficacy is around 55 lm/W. The GLS lamp is intensely bright (it hurts the eyes to look at it directly), while the 600 mm (2 ft) 20-W fluorescent tube gives a gentle light which can be looked at without great discomfort. Thus, neither the lumen output nor the efficacy tell us anything about how *bright* a lamp is.

A pocket torch is intensely bright, but we could not light a whole building with it. Fluorescent tubes, which are not very bright, can be used to light huge buildings to any desired level of illuminance. The difference is that the torch, although bright, does not emit many lumens, while the fluorescent tube, though not very bright, emits a greater number of lumens. The pocket torch concentrates its few lumens in a small area, while the greater number of lumens from the fluorescent tube are distributed over the considerable area of the tube, and therefore the tube is less bright. Thus, to gain an appreciation of brightness, we need to know the area of the emitting body as well as the number of lumens emitted.

3.1.3   The lumen is derived from the *candela* (abbreviation cd), the SI unit of luminous intensity. The candela is the physically defined successor to the now obsolete 'standard candle'. The concept is of a small source, having a uniform luminous intensity of 1 cd in all directions, placed at the centre of a sphere, so that its luminous flux is distributed over the inner surface of the sphere. If the sphere is of unit radius, as there are 4π steradians in a sphere, each unit area of the sphere will receive one unit of luminous flux. By definition, if the unit radius is 1 m, each unit area receives 1 lm; $1 \, \text{lm/m}^2 = 1 \, \text{lx}$.

3.2   Unit of illumination (illuminance): the lux

3.2.1   The word 'illumination' means the act or process of illuminating something, and should not be used to describe the quantity of light received by a plane or surface, for which the proper term is *illuminance*. Quantity of light flowing on to a plane or surface is measured in units of *lux* (abbreviation lx). The lux is analogous to a flow of 'a litre of water per minute per square metre', i.e. it is the rate of flow of luminous energy on to a unit area. Thus, $1 \, \text{lx} = 1 \, \text{lm/m}^2$. An appreciation of the magnitude of the lux may gained by studying Table 1.

Table 1   *Typical illuminance values*

Examples	Illuminance value (lx)
Tropical sunshine	100 000
English June sunshine	80 000
Overcast day, outdoors	5 000
'Bad light stops play' (outdoors)	1 000
Well-lighted office	500
Well-lighted domestic room	200
Working area, building site	50
Good main road lighting in city	15
Typical side-road lighting	5
Minimum design illuminance, security	1
Clear moonlight	0.2
Minimum design illuminance, emergency lighting	0.2
Starlight on clear night	0.01
Cloudy moonless winter night	0.001

If the foot rather than the metre is taken as the unit of linear distance, the unit of illuminance is the *foot candle* (abbreviation fc), a unit formerly used in the UK. Thus, 1 fc = 1 lm/ft². There are 10.76 ft² in 1 m², so conversion between the two systems is simple; 10.76 lx = 1 fc. (The footcandle was dropped from use in the UK in 1968 when the lighting industry adopted metric units.)

3.2.2 We can measure illuminance with an instrument called a *lightmeter* or *luxmeter*. This is not the same as a photographic exposure meter which responds to brightness, not to illuminance. A lightmeter is a device which converts the energy of light falling on its sensitive cell into an electric current, which in turn is measured by a milliammeter calibrated in lux (appendix E).

3.3   Units of luminance; Cd/m² and apostilb

3.3.1   The measurable brightness of a source or object is termed its *luminance*. In lighting work, it is better to avoid the use of the word 'brightness' as it has several meanings, and may be confused with *luminosity*, which is the subjective sensation of brightness and is not related to the absolute brightness or luminance. For example, a candle flame seen in sunlight seems to have little brightness, yet that same flame, seen in a darkened room after an observer has experienced a period of total darkness, will appear to be very bright. Its true brightness (luminance) has not changed, but its luminosity (perceived brightness) has changed markedly.

3.3.2   The SI unit of luminance is *candela per square metre* (abbreviation Cd/m²). Considering an illuminated surface, if we can assume that this is perfectly matt (i.e. it is a perfectly diffusing 'Lambertian' reflector), the luminance will be the product of the surface illuminance (lux) and its reflectance. Its emitted (reflected) flux can be measured in lumens per square metre – a measure more properly termed 'luminous exitance' for which the use of the unit lux would be inappropriate. We can say that a uniform diffuser emitting 1 lm/m² has a luminance of one *apostilb* (abbreviation asb). The relationship between the apostilb and the Cd/m² is given by:

$$1 \text{ asb} = \frac{1}{\pi} \text{ Cd/m}^2$$

The apostilb is a convenient unit, but it is not used in authoritative documents as it is not an SI unit.

3.4 Glare effect

3.4.1 Glare causation

The general nature of glare is discussed in section 1.5. In section 9.7 we see an example of how the glare experienced on passing between two zones of markedly different illuminance may be reduced by providing a zone of intermediate illuminance – a principle that may be applied generally. Our perception of brightness (luminosity) is dependent on our adaptation level (section 3.3.1), and hence our susceptibility to glare is also related to our adaptation level. Glare sensation may be reduced by:

- reducing the ratio of the brightness contrast between any two parts of the visual field; and
- arranging that subjects are not required to observe anything of substantially different luminance to that to which they are adapted without allowing adequate time for further adaptation.

Glare sensation, in both exterior and interior locations, has the following main causative factors:

- the intrinsic brightness of the glare source (cd/m^2);
- the projected area of the glare source presented to the eye, and, taking account of the distance, the angles subtended by this area at the eye;
- a positional factor to take account of the fact that a glare source will be more troublesome to vision if it lies close to the direction of view;
- the brightness contrast ratio between the glare source and its background;
- the time needed by the subject to adapt to a different field luminance.

3.4.2 Glare index – interior locations

The subjective sensation of glare cannot be measured, but it may be evaluated and represented by a non-quantitative number that expresses the degree of discomfort. The British Glare Index System, which is defined in CIBSE TM-10,[17] is based on research carried out by the former Illuminating Engineering Society (UK), and was developed from statistical studies of glare sensation experienced by a large number of subjects visiting some standardized installations. Observations were made in what lighting engineers came to call 'the four just men': the glare had to be described as 'just imperceptible', 'just perceptible', 'just tolerable' or 'just intolerable'.

For interior lighting installations, in which luminaires are arranged in regular patterns, the TM-10 method enables the calculation of a *glare index* for any real or proposed lighting design layout, and permits comparison of this with the *limiting glare index*, i.e. the recommended maximum glare index for that location. The glare index takes account only of the *direct glare* produced by the installation, i.e. that due to the luminance of the luminaires (section 9.3).

3.4.3   Glare index – exterior locations

No system of calculating glare indices for exterior lighting installations has been developed. In the absence of any official recommendations, the arbitrary glare scale in Table 2 may aid assessment and comparison of existing installations.

Table 2   *An arbitrary scale of glare indices for exterior lighting*

Glare index	Description of glare sensation	Examples
50	Unbearable – totally disables vision	Facing main-beam headlights on a dark road; good glare-lighting
40	Bearable for short periods – vision is partially disabled	Typical night driving on unlit roads
30	Just bearable – some discomfort	Typical industrial area lighting
20	Acceptable – there may be some discomfort	Good area lighting
10	Just perceptible – hardly any discomfort	Good sports field lighting (spectators)
0	No perception of glare	Twilight, facing away from sunset, with an overcast sky

3.5   Inverse Square Law of illumination

3.5.1   The Inverse Square Law applies only to a source which is virtually a point-source (i.e. with negligible area in relation to the distance over which the calculation is performed), in circumstances where there is no reflection (e.g. from nearby objects or the ground) and no scattering (due to atmospheric moisture, dust etc.). The relationship is:

$$E = \frac{I}{d^2}$$

where E is the illuminance (lx), I is the source intensity in the direction under consideration (cd), and d is the distance (m) between the source and the point of measurement. The area illuminated within a given angle increases as the square of the distance, and the illuminance (lux, lumens per unit area) decreases proportionally (Figure 3.1). If the incident beam is not normal to the plane of measurement, correction for the cosine effect must be applied (section 3.6).

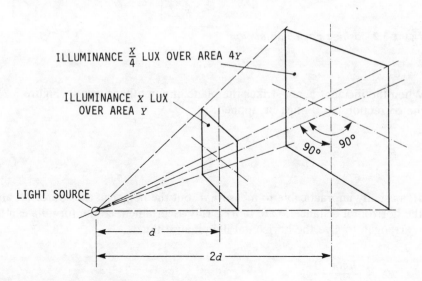

Figure 3.1 *Inverse Square Law of illumination.*

3.5.2 Brightness does not diminish with distance, but is intrinsic to the source. (The brightness of the moon as perceived on Earth is exactly the same as that which is experienced by astronauts on the surface of the moon.) However, in practice, there may be attenuation due to scattering and absorption due to atmospheric moisture, dust etc. With increasing distance, a source will apparently cause less glare as its apparent size diminishes.

3.6   Cosine Law of illumination

The relationship between illuminance, intensity and distance given in section 3.5.1 applies only to illuminance on a plane normal to the incident beam.

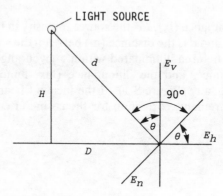

Figure 3.2 *Cosine Law of illumination.*

When the incident beam strikes the plane at any other angle θ (Figure 3.2), the correction $\cos \theta$ must be applied:

$$E = \frac{I}{d^2} \cos \theta$$

It is usually impracticable to measure d, but the height of the source H, and the horizontal distance D are more easily ascertained, so the formula can be rearranged to give the horizontal illuminance thus:

$$E_\mathrm{h} = \frac{I \cos^3 \theta}{H^2}$$

or

$$E_\mathrm{h} = \frac{I}{D^2 + H^2} \cos \theta$$

The illuminance normal to the beam can be calculated from the expression:

$$E_\mathrm{n} = \frac{I \cos^2 \theta}{H^2}$$

The vertical illuminance can be calculated from the expression (Figure 3.2):

$$E_v = \frac{I \cos^2 \theta \sin \theta}{H^2}$$

Practical methods for calculating the above relationships include the use of nomograms auch as those given in *Interior Lighting Design*.[8] Design may also be performed by the use of isolux diagrams derived from such calculations (section 9.2).

3.7 Efficacy of lamps

3.7.1 The efficiency of a machine is the ratio of the output power to the input power, and is usually expressed as a percentage. The true efficiency (thermal efficiency) of any lamp must be 100% if we consider that the heat energy dissipated plus the light energy radiated must be equal to the input power. The power output in radiated light is only a small percentage of the input power, and therefore is not a useful guide to the performance of the lamp. For this reason, the performance of lamps is assessed not by their efficiency but by their *efficacy*.

3.7.2 The efficacy of a lamp is the number of lumens emitted by the lamp per watt of input power. This is expressed in *lumens per watt* (lm/W). Efficacies of lamps are commonly quoted in manufacturers' catalogues as 'lamp efficacies', i.e. the lumens per watt at the lamp, and do not take account of energy consumed in the associated control gear. In general, larger powers of lamps have higher efficacies than smaller powers of the same type of lamp.

One needs to know the lumen output of lamps in order to carry out lighting calculations, but statements and claims about lamp efficacy can be misleading. When making comparisons of lamp performance between types of lamps or between alternative suppliers of a particular type of lamp, it is preferable to compare the 'lamp circuit efficacies', i.e. the number of lumens emitted per watt of total input power (including energy dissipated by the control gear).

The rated or claimed lumen output of a lamp depends upon the method of measurement. For example, the output may be stated as:

- *initial lumens*;
- *100-hour lumens*;
- *2000-hour lumens* (Lighting Design Lumens);

- *average-through-life lumens* (against a stated claim for technical life of the lamp, which may range from 7500 h up to 20 000 h);
- *terminal lumens* at a stated time for end of life.

The 'average-through-life' figure can be misleading, because the designer or user may decide to relamp after a shorter or longer period of use than the nominal or claimed lamp life. Thus, a shorter period will raise the lumen figure, and vice versa. Such adjustments of the relamping time will affect the cost-in-use of the lighting. If the period of lamp use is extended, the cost per day of use will fall very slightly, but the user will get a lower illuminance, and vice versa. There is room for adjustment, and this might be taken into account when compiling the Outline Lighting Specification (sections 15.2.1 and 24.2).

Clearly, the accuracy of lighting calculations will depend on having accurate data on lamp outputs, and having an accurate picture of the manner in which the lamp lumens depreciate throughout the life of the lamp.

The lighting industry is going through a time of commercial change and rapid technical development, and, as can be seen from appendix B, changes are under way in many definitions and conventions relating to lamp performance and lighting. In earlier books by the author it has been possible to tabulate the lumen outputs and efficacies of lamps by type and power, but, with so many changes occurring in lamp manufacturers' products, this is not currently practicable. It is suggested that the reader apply to lamp manufacturers (appendix H) for data on their products.

3.8   Lumen maintenance

3.8.1   The light output of a lamp is not constant, but declines through its life. Thus, any statement as to its efficacy must relate to its nominal output at a certain point in its life or to an average value throughout life (section 3.7.2). High intensity discharge (HID) lamps and fluorescent tubular lamps continue to light long after their lumen outputs have fallen below an economic level (Figure 3.3).

3.8.2   The determination of nominal lamp life is not universally agreed. In the USA, a lamp life of 20 000 h or greater may be claimed, but US lamps last no longer than UK lamps, which will also function for such periods. In the UK it is not considered economic to run lamps beyond their nominal lives of 7 500 or 10 000 h (except, perhaps, for SON lamps in some exterior lighting or roadlighting applications).

Bulk lamp replacement at the calculated nominal end-of-life point usually leads to the most economic operation. If lamps are run for longer periods

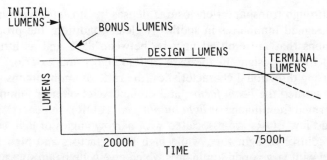

Figure 3.3 *Typical depreciation of lumen output from a fluorescent or HID lamp.*

before relamping, then, in the early part of lamp life the illuminance provided may be significantly in excess of the design illuminance, and/or the achieved illuminance in the latter part of the lamp life will be significantly below the design illuminance. This holds true whatever nominal life is claimed for the lamps (section 23.4).

3.8.3 The lighting designer or user may change the balance of the interrelated values of lumen output and lamp life. For example, in a situation where the cost of relamping is very high, it could be economic to install a higher level of initial lumens and extend the period between relampings, even at an increased running cost per hour due to the falling efficacy of the lamps as they age. The cost of high-frequency control gear with dimming facility is falling, and this control gear can be designed to reduce circuit watts when the lamp is dimmed. Using this control gear allows the period between relampings to be extended by installing a higher nominal lighting level, and negating the excess light output during the earlier part of the lamp life by dimming.

3.8.4 It is too early to make positive statements about the economics of the new Type QL induction lamps (section 5.12), but it is predicted that these lamps may last up to eight times longer than other HID lamps. It remains to be seen if the factors of cost, lumen maintenance, life (and life of the electronic components) etc will result in considerable economies as is hoped. Where the cost of access for relamping is high or involves danger or difficulty, the higher initial cost of these lamps may well be justified.

3.9   Utilance and utilization factor

3.9.1   Utilization of lamp lumens

In any luminaire, not all the lumens emitted by the lamp are projected out of the luminaire. Light is absorbed by the interior reflecting surfaces, and in its passage through transparent enclosures, lenses, prisms etc. Even in the case of well-designed luminaires in spotlessly clean condition, the proportion of lamp lumens that are emitted may be between 85% and as little as 20% according to the design. In the case of floodlights and exterior luminaires having strong directional characteristics, the ratio of lamp lumens to emitted lumens is termed the *beam factor*, and in the case of interior luminaires, this ratio is termed the utilance or *light output ratio* (LOR). In general, low beam factors and low LORs are associated with narrow angles of light spread and accurate control of light flux, while high beam factors and high LORs are associated with large open luminaires which give wide beam spread and only limited directional control of light output.

3.9.2   Beam factor and utilization factor of exterior floodlights

Although it is desirable to employ a luminaire with as high a beam factor as possible for the degree of control required, beam factor is not the only factor determining the economic and efficient use of lumens or capital cost. Of the lumens emitted by the luminaire, only a proportion are directed usefully to the target surface (Figure 3.4). Some light may be wasted by being projected into the sky or at such angles that it is not usefully employed.

If we express the number of lumens we can identify as having arrived on the target plane as a proportion of the lumen output of the lamp, we will find values of something between 50% and 20% for typical installations (Figure 3.5). This ratio may be termed the utilization factor (UF). The higher the UF, the greater will be the economy in the use of capital and energy, though this

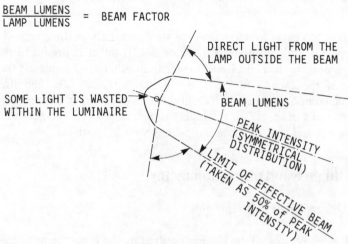

Figure 3.4 *Beam factor of a floodlight luminaire.*

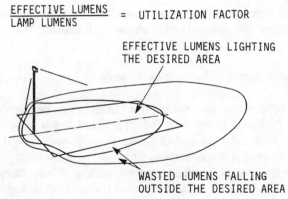

Figure 3.5 *Utilization factor of a floodlight luminaire.*

statement holds good only if the light flux distribution is appropriate to the application.

An exterior luminaire in a particular application, at a particular mounting height and angle of tilt, which produces a high utilization factor would only be the best economic choice if it distributed its lumens so as to produce an acceptable degree of uniformity of illuminance, and an acceptable degree of control of glare. For example, a luminaire which placed a high proportion of its lamp lumens in one bright patch while leaving the rest of the area underlit, might have a high UF, but would produce a most unsatisfactory lighting installation. Similarly, a luminaire which was effective at placing its lumens in the desired area, but which at the same time produced considerable discomfort glare, would probably be less effective in lighting the area than one having a lower UF but which shielded its bright parts at normal angles of view.

There are poorly-designed area-lighting floodlights which dump a high proportion of their output on the ground immediately beneath them, or dissipate the light at upward angles or beyond the edge of the target patch, and fail to project light satisfactorily to the more distant parts of the target area. Such luminaires may have high beam factor, but they are less effective at lighting the area than a slightly less 'efficient' luminaire which controlled and directed the lumens in desired directions – even if a somewhat higher proportion were lost. We see, therefore, that, while beam factors and UFs are helpful in selecting floodlighting equipment and performing lighting calculations, they must not be relied upon without checking that the luminaire has a polar distribution suitable to the application.

A sensible use of the UF concept in exterior lighting design is to employ it in preliminary calculations to determine the approximate number of luminaires required, and then to use isolux diagrams or point-by-point calculations to determine their best disposition and aiming to produce a satisfactory degree of uniformity of illuminance (chapter 15).

3.9.3 Utilization factor of interior luminaires

In the case of interior luminaires, the LOR may be divided into its components of lumens emitted in the upper and lower hemispheres, the upward and downward light output ratios (ULOR and DLOR). The UF, representing the proportion of lumens from an overhead array of luminaires which will reach the working plane, may be calculated from the photometric data of a luminaire. The UF takes account of the absorption and reflection of light at the surfaces of the room.

Standard tables of UF for luminaires having various distribution characteristics have been determined. The UF is incorporated into calculations of illuminance by the Lumen Method of calculation (chapter 9).

4 Lighting and colour

4.1 Colour-appearance and colour-rendering

4.1.1 For electric lighting to illuminate an object so that its colours are recognizable and reasonably faithful to actuality, the spectral composition of the light must approximate to that of daylight. Lighting that is deficient in colour-rendering property, or is of an unfamiliar and unacceptable colour-appearance, can adversely affect the visual performance of workers and their task performance. Colour is an important factor in the decor of workplaces (section 4.2).

For many industrial tasks, some deficiency in colour-rendering of the light from the lamps will not be a serious handicap to efficient work, but reasonable colour-appearance of the lighting is important. Accurate colour judgements cannot be made if the colour properties of the illuminants are not standardized. The techniques of colour matching (sections 4.5, 10.5) must take account of the phenomenon of colour adaptation (section 1.8).

4.1.2 The *colour-appearance* of a lightsource is the colour we perceive the luminous area of the lamp to be when we look at it directly, or the colour of a white object seen in its light. The colour-appearance of a lamp may be described using the same adjectives we use for describing temperature, e.g. 'cool', or 'warm'. Cool colours are bluish, while warm colours are those at the red end of the spectrum. Colour-appearance is difficult to define because the subject becomes adapted to the colour of the light and soon ceases to discriminate accurately.

4.1.3 *Colour-rendering* is the property of the light from a source to reveal the colours of objects. The accuracy of a lamp's colour-rendering property is readily tested within the limits of accuracy of the human eye by trying to match colours in its light. Colour-matching tasks may involve mixing pigments to try to reproduce the colour of a sample; or the task may be to try to select pairs of identically coloured samples from a number of samples.

In light of poor colour-rendering, an apparent match may be found to be false when the task is re-examined under light of better colour-rendering (termed a metameric match, i.e. a match that holds good only under a particular lightsource other than a lightsource of high colour-rendering properties). A true colour match remains a good match when examined under other light sources of different spectral composition.

4.2   Colour in the workplace

The correct colour treatment of working interiors is a matter that is often neglected. The colours of the surfaces in an interior are second in importance only to the lighting in their contribution to the visual comfort of the occupants and their satisfaction with the environment. The colours of the room surfaces also affect visual performance of tasks involving fine colour-discrimination (section 4.5).

4.2.1   The environmental effects of colour in industrial interiors are far reaching, for the general colour with which the eyes are presented helps us to form our psychological attitudes to the space we occupy. Well chosen colours, in combination with good lighting, can make a useful contribution to the wellbeing of the occupants, stimulating morale and promoting activity.

Modern practice in decor tends to the use of large areas of white and grey for the building interior surfaces, with very little relief by use of strong colour. Where such colour schemes are used in areas having very well diffused lighting (e.g. by uplighting), the effect may be to create a 'bland field'. In a bland field, the eye finds no dominant feature upon which to focus, and some subjects may find the effect most disturbing, to the extent that they may find it impossible to work happily in the room, and they may complain vociferously about the lighting, even though it may be adequate in quantity (section 1.7).

The eye craves something interesting to look at, and some object of special regard upon which to dwell. Bland field interiors are unnatural, and in a room having surfaces that are of high reflectance and pale colour, some subjects become conscious of the process of vision, particularly noticing the 'floaters' in the anterior chambers of their eyes, and developing a sensation of unease – even nausea. This unpleasant condition can be relieved by providing one or two small areas of saturated colour in the visual field, perhaps by having some mural paintings or other decoration, or by the use of one or two coloured spotlights directed on to wall surfaces or lighting interesting objects.

4.2.2   It is important to distinguish between the illuminance provided by natural or electric lighting in an interior, and the subjective sensation of brightness experienced by the occupants. The brightness produced by the reflection of light from a surface is related to the reflection factor of the illuminated surface and its colour as well as the incident illuminance. An interior decorated throughout in dark colours will reflect little light, and it will appear gloomy even if provided with generous illuminance, and especially if the lighting is provided by strongly directional downlighters.

4.2.3   When choosing colours for the décor of an interior, the effect of reflectance factor on the brightness appearance should be considered. The lighting designer may recommend the adoption of preferred ranges of reflectance for the walls, ceiling, flooring and furniture (especially the bench-tops and table-tops). The *CIBSE Code*[1] gives guidelines for devising luminance distributions which will be comfortable to the eye and tend to minimize glare.

4.2.4 If the reflectances of paints are not shown on the containers, they can be obtained from the paint manufacturers. Reflectances are expressed as decimal fractions, e.g. 0.5 etc., and these can be related to the reflectances of paints if the *Munsell colour numbers* of the paints are known. In paint descriptions, the Munsell colour number follows the group letter in the Munsell designation. The approximate relationship between reflectance and Munsell value (V) is:

$$\text{Reflectance} = \frac{V(V-1)}{100}$$

Refer to the *CIBSE Code*[1] for more information on this matter.

4.2.5 The colours of décor contribute to the visual performance of the occupants of an interior. By helping to form a pleasant and sympathetic environment the colour scheme will contribute to creating good morale and, indeed, the promotion of the good health of the occupants. It is not only for aesthetic reasons that it has been necessary to create a British Standard for colour co-ordination in buildings[19] for, without some harmonization it would be difficult to find suitable coloured finishes for building components.

There are a number of British Standards on colours for building components, including cladding and floorings. Colours are allotted significance in a method of coding pipelines in factories[20], and a further series of colour significations is of great importance to safety in identifying informational, prohibitory and mandatory signs in factories (Table 3), while another series of colours signifies voltages (Table 4). These matters are of

Table 3 *Colour code BS safety colours*

Colour			Significance
Red	04E53	'Flame'	Fire equipment and alarms
Yellow	08E51	'Gorse'	Where accidents are likely to occur
Green	14E53	'Neptune'	Escape routes
Blue	18E53	'Gentian'	Safety instructions

importance because: (a) colours of décor must not be confusable with significant colours (e.g. Tables 3 and 4); and (b) the lightsources used must have suitable colour-rendering to enable these colours to be readily recognized.

Table 4   *Colour code BS 4343/CEE17 voltage colours)*

Colour	Voltage
Violet	25
White	50
Yellow	110/130
Blue	220/240
Red	380/415
Black	500/750

4.2.6   In exterior lighting, there is a general preference for lightsources that are not excessively 'cool' in colour appearance. For certain outdoor tasks the sources must have a reasonable colour rendering, for example, to enable colour-coded warning and informational signs to be recognized (section 4.2.5), or to enable colour codes on materials such as metal stocks and structural steelwork components to be recognized in construction and engineering tasks. Only rarely is it required to provide light of such colour quality that will enable close judgement of colour to be made in outdoor activities (section 4.3).

4.3   Colour and visibility out of doors

Because our eyes evolved in natural light, light which approximates to the colour properties of daylight is most acceptable to us. Modern people have adapted to the use of electric lamps which generally do not produce light closely similar to daylight, and – in suitable applications – can accept sources such as high-pressure sodium lamps (section 5.10) which produce significant colour distortion. In certain situations we are even able to accept the poor light of low-pressure sodium lamps (section 5.9) which have virtually no colour-rendition at all. This is possible because we experience the phenomenon of 'colour adaptation' (section 1.8) and may not be aware of distortion or absence of colour unless our task demands specific colour decisions.

In section 4.2 we discussed the effect of the spectral composition of light in typical indoor workplaces. For certain outdoor tasks (e.g. fine painting, graded terrazzo work) it is essential to have light of good colour-rendering, and this may be provided more economically by local task lighting rather than by illuminating the whole working area with good colour-rendering sources. The better rendering illuminance must be of such magnitude as to 'swamp' the ambient lighting.

The effect of lighting on visibility out of doors is discussed in section 2.6.2, where it is noted that if an object or scene is presented to our eyes in a blurred or indistinct manner, our time for perception is increased (section 1.9.1). A deficiency in colour-rendering property may have the same effect, while our rapid comprehension of any visual scene is enhanced by the use of good colour-rendering sources. This could be a justification for using lamps of good colour rendering (such as metal-halide) for outdoor security lighting installations.

4.4   Colour properties of light sources

4.4.1   If the spectrum of a lightsource does not contain levels of energy of particular wavelengths, both its colour-appearance and its colour-rendering are affected, and with them the accuracy of our colour-perception (section 4.1). Colour-temperature is a measure of the colour properties of a lightsource, but is only an accurate description if the lightsource emits a continuous spectrum (which most practical lightsources do not). Nonetheless, it is a useful indicator of the colour properties of a source.

Colour temperature is expressed in degrees Kelvin (symbol K). There is, for example, a considerable difference in colour-rendering and colour-appearance between a 'White 3500 K' fluorescent tubular lamp and one described as 'Daylight 4300 K', the latter (being of higher K) appearing more blue in direct comparison with the former. The higher the K value, the colder or bluer the colour-appearance, and – for lamps of continuous spectrum – the better is the colour-rendering. Low illuminances obtained from cooler-coloured sources tend to look dismal. For illuminances below about 300 lx it is advisable to use light-sources below 4000 K. Conversely, high illuminances from fluorescent tubes of warm colour appearance (i.e. low K) can be overpowering, and may give the impression of creating much heat, but lamps of better colour-rendering and of cooler appearance (i.e. high K) produce more acceptable installations for illuminances over 1000 lx.

4.4.2   Generations of engineers accepted the statement that 'better colour-rendering sources are always of cooler colour-temperature', but with the arrival of triphosphor fluorescent lamps, this is no longer true. Because the light outputs from discharge lamps do not have truly continuous spectra, a colour-rendering index (CRI) system has been devised, and this is explained in the *CIBSE Code*[1]. Useful though it is as a guide to choices between possible types of lamps for a particular application, the CRI is not very accurate, and two different types of lamps having the same CRI are not necessarily interchangeable for the same critical duty. CRI is represented by the symbol Ra.

It has been the object of considerable research to try to discover why subjects may prefer the lighting effect of certain sources, and the theory is advanced that they prefer a lamp because it gives them visual clarity, but this must not be confused with the colour matching properties of the source. Some workers believe that this visual clarity preference is largely based on the ability of lamps to render human skin colours attractively.

The factor termed 'visual clarity' is sometimes used in relation to critical colour-matching tasks, but what is meant here is the clarity achieved by the use of polarized light (section 2.9) to reduce the formation of veiling reflections on glossy surfaces.

4.4.3   For non-critical applications of lightsources, e.g. for lighting ordinary interiors pleasantly, and for the recognition of colours (but not necessarily their accurate matching), a colour-preference index (CPI) may be utilized, this being determined by a similar procedure to that employed for determining the CRI. There are thus three current measures of lamp colour:

- colour temperature (K), giving a measure of the coolness or warmth of the source colour;
- colour rendering index (CRI), expressed in Ra, which describes the ability of the lightsource in aiding the matching of colours;
- colour-preference index (CPI), which is based on subjective preference and may have no relation to either K or CRI.

4.4.4   Many factory installations employ HID (high intensity discharge) lamps for general lighting at mounting heights of 4 m or more. If the colour requirements are not demanding, SON or MBF lamps may be used; MBI lamps have better colour performance for this duty. Below around 4 m mounting height, the commonly used sources are fluorescent tubes. If there is a requirement for critical colour vision in a location where practical considerations preclude the use of general lighting by 'better colour rendering' tubes, localized lighting or colour-matching booths may be employed (sections 4.5 and 10.5).

4.4.5   Our eyes have a spectral response curve matched to that of solar visible energy (Figure 4.1). It has long been the aim of lamp makers to produce a lamp with a spectral power distribution to match the human

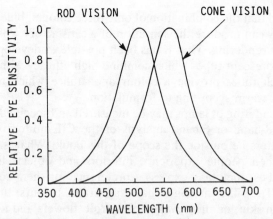

Figure 4.1 *The relative sensitivity of the eye to wavelengths of radiation.*

spectral response curve. The technical difficulties in doing this are great, but a number of practical sources have been developed which, while having spectral power distribution envelopes of quite different contours to that of solar light, give satisfactory performance in many applications. Lamp manufacturers publish such curves for their lamps, but it takes much expertise to interpret them and visualize the colour characteristics they represent.

Quite unusual spectral distributions may give the sensation of white light, or of colours. For example, if orange light (650 nm) and green light (490 nm) are mixed, say, by both illuminating a white object, they appear to produce pale blue light (but without any colour-rendering ability except for those specific orange and green colours (Figure 4.2)). In a similar manner, it has

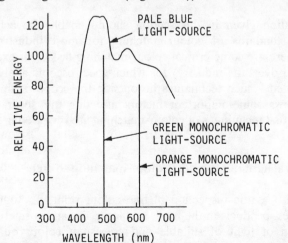

Figure 4.2 *The combination of light from the two monochromatic lightsources produces the same colour appearance (but not the same colour rendering) as the single pale-blue lightsource (after Pritchard[3]).*

been discovered that the combination of only three colours, blue violet, green and orange red, can produce the sensation of a satisfactory white illuminant with fair colour rendering. Thus it has been possible to develop the modern triphosphor fluorescent tubes which combine high efficacy with good colour properties. Such tubes provide a colour appearance which may be cool, intermediate or warm according to formulation.

The colour rendering of lamps may be measured and specified according to BS 1853[21], by 8-band or 6-band analysis or the CIE uniform chromaticity scale; such matters lie outside the scope of this book. What is important to note is that when people experience interiors and objects lit with these three-colour tubes, they may express a preference for them compared with other sources. They like the somewhat exaggerated colours; they prefer the effect on human skin, on meat, vegetables, fruit, flowers and foliage. Under triphosphor light, the complexions of white subjects may seem pinker and healthier; a picture of the sky is rendered bluer than the real sky we see. It is necessary to differentiate between colour-rendering and colour-preference (section 4.4.3).

Caution: purchasers should not allow themselves to be misled by the extravagant and unjustified claims made by certain unscrupulous organizations for the performance of fluorescent lamps with special phosphors or other features (section 8.5.1).

4.5   Colour matching

The information given in this section should enable the achievement of satisfactory standards of colour-matching for most industrial purposes. However, there are some critical colour-matching tasks (for example, in the chemical and dyestuffs industry) for which spectroscopic analysis of colour may be needed – such techniques lie outside the scope of this book. This section reviews some important factors affecting the accuracy of colour matching; actual techniques of colour-matching are reviewed in Section 10.5.

4.5.1   Natural light not the best colour-matching source

For tasks such as the inspection of fine colour print, photographic colour transparencies, products in the food, textiles, paint and cosmetics industries, the provision of light of suitable and consistent colour quality is vital. Contrary to popular opinion, natural light is a poor illuminant for fine colour-discrimination purposes, for it continuously changes in composition according to the season, the time of day and the weather. The spectral characteristics of daylight may not be the same even in two samples of light entering simultaneously through windows on opposite sides of a room.

Traditionally, critical colour tasks were performed in rooms lit only by north-sky daylight. For example, at one time, the grading of hops was deferred annually until the hop samples could be valued under 'a natural cold north light, with the sky unobstructed by heavy clouds or fog'. This involved storing the crop while waiting for these 'ideal conditions' before carrying out the valuation, for the 'wrong' light could considerably alter the appearance and thus the valuation of the crop. This had considerable economic effect, for the hop growers and dealers could not start trading until the grading had been carried out. In the mid-1960s, it was demonstrated that matching carried out under Colour-Matching/Northlight fluorescent tubes at an illuminance of 1500 lx produced reliable results. Modern lighting now produces excellent visual conditions for inspecting the hops, enabling consistent valuations to be made as soon as the hops are brought to the exchange, and appeals against valuations seldom occur.

4.5.2   Lighting affecting colour judgement

Many examples of how the colour quality of the lighting affects accuracy of matching may be given. Under suitable lighting, the fabrics buyer of a London dress house could easily identify twelve distinct shades of black – about twice as good as her performance under the best possible daylight conditions. After the provision of a high standard of inspection lighting, complaints about colour mismatches declined. Previously there had been complaints from clients, for example, that a single panel in a skirt 'appeared to be a different colour at the dance, though it looked all right at home'.

4.5.3   Requirements for good colourmatching

The following sections discuss the factors that should be considered in seeking to provide the correct conditions for accurate colour matching.

The light source

For critical colour-matching work, the choice will invariably be a fluorescent lamp (section 5.5). Note that lamps of higher colour quality tend to be of lower efficacy. Modern lamps based on the triphosphor technology (such as the Graphica 47) are now tending to replace the *BS 950*-based colour-matching lamps.

All lamps exhibit some degree of colour-shift and fall-off of lumen output as they age. To ensure constancy of colour property and minimum variation in illuminance, special care has to be taken in critical inspection processes. Lamp manufacturers usually recommend that lamps used to illuminate critical colour tasks should be bulk-replaced with new lamps (all from the same batch) at planned intervals of 4000 to 6000 hours.

Although these days sporadic early failures of lamps are rare, failure of one lamp at an inspection station using a large number of lamps could necessitate replacing all the lamps in order to ensure constancy of colour. Some users run a few extra lamps on the same circuit as the lightbox; then, if a replacement is required, one can be used from the same batch which has run exactly the same number of hours as those in the array.

Required illuminance for colour matching

The recommendations of the *CIBSE Code*[1] are a sound guide, and it should be noted that the illuminances recommended therein are minima. The eye cannot achieve its potential performance in colour discrimination unless it is light-adapted. For practical purposes, an illuminance of around 1000 lx is the minimum required. Provided glare can be limited, there is virtually no upper limit to the illuminance at which a subject may find visual comfort and best achievement, for we can adapt to sunshine illuminances of 80 000 to 100 000 lx.

In practical tests, a large proportion of subjects judged an illuminance of 2000 lx to be satisfactory (Figure 4.3), to which level our eyes become fully adapted very quickly. It was thought that the satisfaction of the test subjects stemmed from the greater clarity of vision that occurs at this order of

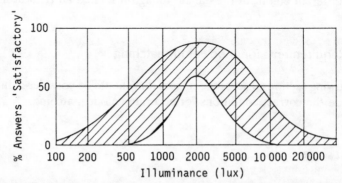

Figure 4.3 *Preferred illuminances for working interiors. Data from ten scientific investigations lie within the hatched band. In each investigation the subjects were asked to state if the illuminance in the test position was 'satisfactory' as judged visually by them. The recommendations of the CIBSE Code[1] take account of such preferences.*

magnitude of illuminance when the eyes register colours fully. It was also believed that the tail-off of those satisfied with illuminances above 2000 lx may have been at least partly due to increasing discomfort from glare under the test conditions.

Trials of lighting methods under industrial conditions do not usually produce convincing results (because of distractions to the experiment and difficulties in controlling conditions). However, trials of lighting for colour vision often indicate that somewhat higher illuminances than those recommended in the *CIBSE Code* do, in fact, bring about better performance of colour work. Some of the levels quoted in the 1984 edition of the *Code* are given in Table 5. If the task illuminance is required over only a relatively

Table 5 *Examples of recommended illuminaces for tasks involving colour discrimination*

Task or location	Standard Service Illuminance (lux)
Bakeries	
Decorating, icing	500
Boot and shoe factories	
Cutting tables and presses	1000
Carpets	
Inspection	1000
Dye works	
Dyehouse labs, dyers' offices	1000
Final inspection	1500
Furniture factories	
Veneer sorting and preparation	1000
Leather working	
Grading, matching	1500
Paint works	
Colour matching	1000
Printing works	
Printed sheet inspection, precision proofing, retouching, etching	1000
Colour reproduction and printing inspection – colour and registration	1500

The above examples are taken from the *CIBSE Code*[1] 1984 edition. The illuminances quoted are for the bench or working plane; in some cases, additional directional local lighting is also needed.

small area, the cost of providing an illuminance of the order of 3000 to 5000 lx is not great, and is an experiment worth trying if the task is critical.

Time for colour adaptation

Adaptation from one illuminance to another takes an appreciable time. If an inspector moves from an area of the factory lighted to a few hundred lux and enters an inspection area or inspection booth lighted to several thousand lux, he or she must allow time for the eyes to adapt to the higher illuminance before starting to exercise critical colour judgement. The time for adaptation varies with personal characteristics and age, and on the magnitudes of the first and second illuminances (Figure 4.4). A subject can only achieve potential performance when well advanced into light adaptation (photopic vision). There are detectable improvements in colour discrimination in some subjects (particularly older ones) up to 5000 or even 10 000 lx if glare is well controlled and if adequate time is allowed for adaptation.

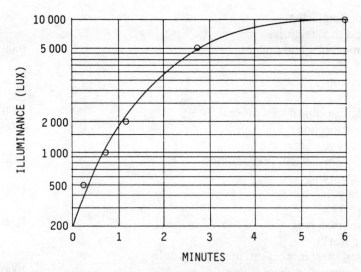

Figure 4.4 *Time to adapt to a higher illuminance. Subjects can only achieve their potential performance in colour discrimination when well advanced into light adaptation. Adaptation to 1000 lx probably represents the minimum standard for industrial colour matching. The time to adapt from a lower to a higher illuminance (from author's own experiments) are indicated on the curve. Thus, to adapt from 1000 lx to 5000 lx takes about 1¾ minutes, and from 500 lx to 5000 lx takes about 2¼ minutes).*

Avoiding colour-reduction

A simple experiment will demonstrate a factor that complicates colour-matching work, namely the phenomenon of *colour-reduction*. Place a small piece of neutral-grey paper or felt measuring about 10 mm square on a sheet of strongly coloured material, say bright red, which is illuminated to say 500 lx or more, and most observers will believe that the small grey sample is green in colour. Repeat the experiment, but this time placing the small grey sample on a green background sheet, and this time most observers will conclude that the grey sample is pinkish in colour. These results are due to the eye tending to 'see' the complementary colour of that to which it is adapted, that is an *after-image* is superimposed on the grey sample so that the illusion of it being coloured is created. The aberrations can be even more confusing. If the grey sample is replaced with a pale blue sample, on the red background the sample will appear to be green, while on the green background it will appear to be a pale tint of purple. If placed on a background of dark blue, it may appear to be white!

The experiments described are extreme examples, but the colour-reduction effect will be met in practical colour-matching work if the background to the visual task is so strongly coloured as to bring about a degree of adaptation to that colour. For that reason, in areas where fine colour work is performed, care has to be taken to restrain the strength of background colours. Plain white interiors may not be satisfactory to work in because of the 'bland field effect' (section 1.7.3). It is good practice to restrict the colours of objects and décor in colour-matching areas to colours not stronger than Munsell Chroma No.1.

Colour degradation of ambient light

In addition to the colour reduction effect just described, impairment of performance is likely to occur if the décor is so strongly coloured that light reflected from it becomes adulterated, even though such coloured surfaces are not within the visual field of the inspector.

Light going directly from overhead general lighting luminaires to the working plane is termed the *direct component*, that which arrives there after reflection from the room surfaces is termed the *indirect component*. In installations having a high indirect component, the light arriving at the working plane may be degraded as regards colour because of reflections, a matter of importance in colour-matching work. Luminaires of higher BZ classification will tend to produce a higher indirect component, and are therefore more likely to introduce some colour degradation of light reaching the working plane in a room having strongly-coloured décor.

In a case seen by the author, the results of matching fine colour prints were below the standard required, yet the light on the task (1000 lx from 'Artificial

Daylight' tubes to *BS 950* Part 1) was satisfactory, the benchtops were of suitable colour, and good practices were being observed. Investigation showed that the ambient light was being degraded by reflection from the pale green plastic floor tiles. In an experiment, the floor between the benches was covered with sheets of white paper, when the required standard of colour judgement immediately became possible. Incidentally, increasing the reflection factor of the floor greatly lightened the appearance of the walls and ceiling, and raised the illuminance on the bench tops by about 10%.

Use of diffusers

In another case investigated by the author, the standard of accuracy in the inspection of coloured plastic objects was not consistent, with the mysterious effect of different results being achieved from the same samples on different days or by different inspectors. This turned out to be due to specular reflections of the luminaires appearing on the surfaces of those colour samples which were more highly polished than others, the effect varying with the position of the test in relation to the overhead luminaires. Again, a simple experiment proved the theory right, for when the overhead bare-tube luminaires were temporarily covered with tracing paper to make them more diffusing, the inconsistency in results ceased. Later the luminaires were fitted with proper diffusers.

Lamp temperature

Another cause of inconsistency in colour performance that has been observed is that the colour performance of fluorescent lamps varies with temperature. In an attempt to overcome the problem, one user had stipulated that the lamps should be switched on 30 minutes before commencement of work, so the tubes could reach their normal operating temperature. But costly mistakes in the matching of screen-printed silks occurred, which on investigation were found to be due to the formation of a pocket of hot air below the ceiling (the luminaires being mounted higher than the window heads), resulting in the tube wall temperatures rising some 10°C with discernible colour shift. Improvement in ventilation, and lowering the luminaires 300 mm, overcame the problem completely.

Fluorescence of the inspected material

There may be difficulties in achieving perfect colour matching of samples because the UV content of natural light or colour-matching lightsources causes fluorescence of some component within the fabric of the examined material. For example, in the matching of paper or print there may be chemical substances in the inks or dressings in the paper that fluoresce under the UV component in daylight or from colour-matching lamps and thus bring about a change in the perceived colour. Examples of other materials that may exhibit this effect include banknotes, stamps, security documents, certain foodstuffs, fabrics and paints, as well as plastic coatings on sheet steel and anodic finishes on aluminium. The effect is particularly confusing when one is trying to match materials of the same nominal colour which contain chemically differing dyestuffs (as may occur in the paper, textiles and carpet industries). The difficulty may be reduced by the use of UV filters over the lamps.

Dichromaticity

A complication in devising lighting to suit difficult colour tasks is that many substances exhibit *dichromaticity*, i.e. they have different colour characteristics under different conditions of lighting and viewing. This may take place either in the transmitted or the reflected mode. For example, gold appears as its familiar colour when viewed by direct light, but a thin sheet of gold transmits only green light, and thus appears to be green. Liquids may also exhibit this phenomenon: bottles of some chemical solutions (and some wines) appear to change colour when viewed against a bright light.

An example of dichromaticity in the reflected mode is silverside of beef which looks mainly red when viewed normal to the surface of the cut, but when viewed at a glancing angle the 'silver' aspect is more noticeable. Certain printed papers, particularly material that has been lithographed, e.g. some postage stamps, appear to change colour when viewed at glancing angles. When a semi-specular material is examined at normal angles, its true colour dominates, but at flatter angles its appearance may be modified by colours reflected in it from other objects or from the lightsources. The effects of dichromaticity in reflection may be reduced or negated by the use of polarized light (section 2.9).

Part 2

Generation and control of light

Part 2

Operation and control of plant

5 Lamps

5.1 Introduction

This is a time of continuing and rapid development in the technology of electric lamps of all kinds. There is now such a wide variety of lamp types, which are available in various ratings and with differing performance features, that the specifier or user may have difficulty in choosing the most suitable type of lamp for the application. Buyers and specifiers of lamps in the various markets have differing requirements and objectives. They use the same lamps, but the criteria by which they judge the qualities of the lamps may differ.

Because some types of lamps are employed in a wide variety of installations, we can no longer talk about 'industrial' and 'commercial' lamps. The same metal-halide lamps and high-pressure sodium lamps are now used in such diverse applications as industrial interiors, shops, offices, eating places, outdoor area lighting, security lighting, building floodlighting and roadlighting.

As well as development in lamps, there is currently much progress in improving ballasts (control gear); for example, in ignitors which give rapid re-strike of high-intensity discharge (HID) lamps, and in high-frequency ballasts for fluorescent lamps. It is important to note that poor lighting performance or short lamp life will almost certainly result from attempting to use lamps on ballasts with which they are not fully compatible.

In general it is safe to use standard fluorescent tubular lamps from other makers in existing luminaires, but, because lamps are not completely standardized between makers, lamps from different suppliers having similar general description may not be identical; they may differ, for example, in lumen output, colour property, claimed life or restarting characteristics etc. In the case of HID lamps (especially metal-halide lamps), they may differ in their match to the control gear, which affects both life and performance. Thus, it is always a wise policy to relamp existing installations with lamps from the same maker as those originally supplied unless true compatibility of the alternative lamp to the circuit is guaranteed.

There have been improvements in HID lamps in recent times, and one learns that SON range lamps are giving a high degree of satisfaction to most users – in terms of consistency of output, life and lumen maintenance. As regards metal-halide lamps, reports from users still contain references to a marked change of colour of the light through the life of these lamps, so that sporadic replacements stand out. The message seems to have got through to users and installers that it is unwise to change to a different brand of lamp when relamping MBI/HQI installations.

The industry forecast is that high-frequency (HF) control gear will completely replace inductive gear within a few years. Initially available for standard tubular fluorescent lamps, HF control gear has now been developed also for HID lamps. With the falling costs of electronic circuitry, it is likely that 'pop-'em-in-the-socket' HID lamps will become available complete with integral throw away control gear, like some of the present compact fluorescent lamps.

The quality of UK lamps continues to rise. It will be in the interests of lamp makers, distributors, installers and users for all lamps to be produced to quality assured (*BS5750*/IEC9000) standards. Some recent allegations about poor quality lamps have proved to be unjustified. For example, a major lighting company put out a helpful warning about some lamps of the 2D type which had been assembled in China. Another lamp maker was recently asked to exchange some defective HID lamps, but these proved to be imitations of their branded lamps, imported from the Far East.

The types of lamps used in industrial interior and exterior lighting installations are discussed in the following sections of this chapter, each section concluding with an indication of the efficacy and salient features of the particular type of lamp.

5.2   Incandescent lamps (GLS)

Incandescent (tungsten-filament) lamps in common use are known as *general lighting service* (GLS) lamps. Despite the availability of low-voltage tungsten–halogen (LVTH) lamps (section 5.3) which are superseding pressed glass aluminized reflector (PAR) lamps and mushroom reflector lamps, and the popularity of compact fluorescent lamps (section 5.6), GLS lamps are still widely used for home lighting, and sales of decorative filament lamps such as spherical lamps and candle lamps are also stable or increasing. However, the use of incandescent lamps (other than LVTH) in industry is small, and declining.

5.2.1   Construction

Filament lamps operate by the passage of electric current through a filament of tungsten, the temperature of which is raised to incandescence. Figure 5.1 shows the relationship of the lumen output and life to the applied voltage. With efficacies in the region of only 10–12 lm/W and a life of only 1000 h,

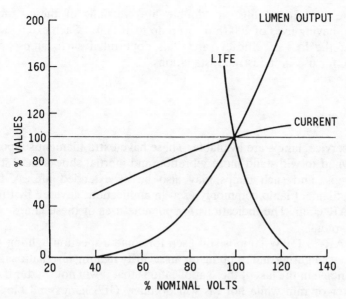

Figure 5.1 *Electrical characteristics of a tungsten-filament lamp.*

these lamps do not have an important place in lighting for industry, commerce or security. Their low efficacy and short life results in a high cost-in-use and high maintenance costs. The filament in an incandescent lamp may be of single-coil or coiled-coil construction, the latter yielding slightly better efficacy.

5.2.2 Operation

Tungsten-filament lamps will operate on a.c. or d.c., and are unaffected by variations in frequency of supply, provided the periodicity is not so low that visible flicker is caused. At 50 Hz the flicker effect is generally imperceptible, and is less in the case of higher powered lamps and lamps designed for operation at reduced voltages. In general, lamps designed for lower voltages are more robust than lamps of equal power designed for a higher voltage. In both cases, the filament is thicker, and thus the thermal inertia is higher, i.e. the lamp has a lower 'negrescence'. All tungsten-filament lamps have the characteristic of instantly relighting on restoration of an interrupted supply.

5.2.3 Life

GLS lamps fail catastrophically at end of life. The 1000-hour life is the average per batch, i.e. some lamps fail before this time, some survive longer.

'Long life', 'extended life' or 'double life' versions of these lamps are available, having lives of 2000 h or even up to 16 000 h. Such extended life is gained at the loss of efficacy, and it is doubtful if such lamps are ever economic for use in industrial installations.

5.2.4 Types

'Rough service' lamps are available. These have extra filament supports and are designed to withstand more vibration and inertial shocks than standard GLS lamps, and such lamps may also be of extended life. A type of tungsten-filament lamp commonly used in applications covered by this book is the PAR lamp. The indications for possible uses of these lamps occur in other chapters.

The PAR-38 150 W lamp is still used for some inspection lighting and for outdoor task lighting and 'topping up' in security lighting installations, having the advantage in the last of these applications that it will not shatter if doused with water or rain while hot (as will ordinary GLS lamps and blown-bulb reflectorized lamps). In task lighting, the PAR-38 lamp is preferably used at 110 V or 100 V for increased electrical safety. For task lighting and display lighting applications, the PAR lamp is gradually being replaced by LVTH lamps (section 5.3) and compact fluorescent lamps (section 5.6). See also the reference to 'Electronic PAR lamps' at the end of section 5.3.2.

5.2.5 Economics

In comparing the economics of tungsten-filament lamps with other types of lamps, the high running cost due to low efficacy and the need for frequent replacement of lamps are only partly offset by the low cost of the lamps and the fact that no control gear is employed. However, even with the availability of compact fluorescent lamps (section 5.6), GLS lamps still might be preferred in a situation in which usage did not exceed a very few hours per week.

5.2.6 Summary

GLS lamps are available in powers of 40 to 2000 W, and have efficacies in the range 10 to 18 lm/W. Standard life rating is 1000 h, but other ratings are available. Restrike is instant. The smaller wattage lamps are useful for topping-up purposes, but short life and low efficacy makes them unsuitable for general use.

5.3   Low-voltage tungsten-halogen lamps (LVTH)

5.3.1   The halogen cycle

Tungsten–halogen (TH) lamps are filament (incandescent) lamps of special construction, consisting of a filament in a tube of fused silica or quartz, with a halogen gas filling.

In all filament lamps, molecules of tungsten boil off the filament in use, and this results in the thinning and eventual failure of the filament, plus blackening of the envelope. In TH lamps, the errant molecules of tungsten have an affinity for the halogen molecules in the gas filling and form into tungsten-halogen molecules which are attracted back to the filament, the elevated temperature of which causes the tungsten and halogen elements to separate; this results in the halogen molecule returning to the gas mass, while the tungsten molecule reforms as part of the filament.

If the tungsten molecules could be persuaded to return to their points of origin, the filament would remain substantially uniform in thickness and thus would have infinite life. In fact, the tungsten molecules return at random, so that eventually there is some thinning of the filament and the lamp fails. The effect of the halogen dosing is beneficial in that it enables the filament to be operated at a higher temperature than that of GLS lamps, resulting in a higher efficacy, while the rated life is extended by a factor of 2 to 5 according to the type of lamp.

As for all filament lamps, the light output of TH lamps relates to the applied voltage, and therefore they can be dimmed. However, if they are dimmed to the point where the filament temperature falls so that the halogen cycle does not operate fully, TH lamps will soon blacken and will fail prematurely.

5.3.2   Low-voltage tungsten-halogen (LVTH) lamps

Miniature LVTH lamps with dichroic reflectors are being put to work in many new applications. As well as being employed for display lighting and for shop and office installations, they are now used in some technical applications. The concentrated beams from these small lamps enable illuminances in excess of 5000 lx to be directed to small areas or into cavities, as may be required in visually demanding tasks and for engineering inspection (chapter 10). Low-voltage track systems for these lamps may provide a neat and safe way of applying them in some industrial applications.

Their compact size, and the controlled light beams from these lamps, make them suitable for some localized lighting tasks, particularly where the lamps must be mounted well away from the actual task to reduce the amount of heat introduced to the work area and to reduce risk of lamp breakage (section

9.4). In industrial clean-rooms, the small size of LVTH lamps enables them to be housed in streamlined luminaires which do not disturb the laminar air flow.

There is a wide variety of powers and designs of LVTH lamps. Generally, those types with axially disposed filaments provide better accuracy and symmetry of beam. Lamp patterns are available with either open or sealed dichroic reflectors, and with front glasses which may be either plain or optically configured, lenticular or diffusing, with coloured front glasses or coloured dichroic reflectors, or arranged to filter out specific wavelengths – especially UV.

These lamps are generally fitted with standard GU5.3 bases in which the lamp is connected and retained by its two pins. Also available is a 'twist and lock' base (Thorn patent) designated GU7, which has robust pins which easily find their keyhole entries and offer better electrical reliability.

There are currently many developments taking place in the design of LVTH lamps, resulting in better optical control by virtue of improved reflectors and front glasses. Other recent technical improvements include a cooler pinch where the conductors pass through the quartz envelope, and improved lens coatings and dichroic reflectors which reduce the amount of heat transmitted in the beam.

Transformers for operating LVTH lamps may be integral with the luminaire, or may be mounted externally. Some track systems incorporate the transformer. Transformers may be of the inductive type, or may employ solid-state technology. LVTH lamps are sensitive to voltage which greatly affects their life. Where a group of lamps is operated from a single transformer, if the voltage regulation of the transformer is not accurate, failure of any lamp will cause a higher voltage to be applied to the remaining lamps and may lead to their premature failure. Lamp life may be shortened by frequent switching, for a tungsten filament has a slightly lower resistance when cold, resulting in an inrush current which will in time deteriorate the filament. There are 'lamp conserver' devices available which are claimed to overcome this effect and result in longer lamp life, but the cost of these, added to the cost of the normal transformer, may be greater than the cost of a solid-state transformer.

Solid-state transformers generally offer better voltage regulation with variation in load, they have more stable output with variation in mains voltage, and they have the capability of limiting inrush current to a cold lamp. Their capital cost may be justified by the better lamp life and the lower maintenance costs that result from their use.

There are plenty of attractive and efficient luminaires for applying these lamps in display and decorative lighting. The problem for users may be to find robust luminaires for these lamps which have enclosures giving the degree of protection necessary to suit the growing number of industrial applications. One can visualize applications of these lamps for tasks in flammable atmospheres, but there are apparently no luminaires designed for them with suitably rated enclosures. Perhaps we shall soon see a new generation of luminaires for these lamps with IP65 protection and for Zone 2 atmospheres.

The integral reflectors of these lamps comprise multiple layers of dichroic material which enable about two-thirds of the lamp heat to be radiated out of the back of the lamp, and the safe dissipation of this heat must be ensured. The bulb wall temperature of LVTH lamps is high, and therefore they may not be suitable for use in luminaires designed for other sources. Care must be exercised in the use of any type of lamp with a directable beam, including all types of LVTH lamps, for the beam is hot enough to ignite flammable materials, especially at short beam throws (for a discussion of fire risks from luminaires, see section 6.5.6).

One form of LVTH lamp is the so-called 'Electronic PAR lamp', which is not a PAR lamp at all, but is a LVTH lamp in a PAR-lamp enclosure, complete with its integral electronic transformer.

5.3.3 Summary

LVTH reflector lamps are available in powers of 35, 50 and 75 W, and the usual claimed life is 2000 h. The lamps have considerable potential for industrial use in special applications of 'remote local lighting' and inspection, but few suitable types of luminaires for industrial use are yet available in the UK.

5.4   Linear tungsten-halogen lamps (TH)

5.4.1   Mains-voltage linear tungsten–halogen (TH) lamps also operate on the 'halogen cycle' described in section 5.3.1. For the halogen cycle to work, linear TH lamps must be operated within 4° of the horizontal, or the halogen will migrate to one end of the lamp and early failure will result. The envelopes of these lamps should not be touched with the ungloved hand, or fats from the skin will migrate into the quartz and cause blistering of the envelope and early lamp failure.

The critical part of all TH lamps is the 'pinch' or seal where the conductors pass through the quartz, and this must be kept below a limiting temperature to prevent early failure. Temperature regulation of the pinch is achieved by the thermal conductance properties of the lampholders and the heat dissipation characteristics of the luminaire body.

Linear TH lamps are used mainly for exterior applications. There are few indoor applications for these lamps apart from lighting barns and farm outbuildings, i.e. for lighting which is used only for a few hours per year. A major market for the larger TH lamps is for temporary lighting on construction sites (section 16.1).

Lighting employing TH lamps tends to have low capital outlay, but incurs a greater energy bill than would result from using HID lamps. TH lamps are durable, not very sensitive to vibration, and need no control-gear.

Low-wattage TH floodlights are widely employed for small exterior lighting installations where the hours of use are short and where good directional control of light is required. Because they come to full light output immediately on being switched on, they are suitable for domestic security-lighting applications where occupancy-sensors are employed (chapter 17). In such applications they have the advantages of providing a greater lumen package than do compact fluorescents, and of having strong directional qualities. Their linear form, when used in suitable reflectors, enables the creation of wide flat-topped beams which are required for many outdoor area lighting and floodlighting applications (chapter 15), and for security lighting (chapter 17).

5.4.2   Summary

TH linear lamps are available in powers of 200 to 2000 W, with efficacies in the range 16 to 22 lm/W. Their life is 2000 h, and restrike is instant. It is important that these lamps are operated horizontally or short life results. They should be used for short- to medium-distance floodlighting. Their relatively low efficacy makes them unsuitable for large-scale use.

5.5   Fluorescent tubular lamps (MCF)

5.5.1   Description

A fluorescent tubular lamp consists of a sealed glass tube (the *envelope*) with an electrode and a heater filament at each end. Each heater filament is connected to a two-pin lamp-cap for external connection. The envelope is filled with a low pressure inert gas containing mercury. The inner wall of the envelope is coated with a chemical powder (the *phosphor coating*) which has the property of fluorescing when subjected to ultraviolet radiation. In operation, a diffuse arc is struck through the gas between the electrodes. In theory, the electrical resistance of an arc diminishes with increasing current until burnout, so an impedance must be connected in series with the lamp. In use, the lamp is connected to an external control circuit or *ballast* which, in conventional or 'inductive' circuits, consists of an iron-cored coil inductance (the *choke*) and a starting device. The circuit usually contains a capacitor for power factor correction.

5.5.2 Starting devices

The starting device may be a thermal starter switch. This is a thermal-delay switch which, after a short time to allow the heater filaments to heat the electrodes to emission temperature, opens to cause an inductive voltage pulse which triggers the lamp to start.

Starting methods that are now becoming obsolete include the use of an inductive device to create a voltage pulse to start the lamp without a starter-switch (*leaky transformer*), and resonant transformer circuits.

Thermal starter canisters generally have a technical life that is longer than one lamp life but less than the luminaire life, and therefore must be replaced periodically. Electronic starter canisters are available (e.g. the Thorn 'Fluoropulse') which last the life of the luminaire, and automatically adjust the cathode pre-heat time to the lamp power. These have a cutout to prevent a faulty lamp 'cycling' (flashing on repeated attempts to start).

Solid state control gear consists of an electronic circuit which produces an analogue of the performance of a choke and starting device, and this is operated at high frequency (typically 28/40 kHz). It seems likely that such *high-frequency ballasts* will entirely supplant other ballasts in the next few years.

5.5.3 Operation

Operation of a fluorescent tubular lamp is initiated by a small current that passes through the heater elements. The elements are coated with a rare earth and, when their temperature is raised, ions 'boil off' and form a cloud around the electrodes. After a brief heating period, the starting device operates. In the case of a thermal starter switch, this opens and reduces the current flowing in the choke, with the result that there is a collapse of the magnetic flux of the choke. The collapsing flux cuts the turns of the choke, inducing into them an electromotive force, with the result that a pulse of higher voltage is impressed across the electrodes. The voltage pulse is sufficient to break down the resistance of the gap between the electrodes and drive ions through the mercury vapour so that a diffuse arc is established. The arc is rich in ultraviolet energy, and this irradiates the phosphor coating within the envelope. It is from the fluorescence of the phosphor coating that practically all the luminance of the lamp emanates.

5.5.4 Summary

Fluorescent tubular lamps are manufactured in a wide range of phosphor colours and in a range of powers from 8 W to 125 W. Efficacies range from about 20 to 75 lm/W for the high-efficacy colours, and somewhat lower for the better colour-rendering colours. The life of lamps of length 1.2 m and longer is 7500 h, and of shorter ones is 5000 h. Restrike is instant. The lamps are suitable for interior general lighting with mounting heights up to about 5 m. They have limited applications in outdoor lighting; cold conditions can significantly reduce light output.

5.6   Compact fluorescent lamps (CF)

5.6.1   Types

It is not surprising that some users find the current plethora of types of compact fluorescent (CF) lamps confusing. Currently we have several series of CF lamps, including a type designated type L; these are either 2-pin 2-arm or 4-pin 2-arm. There is also a 2-pin 4-arm version which is designated Type PLC. There are the self-ballasted lamps, for example, the type SL (jam-jar shape) and type SLD (spherical), and the EL (electronic ballasted) lamps. The GE-Thorn range of '2D' compact fluorescent lamps have a unique configuration, and are supported by an appropriate range of Thorn luminaires. This product range now has competition from 'flat configuration' 4-arm lamps which have recently been introduced.

This is a fast-developing family of CF lamps. A reflectorized version of the EL lamp appeared recently. An overlamp reflector to use with type PLC (4-arm) compact fluorescent lamps is available. A range of coloured PL lamps has been launched which, although apparently intended for the leisure market, might be usefully employed in industrial situations for warning and signal applications.

The electronic gear package that is part of type EL lamps is normally discarded with the lamp at the end of life, but relampable versions are coming on the market, in which the gear may serve during four or more lamp lives.

5.6.2   Applications

When CF lamps first appeared, there was much emphasis on their use for relamping existing GLS luminaires to make immediate savings in current costs, lamp replacement costs and maintenance cost. It must be stated that not all types of luminaires designed for GLS lamps are suitable for use with all types of compact fluorescent lamps. Problems of overheating may occur, and relamping with a lightsource of such different configuration may change the light distribution significantly. Now such a wide range of luminaires designed for CF lamps is available, the marketing emphasis is on replacement of existing GLS luminaires with new ones designed specifically for CF lamps, rather than relamping existing equipment.

CF lamps find many applications in industry, for example, for local and localized task lighting, and for corridors and staircases. Out of doors, they are invaluable for lighting paths, and for marking entrances (chapter 15), as well as for 'topping up' shadowed areas in security lighting installations (chapter 17). However they are not bright enough to be used effectively as perimeter glare-lights.

5.6.3 Power factor

Currently, most CF lamp circuits operating on high frequency ballasts have a rather low power factor, typically 0.5, due mainly to waveform distortion. Where CF lamps are used to replace GLS lamps, there is a gain of about a factor of five in lumen output and, as CF lamps may be expected to form only a very small part of the load of typical industrial premises, their effect on the total power factor (and the percentage of wattless current for the premises – which affects the maximum demand charge) will usually be negligible. However, it is understood that new solid-state circuitry is becoming available which will yield a far better power factor, and this will be of benefit in cases where CF lamps form a substantial part of a consumer's load. Domestic users (who benefit financially from replacing their GLS lamps with CF lamps) are not affected at all by the pf, but it will be in the national interest if ballasts with better power factor come into general use to reduce the wattless current in the distribution system, and to reduce waveform distortion.

5.6.4 Summary

CF lamps have a power range of 7–35 W, an efficacy around 50 lm/W and a life 5000 h. Restrike is immediate. Interior applications include lighting small areas, stairs etc. and exterior applications include bulkhead luminaires, topping up small areas etc.

5.7 Mercury-vapour lamps (MBF)

Type MBF lamps are not now often specified for new interior installations, and many old installations of these lamps have been relamped with high pressure sodium (SON) lamps (section 5.10). There are still substantial sales of MBF lamps for replacements in old installations, but they are being overtaken by other sources such as MBI lamps (section 5.8) which emit better light colour and are more economical. Some users prefer MBF for particular installations of building floodlighting and exterior area lighting, but it is believed that MBF lamps will cease to be specified for any schemes in the next few years. For security lighting applications they are generally superseded by MBI lamps, but the 50 W MBF is still sometimes employed for low-mounted glare lights (section 17.3.7).

The mercury-tungsten 'dual' (MBT) lamp has already become a museum piece, and is no longer specified. In this type of lamp, a tungsten filament acted as the ballast to a mercury discharge capsule. Its main use was as a substitute for high-wattage GLS lamps in floodlights in existing installations but, as its light centre position differed slightly from that of a GLS filament, some distortion of the distribution resulted.

5.7.1   Construction

High-pressure mercury-vapour (MBF) lamps differ from fluorescent tubular lamps (section 5.5) in that the greater pressure and temperature of the mercury-vapour arc necessitates its enclosure within an arc-tube of quartz or similar material. Ultraviolet radiation from the arc passes through the quartz envelope and irradiates the phosphor coating on the inner surface of the outer glass envelope. The UV radiation does not pass through the glass envelope in any harmful quantity. The electrical characteristics of this type of lamp also differ from fluorescent tubular lamps in that the electrodes require no heating filaments. These lamps are available in elliptical bulb shape (MBF) and reflector shape (MBFR).

5.7.2   Operation

In operation, after the arc-tube strikes (i.e. current begins to flow through it), but before the lamp reaches sufficient temperature and gas pressure, the emission of UV is limited, and the luminance of the lamp is low and its colour rather blue. After about 6 minutes the lamp reaches design temperature and full light output. If extinguished and then switched on again, the lamp will not restrike until it has cooled sufficiently; it then must 'run up' again – a total process taking up to 10 minutes. The luminous arc-tube contributes to the light output of the lamp. The 'warm' light output from the phosphor coating renders the combined light output rather less blue/green.

5.7.3   Summary

MBF lamps have a power range of 50–2000 W, an efficacy 30/55 lm/W and a life of 7500 h. Restrike takes 5 to 10 minutes. The capital cost of the lamp and circuit is lower than for MBI, but MBF is being superseded by MBI which has lower operating cost because of higher efficacy and better colour performance.

5.8 Metal-halide lamps (MBI series)

Metal-halide lamps are increasingly being used in industrial high-bay and low-bay installations, and are frequently specified for office applications, both in uplighters and in direct luminaires. MBI lamps are supplanting MBF lamps for many outdoor applications including industrial area lighting (where their good colour-rendering facilitates the picking of colour-coded materials), and for security lighting (where their good colour-rendering makes distant intruders more easily visible against camouflaging backgrounds).

Currently there seems to be some uncertainty regarding the lumen ratings of MBI/HQI lamps claimed by some manufacturers. Outputs are substantially affected by the lamp operating temperature.

5.8.1 Types

These lamps are available in elliptical envelopes (MBI), with phosphor coating (MBIF), in reflector form (MBIR), and in linear form (MBIL). The general construction of this series of lamps is similar to that of MBF lamps (section 5.7) except that the arc-tube is dosed with a halogen substance which has the effect of raising the luminous efficacy of the lamp and also significantly improving its colour performance. In some versions, additional colour correction and further improvement in efficacy is gained by the use of a phosphor coating inside the outer envelope. A development of metal-halide lamps is the *compact-source metal-halide lamp* type CSI, which operates at higher colour-temperature than other metal-halide lamps.

The electrical characteristics of MBI lamps are generally similar to those of MBF lamps except that it is necessary to operate MBI series lamps only with the specific ballast and ignitor recommended by the lamp maker.

5.8.2 Applications

In the UK, elliptical bulb metal-halide lamps are being used for exterior floodlighting and area lighting, as well as being widely employed for industrial general interior lighting. For both interior and exterior lighting, the linear lamp form (MBIL) is an alternative lamp with different colour properties to the SON-T lamp (section 5.10) for use in certain types of luminaires for which reflector systems for linear lamps have been designed. Type CSI lamps are specified for applications where very good colour rendering is required, and have been extensively used for television lighting for sportsfields. They are now used in industrial situations where colour closed-circuit TV links are installed.

5.8.3 Summary

The MBI series of metal-halide lamps is available in powers of 75–2000 W. Efficacy is 50–85 lm/W and life is nominally 7500 h depending on conditions of use. Restrike, if fitted with igniter circuit, occurs within one minute, and with other circuits, within 4 to 5 minutes. Good colour rendering and cool colour appearance.

5.9 Low-pressure sodium-vapour lamps (SOX)

5.9.1 Description

Lamps of the SOX type have an arc-tube containing liquid sodium which, after an initiating discharge through neon gas has warmed the lamp and vaporized the sodium, causes the arc to take on its characteristic yellow colour. The emitted light is virtually monochromatic yellow, without colour-rendering property. The run-up and restrike characteristics of low-pressure sodium-vapour lamps are similar to those of other HID lamps.

Low-pressure sodium-vapour lamps are extensively used for lighting motorways and main traffic routes in some countries (and, regrettably, in the UK), the preference for them by specifiers being based on their high efficacy and the belief – apparently not founded on fact – that yellow light penetrates fog better than does white light.

Efficacies quoted for LP sodium lamps are very high, but many lighting engineers question whether lumens of monochromatic light can be equated with lumens of substantially white light as regards visual performance. It is believed that lightmeters designed for use with substantially white light produce misleadingly high readings when used to measure yellow near-monochromatic light which does not match the needs of the human eye. The efficacy figures themselves are considered to be of doubtful accuracy, as it is known that these lamps tend to consume more energy as they age – a point noted by electricity supply companies who provide unmetered supplies to public lighting installations.

5.9.2 Caution

The near-monochromatic light from low-pressure sodium vapour lamps is considered to be an unnatural and unsuitable form of lighting for workplaces, and these lamps are not recommended by the author for any indoor or outdoor industrial lighting, nor for security lighting. The justifications for this statement are as follows:

- The colours of warning, prohibitory, mandatory and advisory signs cannot be recognized, nor can the colour marks on materials (e.g. steel stocks) which normally serve to identify them.
- It is believed that working in monochromatic light for long periods is more tiring than working in substantially white light, and that judgement of distance and speed is less certain in monochromatic light.
- The unfortunate rendering of skin tones can be depressing for morale – especially of female workers.
- When used for car-park lighting, motorists returning to their cars after dark cannot recognize their cars by colour.
- In public lighting and security lighting, a person seen under near monochromatic yellow light (say, at the scene of a crime, or at a road traffic accident) cannot be accurately described by a witness, who will be unable to state the skin colour, hair colour or colour of clothes of any person they have seen, nor the colour of any vehicle.
- In security lighting, tests have shown that light from these lamps does not give the clarity required to spot a dark-clothed person against a dark background at low light level. Low-pressure sodium lamps are barely satisfactory for security lighting in urban situations involving short distance vision, and are totally unsatisfactory in other situations where the background consists of trees, grass etc.

For the foregoing reasons these lamps are not recommended for any industrial application.

5.10 High-pressure sodium-vapour lamps (SON)

5.10.1 Although the lamps in this series (SON, SON-T, SON-R etc) share the same activating metal in the arc as low-pressure sodium-vapour lamps (SOX) (section 5.9), they are very different in performance. The arc-tube is operated at higher pressure and temperature, and may be 'dosed' with other elements. These lamps do not emit near-monochromatic light, but give a pleasant golden-hued light of quite good colour-rendering which is satisfactory for many indoor and outdoor industrial tasks where fine colour discrimination is not required. The modern 'White SON' (SONDL) is superior to the MBF series for general industrial use, a serious competitor to the MBI series (having been used successfully for lighting canteens), and has proved to be a most satisfactory source for security lighting installations. A high-output version (SONP) is available.

5.10.2   As for other HID lamps, SON series lamps must be operated with an iron-cored ballast in series (or a solid-state HF control gear) to provide the right starting conditions and to limit the arc current. The arc-tube is incandescent in use, its emanations contributing to the light output. The general electrical properties are the same as for other HID lamps, but SON series lamps also require an ignitor circuit to inject a high-voltage/high-frequency pulse during starting. Use of an ignitor enables the lamp to restrike quickly after being switched off. The usual form of this lamp is linear, and both single and double-ended lamps are made, and a reflector-bulb version is also available.

5.10.3   Summary

High-pressure sodium-vapour lamps have a power range of 50–1000 W, and efficacy of 55–110 lm/W. A life for most ratings is 8000 h or more; claimed life for special ratings is 10 000 h. The lamps restrike with an ignitor in less than one minute; with other circuits, 5 minutes or longer. In one form the lamp may be switched on and off with instant restriking and – in sophisticated circuits – may be dimmed. This range of lamps is versatile and will satisfy many industrial interior and exterior lighting needs where high colour discrimination is not required.

5.11   Special lamps

The commonly used lamps for interior and exterior lighting applications are discussed in the preceding sections of this chapter. For a few applications it may be necessary to consider the use of 'special' lamps including the following.

5.11.1   Xenon lamps

These are high-pressure high-intensity discharge lamps employing the rare gas xenon in the arc-tube. The high efficacy and instant full output gained by electronic HF control gear is usually associated with xenon filling. Xenon tends to raise the arc colour-temperature, a xenon discharge being of blue/white of very good colour-rendering. The value of xenon is realized in lamps emitting very high lumen outputs (say over 250 000 lumens per lamp), or where it is desired to flash the lamp for very brief intervals (say down to a few milliseconds), or come to full light output virtually instantly on switching on.

Xenon lamps are used in special applications such as helicopter-borne floodlighting, and for signalling purposes.

5.11.2 Krypton lamps

Krypton is a rare gas that is used as a gas filling for small-power tungsten-filament lamps, which enables a high filament temperature and thus a high lumen output to be achieved with long lamp life.

Krypton-filled filament lamps are used in miners' cap lamps, for signalling lamps, and for some emergency-lighting luminaires.

5.11.3 Neon lamps

Neon arc discharge has a characteristic red colour. As well as being used as the 'starter gas' in fluorescent tubes and discharge lamps, neon is used as a gas filling in very low-power glow-discharge indicator lamps. In association with high-voltage high-frequency pulse generators, large-power neon lamps can be flashed to full brilliance for very short periods, e.g. for signalling purposes as in the long-distance signalling beacons used on airfields (chapter 20).

5.12 Induction lamps (QL)

5.12.1   A new development in lamp technology is the type QL electrodeless *induction lamp* which has been the subject of research by a number of lighting companies for some years. The induction lamp employs a scientific principle discovered by J.J. Thompson about 1900, and differs from other types of discharge lamps in that it has no electrodes. In this type of lamp, a high-frequency (2.65 MHz) oscillating magnetic field is induced in a low-pressure metallic gas within a lamp envelope by means of an induction coil (Figure 5.2). The magnetic field emanates from an antenna which comprises a coil wound on a ferrite rod. The magnetic field ionizes the metallic gas (usually mercury), causing its molecules to emit ultraviolet energy which, as in conventional discharge lamps, is converted to visible light by a layer of phosphor lining the inside of the envelope.

Problems in the development of this lamp have been in the miniaturization and cheapening the electronics, in dealing with the heat that is generated in the lamp, and in dealing with the risk of radiofrequency interference. Patents exist for methods of piping the heat within the lamp to enable its dissipation.

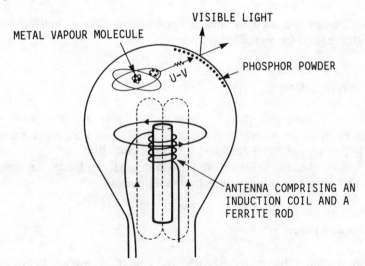

Figure 5.2 *Type 'QL' induction lamp in which the phosphor powder is irradiated with UV to produce visible light. The UV is generated by activation of metal-vapour molecules by magnetic energy from the antenna. (Acknowledgement to Philips Lighting).*

5.12.2   The life of induction lamps is likely to be determined by the durability of the electronic driver rather than of the lamp itself. As they have neither filament nor electrodes (the components which largely determine lamp life in other types of lamp) it is expected that induction lamps will have a technical life up to six times longer than that of other discharge lamps. Lamp life of up to 60 000 h is forecast, over which time the depreciation of lumen output may be of the order of 20 or 30%.

Induction lamps are likely to be around four times the cost of equivalent metal halide lamps, but in typical applications, relamping may not be needed more frequently than every eight years. It is visualized that induction lamps will find their applications in situations where relamping is costly or difficult, e.g. in areas where hazardous atmospheric conditions are constantly present, or where it is necessary to shut down major plant in order to gain access to the luminaires.

The question of electromagnetic compatibility of the QL lamp has necessitated a great deal of research. Equipment such as the lamp system can give rise to radiated electromagnetic fields as well as conducted (mains borne) interference. Simple techniques are claimed to have been adopted to reduce the conducted interference, but there still seems to be some concern regarding radiated interference. Metallic luminaires provide some shielding, but it is believed that screening of the lamp by earthed metal mesh may be necessary, particularly in applications where sensitive electronic, radio or computer equipment is in use, and in television and recording studios, laboratories etc.

The first lamp of this type, launched by Philips Lighting in 1991, is an 85-W 5500-lumen pattern, available in warm and cool 'white' colours derived from triphosphor fluorescent powders. According to press reports (June 1992), a pattern of induction lamp has recently been developed in the USA by Intersource Technologies of Sunnyvale, California, in collaboration with Diablo Research and the American Electric Power Company. This lamp, called the 'E-lamp', appears to have the same salient features of construction as described above, and is said to have a projected life of 10 to 20 years. It is understood that 'E-lamps' will be available in the USA and UK in 1993 at a price of $10 to $20 each. Other companies believed to be in the race to capture a share of this important new lamp market include Matsushita, GE-Thorn and GTE-Sylvania.

5.12.3 Summary

The advantages of induction lamps can be summarized as follows:

- Virtually instantaneous starting (less than 0.5 s) at full light output;
- starting characteristic maintained hot and cold (down to -20°C);
- may be dimmed electronically;
- physically small;
- efficacy of smaller powers comparable with compact fluorescent lamps, and may compete with high-efficacy metal halide lamps in larger powers;
- any required colour characteristic may be produced by use of appropriate phosphors;
- light output colour does not change when dimmed;
- no discernable flicker or stroboscopic effect;
- constant light output during mains fluctuations;
- claimed to comply with current EEC limits for electrical interference if screened and earthed to the maker's instructions;
- long technical life.

5.13 Cold light generation

There are a number of ways of generating light without generating heat. These technologies may become of greater importance in the future.

5.13.1   Chemiluminescence

Chemiluminescence is the production of light by chemical action. A useful application of this principle is the range of 'Cyalume'® lightstick chemical lights available from Cyanamid of Great Britain Ltd (appendix G). These are used for warning purposes in emergencies, and as signal flares for distressed persons in the sea, or on mountains etc. They can serve an important role by providing a small amount of light for the safety and comfort of persons trapped in the dark, e.g. entombed miners or tunnellers, or submariners marooned on the ocean bed.

Each chemical light is a sealed plastic tube about 150 mm long and about 12 mm in diameter, which is filled with a liquid. In the liquid there is a small sealed glass tube containing another liquid. To activate the unit, the outer tube is slightly bent by hand, fracturing the inner glass tube. This allows the two liquids to mix, and react chemically so that light is emitted without heat. Development of light output is accelerated by shaking the device. Light output is slightly increased by warming the device. No heat is produced and the unit is sealed, so it can be used in the presence of flammable gases and can be operated under water. The chemicals used are nontoxic.

A range of patterns is available, in colours (white, green, orange, red and blue), and various durations and outputs (12 h, 8 h, 30-minute Hi-Intensity, and 5-minute Ultra Hi-Intensity).

The light produced is sufficient to enable emergency work to be done in difficult conditions, including in hazardous atmospheres. Because the devices are small and light in weight, they are easily carried in the pocket for emergency use. It is recommended that they should be carried routinely by divers, oil rig workers, and personnel who must enter drains, tunnels, boilers, shafts etc, where a little light could save a life.

5.13.2   Phosphorescence

Phosphorescence is the production of light from a chemical substance which has been activated by light or other radiation. The common example is the material used for luminous dials and watch faces, and similar materials may be applied to switches and exit signs to enable them to be seen in the dark.

Another application is the use of chemicals which contain a radioactive isotope to activate a chemical substance and cause it to emit light. Exit signs and lane markers etc using this principle are sealed, and can be used in wet conditions and in the presence of flammable gases. They have half-lives measured in years, and will continue to be luminous without having to be exposed to light.

5.13.3 Electroluminescence

Electroluminescence is the production of light in a crystal or other substance when subjected to electrical stress. So far, the idea has had only limited application as in signs in aircraft. There is the possibility that research will reveal ways of making electroluminescence panels produce enough light to be used as practical light sources.

Light emitting diodes (LEDs) work on a similar principle, and these are extensively used as indicators on control equipment, e.g. as charge indicators and 'mains healthy' indicators on emergency lighting luminaires.

5.13.4 Fibre optics

The phenomenon of 'piping of light' is discussed in section 10.3. Fibre optic devices are not light sources in themselves, but are used to deliver light from a remote lamp to where it is needed, without any accompanying heat.

This technology is developing rapidly, and it is possible that quite soon we shall see buildings in which light is piped from a heliostat (a light-collecting mirror system that follows the sun's arc) to possibly hundreds or thousands of points of use throughout the building, each with appropriate light distribution characteristics. When sunshine is not available, natural light would be replaced by perhaps just one or two very large high-efficacy lamps located in the transmitting room on the roof. Unlike present luminaires, the lighting points would radiate no heat into the building, so the lighting would not create an air-conditioning load and could not be a cause of fire. The heat from the big lamp or lamps could be captured and utilized.

It is already possible to replace up to 50 dichroic LVTH lamps with a single fibre-optic system powered by one far more efficient lamp. It may be generally known that fibre-optic systems are used for lighting museums and art galleries (where there is a need for cold, UV-free lighting), but perhaps not generally realized is that display spotlighting, and even complete lighting installations for hotels, offices, supermarkets or industrial plant, are possible with this equipment. For some years, light-piping has been used in engineering workshops to illuminate the interiors of hollow objects having small entrance apertures. The potential uses include many kinds of local lighting in workshops, e.g. on the cutting tools of lathes, milling machines etc., where the absence of electrical connection to the lightsource at the point of use makes the lighting unit safe to use in contact with water, soluble oil etc, as well as in the presence of flammable liquids and gases.

The piping of light must surely be one of the most promising fields for future lighting development[46].

6 Luminaires

The term *luminaire* is preferred to 'lighting fitting' or 'lighting fixture', as the latter terms are used to describe other electrical fitments.

All luminaire designs are a compromise between performance, limitations of practicability of manufacture, and cost. The ideal form of any luminaire has yet to be designed. Very high performance (in terms of light output ratio, optical performance, cut-off etc) can be achieved at greater cost or bulk. For example, large exterior luminaires offer greater wind resistance and therefore require more costly towers or supports, but small luminaires (which may have lower light output ratios) would be required in greater quantities, and thus the energy cost and first cost would tend to be higher.

The quality requirements for all types of luminaires, e.g. constructional features, optical performance, electrical safety and durability, are covered by *BS 4533*[30]. The trend is to miniaturization of high intensity discharge (HID) lamps, for example the MBI/HQI-T lamps which are literally thumb-sized single-ended lamps. Having physically smaller lamps available does not necessarily mean that luminaires will get much smaller, even though the use of lamps approaching point-source dimensions enables better geometrical light control. This is because the smaller the lamp for a given output, the higher is its luminance, and thus the problems of glare control become greater. Smaller lamps demand to be housed in better quality luminaires having a high degree of optical control.

6.1 Functions of luminaires

An appreciation of the functions of luminaires will assist in the specification of suitable equipment. The following functions may be considered, the nature of the construction and of any enclosure being appropriate to the duty and the environment in which the luminaire will function:

- *Mechanical:* to physically support the lamp/s and enclose the control gear, and to provide a means of attaching the luminaire to a structure or support.
- *Electrical:* to provide electrical connections to the lamp/s and the electrical components, and a means of safely introducing the mains connection.
- *Light control:* To provide means of controlling the distribution of light from the lamp.

- *Safety:*
  - to provide enclosure of the lamp/s, control gear and electrical circuitry for protection against electric shock, burns etc.;
  - to protect the lamp and ancillary components from mechanical damage, and prevent the ingress of moisture, corrosive substances, dusts etc.;
  - to provide environmental protection against the effects of a broken lamp or the heat of the lamp and other components, so as to prevent fire and explosion.

6.2   Control of light distribution

The control of light by a luminaire is performed in various ways according to the nature of the light source and the duties of the luminaire. Some or all of the following features may be present:

- to collect light from the lamp and redirect it in required directions according to the function of the luminaire;
- to diffuse the light from the lamp/s by use of prismatic or scattering media so as to reduce the brightness seen by the user, or the brightness of reflections;
- to mask the lamp/s from view at certain angles as required by the desired function, i.e. to provide cut-off.

The functions of light control may be exercised by reflection (section 6.2.1), refraction (section 6.2.2), diffusion (section 6.2.3) or absorption (section 6.2.4) or a combination of any of these.

6.2.1   Control of light by reflection

Reflection of light at specular (mirror-like) surfaces takes place at an efficiency generally higher than 90%. Materials such as 'Silverlux'® plastic sheet which bears a very thin coating of pure silver have a reflection factor of over 98% when clean and new.

In practice, part of the emitted flux from the lamp is from the lamp directly (direct component), and part is reflected one or more times before being emitted (reflected component), the ratio between these two components depending on the shape, particularly the depth, of the reflector. Concentrating reflectors tend to have a high reflected component; dispersive and wide angle luminaires tend to have a high direct component (Figure 6.1).

Specular reflectors tend to project distorted images of the light source on to the ground or lighted surfaces, or they may produce striations of brightness. To prevent these effects, a specular reflector may have a ribbed or stippled surface.

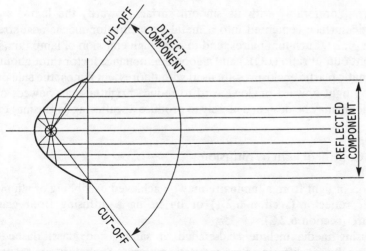

Figure 6.1 *Direct and indirect components of light emission from a typical luminaire.*

6.2.2 Control of light by refraction

Prismatic front glasses to luminaires may have the function of optical control, or may be provided to exercise a scattering function to prevent bright spots and striation. If provided for optical control, prismatic front glasses may act as Fresnel lenses and concentrate the beam, or the reverse. Ribbed prismatic front glasses may be used to condense the beam in one plane (make the beam flatter), or to control the beam to produce extra-narrow or extra-wide distribution in one plane.

Control of light is achieved in some luminaires by total internal reflection at prismatic elements (Figure 6.2A). A moulded-glass reflector/refractor is

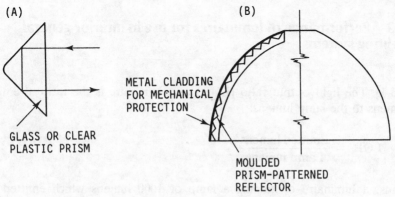

Figure 6.2 *Total internal reflection in prismatic reflectors.*

employed, preferably with its smooth surface towards the lamp, and its prismatic surface cemented into a metal enclosure for mechanical strength (Figure 6.2B). Such luminaires tend to have high emission of lamp flux, i.e. a high light output ratio (LOR) and a good maintenance factor throughout life.

Prismatic plastic enclosures are used with fluorescent lamps, the side prisms directing light upward or downward to reduce brightness as viewed, or the prism detail may be designed merely to produce a scattering (diffusing) effect.

6.2.3   Control of light by diffusion

Diffusion of light from a luminaire may be achieved by ribbing or stippling a specular reflector (section 6.2.1) or by fitting a diffusing front glass or enclosure (section 6.2.2).

Diffusing media include acid-etched or sandblasted glass, flashed-opal glass, translucent diffusing glass or plastic material. Such diffusers reduce the directional properties of the lamp, the resultant polar distribution generally being wider.

Light losses in diffusers may be high; for example, a loss of 25% may be associated with a diffusion factor of only 10%.

6.2.4   Control of light by absorption

Absorption is a costly way of controlling light, and is only employed out of necessity, e.g. to fit 'barn door' external louvres to prevent a floodlight luminaire causing glare in a particular direction. Absorbing louvres, in the form of matt black concentric cylinders are used in some downlighters to screen the light source, and these do not greatly affect the beam intensity in the required downward direction.

6.3   Performance of luminaires for use in interior general lighting systems

6.3.1   The light output ratio (LOR) of a luminaire is the ratio of emitted lumens to the lamp lumens:

$$LOR = \frac{Emitted\ lumens}{Lamp\ lumens}$$

Thus, a luminaire containing a lamp of 1000 lumens which emitted 600 lumens, would have an LOR of 0.6.

The LOR is the sum of the lumens emitted in the upper and lower hemispheres, known as the upper and lower flux fractions respectively. The ratios of these flux fractions to the total emitted lumens give the upper and lower light output ratios (ULOR and DLOR). The relative magnitudes of these ratios give an appreciation of the overall performance of the luminaire in the broad classifications of distribution which are widely used[8], and are summarized in Table 6.

Table 6 *Classification of interior luminaires according to light distribution*

Class	Typical ULOR (%)	Typical DLOR (%)
Direct	0–10	90–100
Semi-direct	10–40	60–90
General diffusing	40–60	40–60
Semi-indirect	60–90	10–40
Indirect	90–100	0–10

6.3.2 The manner in which luminaires distribute their flux affects both the *utilization factor* (the proportion of lumens emitted by the luminaire arriving at the working plane) and the *glare index* (an arbitrary index of the degree of direct glare produced by an overhead system of general lighting), both of which are discussed in section 9.2.

6.4 Performance of outdoor floodlight luminaires

Various systems of classification are used to describe how light is distributed from floodlight luminaires. Designations of beam angle such as 'narrow', 'medium' or 'wide' may not have identical meanings in terms of performance from different manufacturers. While of value in considering luminaires for floodlighting buildings, these designations are of little help in choosing luminaires for area lighting (where the objective is to spread light on the horizontal plane with acceptable uniformity).

6.4.1 Analysis of the needs for area lighting for many industrial applications and security lighting indicates that three basic distributions are mainly needed, designated here as 'X', 'Y' and 'Z'. However, in situations where it is vital to prevent 'pollution by light', i.e. casting of light above the horizontal, there is a strong case for using 'flat glass luminaires'. These have an opaque top housing and a horizontal flat glass, and which project no light above the horizontal, and are designated here as type 'F'.

'X' distribution

This provides a flow of light from the luminaire that is substantially conical. The beam angle is defined as that conical angle at which the luminous intensity of the beam falls to 50% of the peak intensity. A convenient arbitrary classification of the beam angle for such luminaires is: 'N' (narrow), <15°; 'M' (medium), 15–30°; and 'W' (wide), >30°. These distribution types may be referred to as types XN, XM and XW. (Figure 6.3)

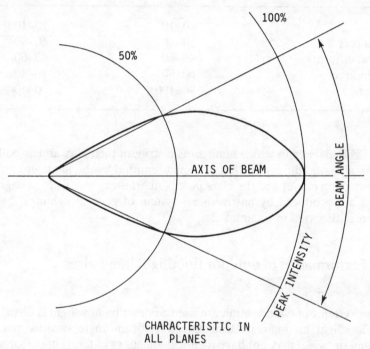

Figure 6.3 *Type 'X' light distribution (substantially conical distribution).*

'Y' distribution

This describes a polar curve that is substantially fan-shaped, i.e. wider in the horizontal plane than in the vertical, and substantially symmetrical in each plane about the aiming axis. Typically, the depth of vertical beam angle will be about half the beam angle in the horizontal plane. The beam angle is defined as that angle at which the luminous intensity of the beam falls to 50% of the peak intensity in the stated plane. Commonly, luminaires in this classification have a beam angle of 70–90° in the horizontal plane, with some up to 120° (Figure 6.4).

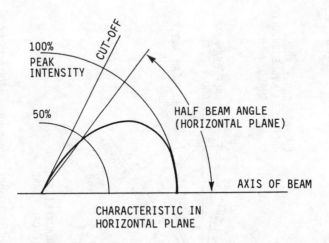

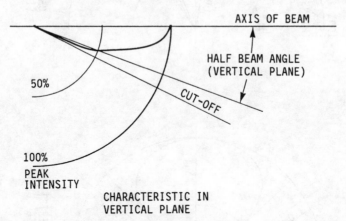

Figure 6.4 *Type 'Y' light distribution (substantially fan-shaped, and symmetrical in the vertical plane and in the horizontal plane).*

'Z' distribution

This describes a polar distribution that again is substantially fan-shaped, i.e. wider in the horizontal plane than in the vertical, but which is not symmetrical in the vertical plane. In the horizontal plane the distribution is symmetrical about the aiming axis; in the vertical plane the distribution is asymmetrical,

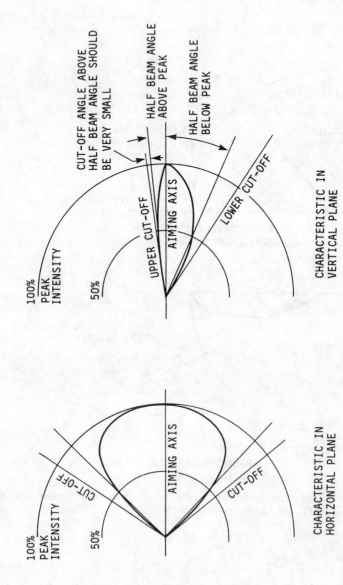

Figure 6.5 Type 'Z' light distribution (substantially fan-shaped and wide-angled in the horizontal plane, but asymmetrical in the vertical plane with fast run-back above the peak).

with a fast 'run-back' above the peak intensity. The beam angle is that angle at which the beam intensity falls to 50% of the peak intensity in the stated plane. Commonly, luminaires of this type have a horizontal beam angle of 70–90°. In the vertical plane, the distribution below the peak gives a half beam angle of about 20°, but above the peak the half beam angle should be as small as possible, commonly 7° (Figure 6.5).

'F' distribution

Type 'F' luminaires cannot be tilted, their bottom flat glass must be horizontal. Thus, in order to get sufficient throw, their use may involve having higher mounting heights than would otherwise be employed. With the use of higher mounting height comes the advantage that the spacing between masts may be extended, and that fewer masts will be needed. However, each mast may carry a cluster of up to 12 luminaires (Figure 6.6).

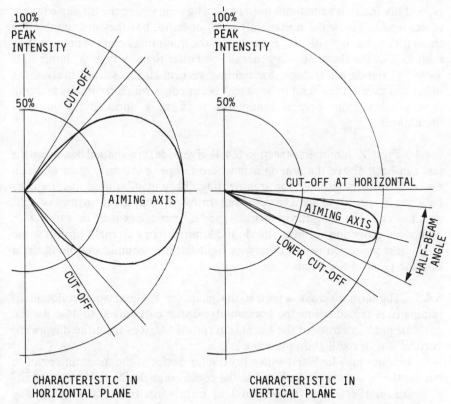

Figure 6.6 *Type 'F' light distribution (flat glass luminaire; wide or medium angle in horizontal plane, but no light projected above the horizontal).*

6.4.2 A type 'X' conical-beam floodlight (section 6.4.1) may emit a substantial proportion of its output outside the defined beam angle because, unless the reflector is very deep, there will be a high direct component (Figure 6.3). To improve the *beam factor* (proportion of lamp lumens in the beam), additional means of light direction may be employed, including prismatic front glasses or cylindrical specular louvres on the axis of the beam. This type of floodlight is used for 'remote topping up' of security lighting (chapter 17), and for the floodlighting of narrow areas from one end. Type 'X' floodlights are also used in the floodlighting of buildings, or for giving extra lighting emphasis where needed.

Type 'XN' and 'XM' floodlights are intensely bright and can cause painful disabling glare if directed towards the eyes of a subject. Type 'XN' floods give especially intense bright beams if fitted with compact-source tungsten-halogen projector lamps.

6.4.3 Type 'Y' luminaires (section 6.4.1) give wide, fan-shaped beams (Figure 6.4). They are sometimes referred to as 'dispersive floodlights'. This type of luminaire is sometimes used in area lighting where the tilt angle (angle of depression below the horizontal) is 35° or more, but they are suitable for small tilt angles, e.g. 10–20°. However, if the mounting height represents less than 25% of the throw, it will generally be better to use a type 'Z' luminaire. Type 'Y' floods are suitable for lighting vertical surfaces from fairly short offset distances. They tend to have soft beam edges which facilitates merging light patches from several luminaires to form a substantially uniform illuminance.

6.4.4 Type 'Z' luminaires (section 6.4.1) give wide, fan-shaped beams with a fast run-back above the aiming axis where the peak intensity occurs, which makes them very suitable for area lighting. They may be used at tilt angles between 5° and 20° to achieve very long throws without great wastage of light into the sky. For long-throw work, type 'Z' luminaires may be fitted with additional reflecting surfaces (inside the luminaire or externally) to lift some of the lumens which would otherwise light the foreground, and direct them into the peak of the beam.

6.4.5 The term *tilt angle* is used to designate the angle at which a floodlight luminaire is angled below the horizontal. Former custom was to describe the angle of peak intensity of the beam in terms of 'degrees from the downward vertical' (as in roadlighting practice).

Luminaires may be fitted with a protractor device at the mounting pivot to enable the beam axis to be tilted at the required angle. Note that under EC Regulations (Section 5 (Floodlights) of EuroNorm EN60958 (1989)), any floodlight mounted higher than 3 m is required to have two independent fixings to its support.

Type 'F' luminaires have no tilt angle, for they are installed with the bottom glass horizontal.

6.4.6 The most widely used type of floodlight for industrial area floodlighting today is the type 'Z', or a type intermediate between 'Y' and 'Z'. Such reflectors work best with a linear source such as the SON-T lamp. For economy in capital cost, sometimes a type 'X' luminaire having a large substantially parabolic reflector is selected, used with lamps such as MBF or MBI, but this does not lead to true economy (especially in medium-sized and small schemes) because of the high *light-loss factor* (lumens cast up into the sky or falling outside the perimeter of the plot). 'Wide light' luminaires, with a beam spread of up to 120° in the horizontal plane are sometimes selected because they permit arrayed luminaires to be spaced wide apart. However, control of light above the horizontal is not as good with wide light luminaires as with type 'Z' luminaires, and therefore sometimes external absorbing or reflecting louvres or 'barn doors' have to be added at increased cost, and with increased likelihood of wind-damage.

6.4.7 Outdoor floodlight luminaires tend to be classified by their photometric performance rather than by their physical construction, but there are significant differences in quality and durability over the ranges available. The cheapest does not always return the lowest cost. For example, high quality luminaires such as prismatic projectors, may return a dividend due to such factors as:

- Good glare control, may be mounted lower than other types, saving in column costs.
- Higher beam factor, fewer luminaires may be required.
- Compact design, may offer less wind resistance and permit use of cheaper supports.
- Good optical control; may be mounted at wider spacings.
- Good engineering construction; may have very long technical life with minimum maintenance.

6.5 Technical features of luminaires

6.5.1 Trends

A current general trend in industrial lighting is towards miniaturization, a trend with somewhat mixed benefits. As lamps and luminaires get smaller, they tend to be brighter, so that the control of direct and reflected glare once more becomes a problem.

One luminaire development that seems to be against this trend is the 'Conductalite' fitting from Simplex Lighting. This employs an MBI/T or SON/T lamp, the light from which is distributed along a reflector of 5.1 or 3.9 m length, using the wedge reflector principle similar to that employed in some illuminated signs. Whether the benefits claimed for this interesting fitting will secure for it a major place in the market is yet to be seen, though it seems to be more suitable for sports halls and some industrial locations than for offices.

However, in office lighting, there is a clear trend away from use of long luminaires; long fluorescent tubes are giving way to 'long compact' fluorescent lamps. This trend is related to the general adoption of 600 mm module suspended-ceiling systems for new and redeveloped office premises, and also for many retail premises. Modern ceiling systems favour the use of compact fluorescent lamps (section 5.6), especially in downlighters. Even with specular 'dark' reflectors there may be problems due to the high brightness of these lamps.

With justification, industry is starting to make use of low-voltage tungsten–halogen (LVTH) projector lamps (section 5.3). They are used for such applications as providing some local/localized light on small areas, where the nature of the working conditions requires that the luminaire be positioned well clear of the task area. Currently there is a need for new designs of robust luminaires for these lamps for use in industrial situations so they may be used safely in rugged, hostile or hazardous environments.

The potential in the 'piping of light' (section 5.13) has yet to be exploited, but it is noted that some fibre-optic devices are now being used in specialized branches of tool-making and instrument-making to direct light into otherwise inaccessible positions, and also to bring back an image.

It may be forecast that high-frequency control gear will become the norm for both fluorescent lamps and HID lamps before the end of the 1990s, and that progress in semiconductor circuitry will soon enable the general use of 'throw-away ballasts' to be discarded with the lamps at the end of lamp life.

6.5.2 Thermal considerations

Discharge lamps can only produce their designed output and give their rated life if operated within the range of their optimum operating temperature. If lampholders are operated above their limiting temperature, the contact springs may become de-tempered, leading to poor electrical contact.

In the case of tungsten–halogen lamps (section 5.4), the design of the lampholder and any associated heatsink is critical to good lamp performance. If the heat conductance is too great (especially in open luminaires used in very cold conditions) the lamp may not reach its optimum temperature (so that the halogen cycle cannot operate) and short lamp life will result. If the heat conductance and dissipation is insufficient, the lamp may give short life because of overheating of the seals where the electrical connections enter the quartz envelope.

Enclosed exterior luminaires can operate satisfactorily down to −20°C, though HID lamps may take longer than usual to run up to full output. Open and ventilated luminaires are not usually successful in sub-zero conditions because frost forms on electrical components causing tracking. For operation at very low temperatures, SON lamps seem the least likely to give trouble, and having an ignitor, usually start reliably under adverse cold conditions.

Although a luminaire may be classified as capable of withstanding a jet of cold water (chapter 7), the water-jetting of hot luminaires is not recommended. Note the caution about water-jetting luminaires given in section 21.2.

Some luminaires incorporate a gasket, 'O'-ring, seal or washer for sealing the luminaire enclosure. If this component is made of an elastomer it may deteriorate or embrittle at elevated or low temperatures respectively. It is recommended that the luminaire manufacturer be advised of the climatic conditions where the equipment is to operate.

Lamp replacements should always be carried out with lamps of exactly the original specification, and − especially in the case of all metal-halide lamps − preferably from the lampmaker who supplied the original lamps. To this end, luminaires may be marked with the replacement lamp details. In the case of TH-lamp and filament-lamp luminaires, it may be possible to mistakenly relamp with a greater power of lamp than that for which the luminaire was designed, and this can lead to failure of the unit, overheating or fire.

6.5.3 Electrical considerations

Internal wiring of luminaires and control gear enclosures should not be replaced without confirming the correct grade and temperature duty. Unless the manufacturer advises otherwise, wiring of normal insulation, i.e. to withstand temperatures up to around 70°C will be satisfactory for wiring into luminaire terminals.

Slight progressive changes in lamp design over the years may affect lampholder performance. For example, the method of fixing the envelope to the lampcap shell in HID lamps has changed, and a ceramic ring is usually visible at the junction of the envelope and cap. If any screw lampholder employed does not have a spring-loaded centre contact, lamps without the ceramic ring could be overtightened and deformed, so that the lampholder makes poor contact at lamp replacement. Always inspect lampholders before inserting a replacement lamp, and specify that the centre contact of screw lampholders shall be spring-loaded.

The preferred method of power factor correction for lighting installations is for pfc condensers to be incorporated into the luminaires or mounted close by, i.e. 'lumped correction' is not recommended, though it may be a little cheaper.

Note the manufacturer's instructions as to the maximum distance any control gear may be mounted from the lamp. Lamps operating with ignitors may tolerate a greater distance between gear and lamp if the ignitor is located at the lamp – a feature that may be of value in lighting hazardous zones with luminaires having remote ballasts.

Because of voltage spikes occurring at switching, mineral-insulated/metal sheathed cable should not be used to connect between lamp and ballast except for short lengths approved by the manufacturer of the control gear.

It is recommended that when performing insulation resistance tests on luminaires, the lamps should be removed, and that any electronic devices (e.g. ignitors) should be removed or short-circuited to prevent damage by, for example, over-voltage or reverse-voltage.

6.5.4   Mechanical considerations

Experience indicates that mechanical failure of luminaires is not a frequent occurrence, provided only that the luminaires are securely mounted and they do not vibrate too strongly under the influence of the wind or due to vibration of the building or structure to which they are fixed. There may be a surprising amount of building vibration associated with the movement of gantry cranes.

For luminaires which must be serviced at height, the hinging of openable parts, and the attachment of parts by safety chains is recommended. Fixings should be captive.

Luminaires which are for use in situations where there may be vibration from the structure or other support, may be supplied by the manufacturer with special lamp steadies. For some situations, a flexible damping suspension device may be used to insulate the luminaire from gross vibration of the support.

6.5.5   Special environments

These are reviewed in chapter 7, including some guidelines for the construction and use of luminaires in clean-rooms and sterile rooms in section 7.4.

6.5.6   Fire risks from luminaires

Luminaires constructed within the constraints of *BS 4533*[30] and relating British Standards, and manufactured to *BS 5750*[35] quality standards, are unlikely to cause a fire hazard if correctly applied within the environments for which they are designed (chapter 7), and not misused or abused (section 7.8).

In particular, attention is drawn to the use of luminaires with directable beams, e.g. crown-silvered incandescent lamps in reflectors, PAR lamps, and low-voltage tungsten–halogen (LVTH) dichroic lamps. The beams from these lamps are hot enough to ignite flammable materials, and so must not cause the temperature of any illuminated surface or object to exceed 90°C. *BS 4533*[30] requires that the minimum permissible distance (in metres) between the lamp and the illuminated surface or object shall be marked on the luminaire with the 'F' symbol, and users should ensure that the lamps are not mounted closer than this distance from any flammable material. It must be remembered that the lamps and luminaires themselves are hot, and may directly ignite flammable materials such as curtains if allowed to come into contact or close proximity. If they are ceiling-recessed, a proportion of the heat from downlighters and dichroic lamps is radiated or convected from the back of the fitting, and this can cause overheating of wiring or other materials in the ceiling cavity.

Part 3

Environmental and safety considerations

7 Lighting environments

7.1 Environmental protection of lighting equipment

7.1.1 The safety of persons on industrial premises may be dependent on the provision of electrical installations and lighting equipment certified as suitable for the hazardous environmental conditions where it is installed. In some cases the equipment may have to withstand two or more adverse factors of the environment. For example, there may be situations requiring the use only of electrical enclosures designed and certified for use in a flammable atmosphere, in which the equipment may also be subject to vibration, or have to operate in the presence of conductive dusts.

7.1.2 While the majority of industrial lighting installations are in normal, dry, clean environments, luminaires and electrical equipment of appropriate types must be selected to withstand adverse environmental conditions such as damp, dust, vibration etc. In this chapter there are numerous references to the necessity of using luminaires and electrical equipment having enclosures appropriate to the environments under consideration, and the reader is referred to the following British Standards for further information:

- *BS 5490*[55] regarding ingress protection (IP classification);
- *BS 5345*[38] regarding use of electrical equipment in potentially explosive atmospheres;
- *BS 6467*[56] regarding enclosures for use in the presence of combustible dusts;
- *BS 4533*[30] regarding mechanical and electrical features of luminaires.

7.1.3 It is important to note that luminaires designed for use in 'normal dry interiors' are likely to rust or corrode if subjected to damp. If such luminaires are stored or installed in rooms which have recently been screeded, plastered or treated with water-based emulsion paint, deterioration from damp caused by condensation may result. Indeed, luminaires installed in any unheated industrial building in the winter months are likely to deteriorate unless specially protected against damp.

7.2   Lighting in high ambient temperatures

7.2.1   In factories operating hot processes, it is common to find that the ambient temperature at luminaire level is considerably above the normal range in which standard lamps, control gear and wiring are designed to operate, often resulting in short lamp life and early failure of components.

In rooms with flat ceilings or well-insulated roofs it is possible for the static air above the highest window opening to be considerably hotter than the air in the 'occupancy zone' up to 2 m above floor level. Such pockets of hot air may be unsuspected until low light output or colour change of fluorescent tubes, or early failures of control gear call attention to them. Similarly, luminaires recessed into a ceiling void may be subject to overheating if the void is not ventilated and the roof is not insulated from solar heat.

In some processes a column of hot air rises which, while not greatly affecting the ambient temperature, may seriously overheat any luminaire upon which it impinges. There are factories, for example in the paper industry, where, because of the heat, it is impossible for staff to ascend to luminaire level when the plant is operating, so that relamping and maintenance of the lighting has to await the infrequent times of shut-down.

7.2.2   These problems may be overcome by specifying the environmental conditions to the lighting provider – an important matter when inviting tenders (section 24.2). Suitable lighting equipment is usually available, though at extra cost, to enable lighting to operate efficiently and safely at all temperatures in which a person can work, say up to 35°C, but this statement should not be taken to mean that standard lighting products will be satisfactory at such temperatures. The optimum wall temperature for standard fluorescent tubular lamps is 38°C, and if the tube runs hotter or cooler than this, its light output will be reduced, in some cases with a change in colour performance.

The wall temperature of HID lamps is generally around 120°C, so that they are not much affected by high ambient temperature if operated in suitable luminaires. However, control gear operated at elevated temperatures will tend to have a short life. In particular, capacitors may fail, and those which are liquid filled may leak with the risk of fire. Most types of HID lamps seem to tolerate fairly high ambient temperatures, but SON lamps, especially when operated in enclosed luminaires, can become unstable. It is therefore wise to consult the lighting provider as to the suitability of any proposed combination of lamp, control gear and luminaire for use in very hot conditions.

7.2.3   A fluorescent tubular lamp can perform well at elevated temperature if its internal gas pressure can be reduced by spot cooling. One method of doing this is to provide a small 'pip' of glass drawn out from the envelope, which being away from the arc is cooler. Another method uses an electric junction (reverse Peltier Effect) to get the necessary cold spot. A widely used method is to employ 'high temperature' tubes which are fitted with a ring of indium which, with other adjustments to pressure and gas composition, enable the lamp to operate at equatorial temperatures. The latter type of lamp may be used to replace standard lamps, but only if the ambient temperature does not exceed the tolerance of the lamp, control gear and wiring.

7.2.4 In hot conditions, the luminaire connections must be such that the cable insulants are not damaged by heat. Typical modern cable insulants intended for normal temperatures will show signs of softening or creepage at 95°C, and some will start to flow at around 105°C.

7.3 Lighting in low ambient temperatures

7.3.1 Most types of HID lamps will start satisfactorily and run up to full output in temperatures down to −5°C, and are more likely to be affected by damp and condensation than by temperature.

Tungsten-halogen lamps may give short life at temperatures below about −5°C if the lamps cannot get hot enough for their halogen cycle to operate (section 5.3.1). In open luminaires, ice forming on these lamps when not switched on can cause problems, and they are also prone to burning of the lampholder contacts under these conditions.

Fluorescent tubular lamps may give trouble in starting at ambient temperatures below 5°C, particularly under damp conditions, and will not give their full light output if the tube wall temperature is substantially below 38°C. In ordinary cold industrial buildings in winter, 'switch start' luminaires will usually start satisfactorily. Fluorescent tubes operating on high-frequency control gear will usually give faultless starting under these conditions. The colour properties of fluorescent tubes may be affected at low operating temperatures.

7.3.2 In refrigerated stores and cold rooms, fluorescent tubes are successfully operated in temperatures down to about −20°C by using fully-enclosed luminaires. To start the lamps they may first be warmed before being installed in the luminaires, and they may then be operated continuously for life. Tubes thus operated may give two to three years continuous burning, perhaps more if there are few voltage fluctuations large enough to cause extinguishing and restriking. A useful tip is not to locate fluorescent luminaires close to the cold-air recirculating fan outlets, for a cold blast can 'blow out the lamps' due to chilling.

For cold-room applications, lamp-makers may supply a special fluorescent lamp in which the gas pressure has been adjusted to aid low-temperature starting. Fully enclosed luminaires are recommended, and under extreme conditions these may be fitted with an auxiliary filament lamp to act as a heater, or small heater elements may be incorporated in the luminaires. Advise the lighting supplier of the application; variants of standard products suitable for this duty may be available.

7.3.3   In cold-rooms operating continuously below 0°C, the humidity is very low, and rust and corrosion is not usually a great problem if the luminaires are of good quality. But, if the same type of luminaire is used in locations such as under loading-bay canopies, or in unheated garage areas and stores, corrosion or rust may occur, as well as starting problems. Luminaires designed for interior use will not give long and safe service under these conditions (section 7.7).

7.4   Lighting in clean-rooms and sterile rooms

7.4.1   Clean-rooms

Clean-rooms are used to provide working conditions of stringent cleanliness for critical industrial processes. They are used for fine engineering work and microelectronics, in nuclear research and engineering, for semiconductor manufacture, in the pharmaceutical industry, and in the preparation of dust-sensitive items such as the selenium drums used in xerography.

To produce a dust-free atmosphere, clean-rooms have total mechanical ventilation with filtered air. To ventilate the room with very low air speed and thus avoid entraining dust particles in the airflow, input and output grilles of large area are used, in some cases consisting of virtually the whole floor and the whole ceiling. Such installations produce laminar flow air movement.

The work performed in such rooms is usually of meticulous accuracy, requiring illuminances of at least 1000 lx. The heat from the lamps is ducted away with the airflow. However, it is not good practice to use ordinary 'air-handling luminaires' (in which effluent air is drawn over the lamps) as dust can be trapped in the luminaires and may be released into the room when the system is switched off. It is preferable to place the lamps behind glass or clear plastic panels in the ceiling or walls, so that access to them is gained from outside the controlled atmosphere zone (Figure 7.1). If luminaires must be positioned within the clean-room, they should be of small cross-section and preferably be of streamlined profile so as to cause minimum disturbance of the laminar air flow.

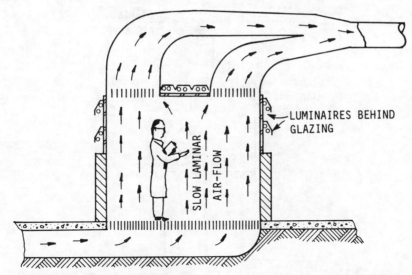

Figure 7.1 *Lighting in a laminar-flow clean-room. Air enters through the large floor grille and leaves through ceiling grilles. Slow air movement ensures minimum entrainment of dust.*

7.4.2 Sterile rooms

Sterile rooms present somewhat different problems to those met in industrial clean-rooms, in that they must not only be kept dust-free, but all internal surfaces must be of surgically sterile cleanliness. Sterile rooms are used for such tasks as the preparation and packing of pharmaceutical products, sutures, surgical dressings etc., and may have to accommodate many operators.

The general lighting and ventilation of the room may be arranged in the same manner as for clean-rooms, though the direction of the air flow is more usually downward. Additionally there may be hooded benches located in the room. Each bench is provided with a positive pressure of sterilized air, so that contamination cannot be drawn into these critical spaces by air entering from the general air volume of the room (Figure 7.2).

7.4.3 Luminaire requirements

For locations where very high standards of cleanliness are required, the following guidelines for the construction of fluorescent luminaires should be complied with:

- The whole luminaire shall be capable of withstanding repeated thorough cleansing with water and detergent and an approved chemical disinfectant.

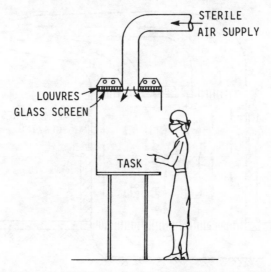

Figure 7.2 *Lighting for a sterile work cabinet in a sterile room or clean-room.*

- Any spaces between lids or diffusers etc. and the apertures they mate with shall be as small as possible to limit the entry of dust. Preferably such conjunctions shall be sealed with impervious flexible seals.
- The heating and cooling of air within an enclosure may cause air to be expelled from and drawn into the enclosure. Such air movements could cause the transfer of contaminated dust from within the enclosure to the surrounding atmosphere. If the enclosure cannot be properly sealed, it should be the objective to ensure that dust that enters the enclosure stays there and does not become recycled.
- Any openings such as between louvres or between the lamp and reflector shall be as large as possible to permit easy cleaning.
- Any seals, glands, packing washers etc. used in the construction of luminaires shall be made from substances which will not support mould growth and which are not hygroscopic.
- Preferred designs for luminaires in these special environments shall have minimum horizontal areas upon which dust particles can settle.
- Luminaires may be specially constructed so they are easy to dismantle for cleaning. Dust-tight luminaires to the relevant sections of *BS 4533*[30] may be constructed with specially smooth and impervious exterior surfaces and other features that make them easier to clean.
- If a sealed fully enclosed luminaire is used in a sterile area, it must be fitted with a breather fitted with a filter of material fine enough to trap all particles of a magnitude to be agreed with the medical or technical experts.
- For critical situations, consideration should be given to the use of a construction permitting the luminaires to be periodically removed from the

clean environment to be cleansed and sterilized elsewhere before replacement, or replaced with new luminaires.

7.5   Lighting with reduced radiofrequency interference

All gaseous discharge lamps (HID lamps and fluorescent tubular lamps) produce emanations in the range of radiofrequencies (r.f.) which can cause r.f. interference (r.f.i.) with communications equipment. Fluorescent tubular lamps operating on dimmer circuits (both variable dimmers and those giving preset outputs), and all lamps operating on solid-state high-frequency control gear are prone to producing r.f.i. Dimmers which use 'wave-chopping' (e.g. thyrister dimmers) are prone to creating r.f.i. and feeding harmonics back into the mains. At the time of writing it is not yet known if there will be any special problems with r.f.i. from the new induction lamps (section 5.12).

Interference in the r.f. bands may be of various types:

- *Radiated interference*. Emanations in the r.f. band may be radiated by lamps, control gear and circuitry, and from ignitors of HID lamps. That during operation of the lamps will be continuous; that from ignitors and starting circuits will be intermittent and confined to the brief starting period. A failed lamp that is cycling (repeatedly attempting to start) will cause intermittent interference.
- *Mains-borne interference*. If the equipment is not adequately isolated by an r.f. filter, frequencies can leak into the mains and affect nearby sensitive equipment.
- *Mains-borne radiated interference*. Equipment which is not isolated by an r.f. filter may inject interference into the connecting cables, and if these cables are not adequately screened with earthed metal, the cables can radiate interference which can affect nearby sensitive equipment.

Interference can vary from 'clicks' due to peaks and inverse peaks generated during switching and starting, to continuous or intermittent interference in the r.f. and audiofrequency ranges. Interference generated by lighting circuits can affect sensitive equipment such as electrical test instruments, televisions, radios, security monitoring devices, computers, and audio and video recording and playing apparatus. Interference from commonly-used types of lighting equipment can affect long-, medium- and short-wave radio reception, and bands I and III which affect domestic radio and television reception. The effects in industry can be serious because of the possibility of interference with microprocessors and computer-controlled equipment.

In broadcasting and recording studios, and in laboratories where delicate electrostatic equipment is used, the greatest problem may be due to radiated emanations from fluorescent tubes. To ameliorate this problem, tubes or the transparent enclosures thereof, may be covered with metalized clear plastic film, or with fine metallic mesh. Such screens must be earthed. All wiring must be metal sheathed and the sheathing earthed. In difficult cases it may be necessary to replace lighting equipment and rewire the premises, fitting appropriate r.f. filters. The subject of r.f.i. is covered by *BS 4533*[30], by *BS 5394*, Part 1 1988[31], and by the *Electromagnetic Compatibility (EMC) Regulations* 1990[64]

Creating r.f.i. is an offence in law, and relating regulations have been published[36]. Further information is available from the following organizations (see appendix G): the British Standards Institution, the Lighting Industry Federation, the Radiocommunications Agency of the DTI (to whom complaints about r.f.i. generated by others may be directed), and the Electricity Association. A publication from the former Electricity Council provided much useful information[33], but may no longer be available, and a publication of the International Special Committee on Radio Interference[32] is also recommended.

7.6 Lighting in dusty or soiled atmospheres

7.6.1   Light which is obstructed by dirt on the lamp or luminaire surfaces does not benefit the user but still has to be paid for. An allowance is made in lighting calculations for an acceptable degree of light loss (section 3.10), the recovery of which would involve uneconomic costs of extra maintenance. Even luminaires designed as 'dust-tight' will tend to 'breath' and, unless fitted with a suitable filter at the breather-hole, will expel heated air as they warm, and draw in contaminants as they cool.

7.6.2   In addition to absorbing light, a layer of dust or dirt on the lamp or luminaire has the effect of diffusing the light, so that beam angles are widened and directional effects diminished. The cut-off may be affected, and a result may be an increase in direct glare from the luminaire.

7.6.3   Airborne conductive dusts (e.g. carbon or fine metallic dusts) entering luminaires can cause tracking and electrical breakdown. Sooty or oily particles will rapidly reduce the luminaire light output. Organic dusts (e.g. flour or spices) are hygroscopic and will support mould growth – it is not unknown for an enclosed luminaire so contaminated to fill up with water extracted from the atmosphere.

Fibrous dusts (e.g. wood dust, cotton linters, wool fibres and asbestos) can settle on luminaires and insulate them to the extent that they seriously overheat. This has been known to cause fire. Airborne microdroplets of fats passing into a luminaire when heated, can solidify when the luminaire is cold. When the luminaire is switched on again, the deposits melt and drip off, acting as a serious potential cause of contamination of foodstuffs.

7.6.4 Unless the luminaire is sealed and fitted with a filter-breather, fitting a translucent cover is not necessarily the best way of reducing light loss due to contamination, for more surfaces between the lamp and the working plane will have been interposed, thereby increasing the opportunities for absorption of light (Figure 7.3). The arrangement shown in Figure 7.3A gives the best performance, and that in Figure 7.3C the worst. However, if the (C) arrangement can be properly sealed, it may give a performance equal to or better than the 'A' arrangement.

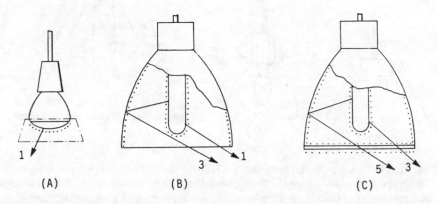

Figure 7.3 *Effect of soiling of luminaires. The numbers on the arrows indicate the number of layers of soiling through which an incident ray has to pass. (A) reflector lamp; (B) lamp in an open reflector; (C) lamp in a reflector with cover-glass (unsealed).*

7.6.5 In conditions of heavy pollution, luminaires of water-jet proof construction (IP64) may be cleansed by hosing, but certain precautions are necessary (sections 7.7 and 23.2.2).

7.6.6 Most contaminating particles can only settle on luminaire surfaces if the speed of the air in which they are entrained is below a critical speed. If it can be arranged that the surfaces which it is desired to keep clean are constantly scoured by an air current faster than that critical speed, the surfaces will remain clean. This is the theory of *through-vented luminaires*. The through-venting feature has been used in slotted-top fluorescent trough luminaires since the earliest days of fluorescent lamps, when they were employed to prevent the *tunnel effect* (the gloomy appearance of a room lit by direct lighting with little light directed to the ceiling); it was noted that slotted troughs invariably kept cleaner than unslotted ones (Figure 7.4A).

In modern luminaires for HID lamps, the through-venting design is employed to keep the reflector clean, the through draft also helping to cool the control-gear (Figure 7.4B). This enables the control-gear compartment to be smaller and closer to the lamp, resulting in cost savings. A further development is the use of convected air through a luminaire to scour the external surface of the cover glass (Figure 7.4C) to prevent deposition of dirt.

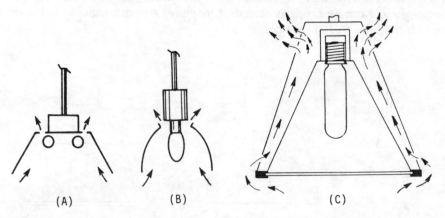

(A)   (B)   (C)

Figure 7.4 *Through-vented luminaires. (A) cross-section of a fluorescent trough luminaire with slotted top; (B) through-vented HID luminaire in which the through draught helps to cool the control gear; (C) sealed luminaire with convection passages, the entering air being drawn across the cover-glass, and the venting air helping to cool the control gear.*

A type of luminaire sometimes employed in the food and pharmaceutical industries is the enclosed *pressurized luminaire*. In this type, the internal air pressure is maintained from an air-line. The luminaire is fitted with a small purging valve, so there is a continuous flow of clean air through the luminaire cavity which helps to cool the lamp and control gear. It seems likely that new types of pressurized luminaires may in the future become a standard method of providing the necessary protection in hazardous zones (section 7.10.3).

Reference should be made to *BS 4533*[30] regarding types of luminaires such as dust-tight, dust-proof, hoseproof and corrosion-resistant.

7.6.8   There are fire risks associated with suspensions of flammable dusts. For example, atmospheric suspensions of materials such as flour, custard powder, cornflour and icing sugar can burn with explosive force. If the density of flammable particles in suspension is such that each particle is within the flame zone of the next particle, once conflagration commences, the flame may be propagated at high speed, creating rapid heating and expansion of air with sufficient explosive force to demolish buildings.

7.7 Lighting in wet and corrosive atmospheres

7.7.1 The constructional differences between *dust-tight* luminaires and *water-jet-proof* luminaires to BS 4533[30] may not be great, but the casing of the latter must be corrosion-resistant and the seals must be capable of withstanding water and the particular corrosive substances to which the luminaire is exposed. Non-metallic luminaires, constructed entirely of plastic materials, can be highly effective for this duty, especially if pressurized with air (section 7.6.6). Metal luminaires coated with plastic may be equally suitable. If luminaires are to be hosed, note the warnings in sections 7.7 and 23.3.2.

The user must be careful to distinguish between the luminaire classifications in *BS 4533*[30]: a *drip-proof* enclosure may not be *rainproof*; a rainproof enclosure may not be water-jet-proof; a water-jet-proof enclosure may not be *watertight*; a watertight enclosure may not be suitable for submersion; and finally, a *submersible* enclosure may not necessarily be watertight, but may be constructed on the diving-bell principle.

7.7.2 Many industrial atmospheres contain steam, and some processes create atmospheres containing entrained droplets of corrosive substances, as in wool-stripping, hide-pickling and leather-making. In the food industry a common air contaminant is microdroplets of acetic acid. In environments where there is wetness and contamination in the atmosphere, it is not only the luminaires which must withstand the hostile conditions; the entire electrical installation is required to offer resistance to these conditions to ensure electrical safety.

In these conditions, there is a potential danger when it is necessary to open up any electrical enclosure, e.g. opening enclosed luminaires for cleaning or relamping. In all cases it is vital to isolate the equipment from the mains, and, if at the time of the operation the atmospheric conditions are adverse, a protective tent or other enclosure may have to be erected. It is best practice to carry out such maintenance when the plant is shut down. In really bad conditions where there is 24-hour working, it may be better to have all luminaires connected to the supply by suitable plugs and sockets so that a faulty or dirty luminaire can be safely disconnected, removed, and replaced with a clean serviceable one.

7.8   Lighting in rugged environments

This chapter deals generally with methods of matching lighting equipment to the environment in the quest for safety and efficient operation. If all who work in industry always performed their tasks with care and without moments of thoughtlessness or irresponsibility, it would not be necessary to include this section 7.8 to call attention to abuses of lighting equipment which can take place.

7.8.1   The reader is invited to provide his own solutions to preventing abuse and misuse of lighting equipment. The author has witnessed the following acts or seen their results:

- Suspending a chain-hoist from an overhead lighting trunking to lift a heavy load.
- Using a bayonet-cap adaptor to connect an improvized electrical supply to an unearthed handlamp having a metal wire cage, the wiring passing out through a window and lying in puddles on its way to the point where the handlamp was being used.
- Using a wall-mounted lighting switch as a foot support while climbing to reach a high object.
- Placing a number of heavy carpet strips over a low-mounted suspended fluorescent trough luminaire (which was switched on) to keep them off the floor while carrying out repairs.
- Standing on the tines of a forklift truck and getting the operator to elevate the lift in order to get access to replace an HID lamp mounted 6 m above the floor.
- Attempting to remove the remains of a smashed MBFR lamp from its socket by wrapping a piece of cloth around it, while the circuit was still live.
- Inspecting the inside of a metal vat (an 'earthy location') by use of an unprotected fluorescent batten luminaire; the luminaire was connected to a nearby 13 A socket outlet by means of bare cable ends pushed into the socket, the shutter of the outlet being held up with a screwdriver pushed into the earth socket, and the earth wire not used.
- Experimenting to see if a 400-W MBF lamp would work in a luminaire intended for a 400-W SON lamp.
- Moving a fluorescent luminaire to a new position, suspending it on two pieces of string, and extending the connecting cable by a piece of lighting flex, the connections being made by twisting the bare ends which were left exposed.
- Unhooking a high-bay HID luminaire from its suspension point, and allowing it to hang down on its PVC/PVC connecting cable.

- Placing a pole across two suspended luminaires and using this to support some wet canvas sheets to dry overnight.
- Connecting a festoon of Christmas-tree lights into a wall-switch on a staircase, and leaving the cover off the switch.
- Attempting to liquefy some petroleum jelly by immersing a 100-W tungsten-filament lamp in it.

All the above abuses of lighting equipment contravene the *IEE Wiring Regulations*[27], and contravene the *Health & Safety at Work Act, 1974*[25] and the *Electricity at Work Regulations*[78], and could be the subject of prosecutions against the persons concerned and the companies on whose premises the offences occurred.

7.8.2 There are rugged environments where there may be abuse of lighting installations. For such locations, lighting equipment must be selected with care, and positioned so as to minimize the opportunity for its abuse. It must be robustly installed, and inspected frequently.

The problem of abuse of lighting equipment within an organization is essentially one of management and training (section 22.1). The person responsible for specifying and maintaining the lighting installation has an important part to play, for if the installation is suitable for its purpose in the first place, and is properly modified to suit locally changed needs as they occur, there should be little motivation for unofficial interference with installations.

As regards abuse of lighting equipment by outsiders (e.g. stone throwing, air-gun attacks), much can be done by choosing and siting equipment carefully. Wall-mounted luminaires in areas to which the public have access should be positioned well above hand-reach height. *Vandal-resistant* equipment may be specified, e.g. luminaires made of or glazed with tough plastics such as Cabulite®, a high-impact acrylic material, or UV-stabilized polycarbonate. Experience shows that luminaires inside fence lines are unlikely to be hit by objects thrown from outside because the required tragectory is of such a height that few vandals have the strength to throw a heavy object effectively. Air-gun attacks on luminaires at night are unlikely to be effective because the glare makes accurate aiming difficult, but luminaires are vulnerable to an air-gun marksman during the day.

7.9 Lighting in windy or vibrating environments

7.9.1 Indoor environments

Suspended luminaires in industrial high-roofed buildings may be subjected to considerable wind pressures when large doors are opened, sufficient to fracture rigid suspension drops. If swinging rigid suspensions or chain suspensions are employed in situations where the luminaires may be subjected to strong internal air movements, lateral staying with rods or cables may be necessary. A better method is to mount the luminaires on rigid trunking, the trunking itself being laterally stayed if required.

Steel-framed skin-cladded industrial buildings may be subject to considerable and sustained vibration due to wind pressures or the operation of a gantry crane. Mechanical shocks can be sufficient to shorten the life of lamps or to cause lamps to fracture in their sockets. Under conditions of vibration, fluorescent tubular lamps have been known to shear their pin connectors and fall from the luminaires. Screw-cap lamps, including HID lamps, if not firmly screwed home, may fall from their sockets.

The first approach to dealing with this problem is to attack the source of the vibration. For example, bolt up tightly any stays on the structure, ensure that gantry crane rails are tightly bolted (or mounted on anti-vibration mounts), and put soft bumper stops on large sliding doors.

Next, consider interposing some heavier supports between the structure and the luminaires. In bad cases, consider interposing anti-vibration mounts (and ensuring that earth continuity of the supports is maintained). Note that the wrong design of anti-vibration mount can actually increase the amplitude of the vibration.

If conditions preclude preventing the vibration or shock, select luminaires which incorporate robust lamp-steadies, and select enclosed types of luminaires or those fitted with mesh screens to prevent detached lamps or parts thereof falling. In the case of premises where food is handled or stored, there is a specific requirement under the *Food Regulations*[34] to prevent parts of a shattered lamp from falling, as well as general safety and hygiene requirements under the *Health and Safety at Work Act*, 1974[23] relating to all premises where persons work.

7.9.2   Outdoor environments

Towers, masts, poles etc for supporting exterior lighting equipment may be damaged by wind, and in falling can cause damage and injury. Accordingly, insurers normally require that the wind resistance of supports shall be designed to withstand the highest wind velocity likely to occur according to 100-year statistics. This requirement also applies to lighting equipment which is supported by buildings and other structures.

The wind resistance of a luminaire is determined mainly by its projected area, and also by its shape. Designs which use advanced reflector design and physically small lamps will have a smaller projected area.

The number of luminaires on a support, and their aerodynamic properties, will help determine whether there will be vibration likely to damage the equipment or cause short lamp life, and this is a matter which should be discussed with lighting providers. The secure fixing of luminaires (and any separate control gear) is a first priority, with each item fixed by two independent fixings. If the luminaire adjusting means are not locked securely, high winds may move luminaires from their set aiming positions. If an integral winch is provided to lower a luminaire, it is wiser not to use it during periods of high wind.

7.10 Lighting in hazardous environments

This chapter has reviewed some salient points about the provision of lighting in areas having a risk of fire or explosion due to the presence of flammable materials, liquids, vapours, gases or fumes.

The relating UK legislation, and the safety requirements under European and British standards, are developing and changing at such speed that any specific information regarding definitions of enclosures and practices, standards and laws etc. given here would soon be out of date. For current information, the reader may refer to the following organizations (whose addresses are given in appendix G): the British Standards Institution for current British and European standards; the Chartered Institution of Building Services Engineers for current publications giving technical information on practices, and possibly for reference to the Institution's Register of Consultants; the Health and Safety Executive for current publications, and for general guidance on legal responsibilities of occupiers in relation to special environments; and the Lighting Industry Federation for current publications and the names and addresses of members of the Federation who are providers of specialist equipment for these environments.

7.10.1 Hazardous Zone classifications

Areas where a hazard may exist due to the presence of flammable substances are classified into three types which are designated 'Zones' in IEC Standard 79–10, thus:

> *Zone 0*: An area where a flammable gas-air mixture is continuously present or is present for long periods.
> *Zone 1*: An area where a flammable gas-air mixture is likely to occur during normal operation of the premises.
> *Zone 2*: An area in which flammable substances are so well under control that a hazard is only likely under abnormal conditions.

Note the correct use of the word 'flammable', meaning 'easy to set on fire'. The traditional word 'inflammable' means exactly the same thing, but its prefix 'in-' suggests the opposite, particularly to persons for whom English is not their first language, and is therefore no longer used in British Standards.

Areas which are not designated as Hazardous Zones are termed 'normal atmospheres', the use of the unofficial term 'Zone 3' being deprecated.

Zones are conceived to exist in three dimensions, not just in the horizontal plane (Figure 7.5). Thus, for example, in an area where there may be heavy petroleum vapour, while only specially constructed protected luminaires and electrical equipment may be used within a certain distance of the ground, equipment of lesser protection might safely be used above that level. In such a case, the Zone 0 might apply to any pits, depressions in the ground and spaces under vessels, from the lowest level up to the height above which it is judged that a gas–air mixture would be unlikely to be present, or would not be present for long periods. Above this there could be a further stratum designated as Zone 1, and yet another above that designated Zone 2.

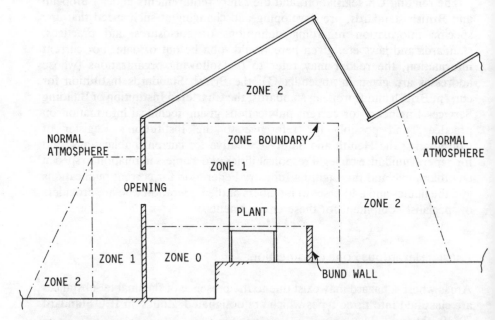

Figure 7.5 *Hazardous Zones. A sketch showing a building in cross-section (not to scale) indicating how zones might be defined in respect of risks associated with plant emitting a flammable vapour that is heavier than air.*

It is the occupier's responsibility (i.e. the responsibility of management) to ensure that these important regulations are properly observed. If, as is common, the personnel of the factory are not qualified to evaluate and declare the nature of the Zoning and the extent of the Zones, expert help must be obtained. Some advice may be given by the Health and Safety Executive inspector, but it is normal good practice to employ expert professional consultants who will issue certificates designating the Zoning. In each Zone, only equipment of appropriate type may be employed (section 7.10.2).

It is important to note that in addition to the hazards due to flammable atmospheres, some atmospheres are also wet or corrosive, so that the lighting equipment and installation components may need to be water resistant and corrosion resistant as well as certified for the appropriate Zone of use.

7.10.2   Protected equipment

Luminaires and electrical equipment for use in Hazardous Zones must be so constructed as to prevent fire or explosion being caused by sparks or by high temperatures as defined by *BS 5345*[38]. The various possible kinds of protection are described in the following sections:

- *Intrinsically safe:* Within a Zone 0, only intrinsically safe equipment may be employed. This equipment is so constructed that it cannot generate a spark or a temperature high enough to ignite a flammable gas-air mixture. Intrinsically safe electrical equipment operates at low voltage, and will have circuits of such high impedance that they are incapable of generating a hot spark. Equipment of this kind includes telephones, bell and buzzer systems, annunciators and instrumentation of various kinds. The only lighting equipment which is intrinsically safe is that operating by 'cold light' (section 5.13).
- *Flameproof enclosure:* Electrical apparatus may be contained within a flameproof enclosure capable of withstanding the greatest temperature rise that is likely to occur, and so constructed that the surfaces of the enclosure in contact with the surrounding atmosphere will not rise to a temperature such as to cause ignition. As the flameproof container is not sealed, contaminated air will enter the enclosure. The enclosure is robustly constructed, and is fitted with wide flanges, accurately machined flat, and with a controlled small gap between them. If the gas-air mixture within the enclosure should ignite, the enclosure is robust enough to withstand the force of the gas expansion. If internal ignition occurs, the gas pressure generated is relieved by escaping through the controlled flange gap, the flanges being wide enough to cool the gases to a safe temperature before they leave the luminaire. Thus the escaping gases would be unable to propagate flame outside the luminaire.

Protected equipment

Luminaires and other electrical enclosures designed for use in Zone 1 and Zone 2 are available. In these Zones, only equipment with appropriate certification may be employed. Luminaires for this duty are characterized by robust construction, and design features to ensure minimum temperature rise in use or under fault conditions.

7.10.3   Pressurized enclosures

Part 8 of *BS 5345*[38] deals with the various types of special electrical apparatus which may be certified for use in explosive gas atmospheres but which do not conform to any of the standard techniques of protection as specified in existing standards. There is increasing interest in the use of pressurized and ventilated luminaires which are continuously fed with a supply of clean air. It would seem that such luminaires could offer considerable advantages over conventional designs of enclosure for use in potentially explosive atmospheres, and some points in this regard are as follows:

- An enclosed luminaire that is continuously fed with clean air under pressure cannot aspirate flammable substances (nor moisture or corrosive material), and thus presents the possibility of operating the lamps and their control gear in an uncontaminated atmosphere within the luminaire.
- It can be arranged that the pressurizing air vents continuously through a relief valve through which the external atmosphere cannot enter. The resulting flow of air through the luminaire can be arranged to cool the lamp and control gear so that the luminaire may have a reduced surface area without undue temperature rise.
- It can be arranged that as long as pressure is maintained within the luminaire, the circuit will remain operable. If the pressure within the luminaire should fall (e.g. because of a fault, or because the luminaire was damaged or had been opened), a pressure switch within the enclosure would detect the fall of air pressure within the luminaire and cause a remote contactor to isolate the luminaire from the electrical supply. (The contactor would be in a normal atmosphere zone, or be housed in an enclosure suitable to the local zone.)
- Because a pressurized luminaire as described would not have to be constructed to withstand the force of an internal explosion, it could be constructed lightly at much lower cost than present conventional enclosures for hazardous zones.
- Pressurized luminaires as described could be fitted with a non-sparking means of connection and disconnection, e.g. a suitable plug and socket, and the air line could also be fitted with a coupler which would permit connection and disconnection without loss of air pressure. With these features it would be safe to enter a contaminated zone and disconnect and remove the complete luminaire for servicing outside the zone, and to replace it at once with a clean serviceable luminaire which was filled with clean air. It would thus become possible to carry out full maintenance and

lamp replacements without having to close down and ventilate the plant area.

- For greater reliability, luminaires in any location could be served with the supply of clean pressurizing air from local air receivers which would maintain the air supply to luminaires at pressure during any outage of the mains supply to the compressor.
- Should there be loss of electrical supply to pressurized luminaires as described, a time delay device could ensure that, on restoration of the electrical supply, the air pressure would be applied to the luminaires for long enough to purge them of any trapped gases before the electrical supply is reapplied to the luminaire.

It must be commented that pressurized fluorescent luminaires have been successfully used in many installations in the UK, usually in locations having a hazard from petroleum vapours, e.g. at garages, service stations and petroleum premises, and in areas where flammable printing inks are used or stored, and such installations have been approved by the Petroleum Officers, HSE inspectors and the insurers. What is being suggested here is that, subject to the necessary approvals, pressurized luminaires could be used to provide safe lighting in all industrial hazardous zones.

If such equipment were to become available commercially, with a choice of luminaire type (perhaps including HID lamp luminaires as well as fluorescents), complete with a range of compressors and air receivers, non-sparking plugs and sockets, air-line accessories etc., there would seem to be no reason for continuing the use of expensive equipment of other types. A standardized system of pressurized luminaires and accessories could do away with the technically difficult activity of determining the limits of zones, and only two zones would need to be considered – hazardous and non-hazardous.

Other benefits would stem from this policy. The same range of pressurized luminaires and auxiliary equipment could be used in damp, dusty and corrosive atmospheres, in wet or corrosive conditions (section 7.7), in clean-rooms and sterile rooms (section 7.4), and in difficult situations, e.g. in the presence of methane, hydrogen, and corrosive gases. It is quite possible that a single type of luminaire enclosure constructed of plastic materials could be resistant to corrosion and moisture, and suitable for use in all these hostile atmospheres.

The approval and general use of such a range of equipment would bring the boon of lowered costs of lighting to a wide range of situations where the present cumbersome and costly explosion-proof types of luminaires must at present be employed, and would encourage the provision of better lighting conditions. Further, if the additional cost of pressurized luminaires over standard luminaires was not great, it would encourage the use of the former in borderline situations where at present there is a tendency to fit unprotected equipment and hope for the best.

While it advocates a rethinking of the problem and suggests possible new practices, nothing in this section should be taken as proposing that any lighting or electrical apparatus should be used in any potentially explosive or flammable atmosphere unless it has been approved for use in that atmosphere.

7.10.4 Economy in lighting hazardous zones

The cost of providing lighting in hazardous zones may be reduced by finding ways of making the hazardous zone smaller, for example, by removing from the zone all activities which are not in themselves hazardous. If there is one hazardous process in a department, it may be economical to move that process to separate premises to permit the rest of the department to be one having a normal atmosphere.

Another approach to economy is to consider applying to interior installations a practice well known in exterior lighting work, in which light is projected into a hazardous zone from non-protected luminaires sited in an adjacent zone of normal atmosphere. This idea can be applied to industrial buildings having roof-lights, where luminaires may be mounted above the roof glazing in normal atmosphere and project their light into the building (Figure 7.6). Such an installation must be properly engineered to a standard

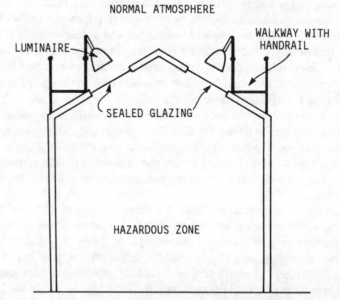

Figure 7.6 *Luminaires designed for exterior use in normal atmospheres mounted above sealed glazing to light an interior which is a hazardous zone (not to scale).*

acceptable to the Health and Safety Executive, to the enforcing officer under the Fire Precautions Act, and to the insurers. It would be required that air leaving or being extracted from the building does not pass near the luminaires. Access to the luminaires must be provided by installing walkways on the roof. The light loss through the glazing can be minimized by arranging that the flow of light is substantially normal to the plane of the glazing. There will be appreciable light loss if the glazing is wire-reinforced and has a figured surface. It might be practicable to replace the existing glass with plate glass having a layer of plastic sheet or plastic film on its underside, or with a protective wire-mesh screen under, to guard against glass fragments falling if a pane should be broken.

The piping of light into hazardous zones from lightsources located in a non-hazardous zone is another option that is likely soon to become more practicable and economic (section 5.13).

7.10.5 Problems during installation

There are likely to be few problems regarding installing proof luminaires and electrical equipment in new premises, or in those in which the plant is not yet in operation, but if the installation has to be carried out in what is already a hazardous zone, the difficulties may be considerable.

For example, it is necessary to ensure that no sparks are created, and, of course, there must be no naked flames and no welding during installation. This will require the use of nut-and-bolt fixings, but will preclude the use of masonry drills and cartridge-operated bolt-fixers. Any powered means of access or transport would have to be fitted with proof-type electrical drives, and, if engine-driven, would have to be adapted and certified for use in the presence of the particular atmospheric contaminant.

No 'live' connections may be made to wiring in a hazardous zone unless the plant is shut down to safety, the area purged and ventilated and a 'permit to work' issued by the responsible engineer[18, 38]. During the time the 'permit to work' is valid for making electrical connections or carrying out hot processes (e.g. welding, soldering), the permit for normal operation of the plant is withdrawn, and only becomes effective again when the responsible engineer has carried out acceptance inspection of the new installation and is satisfied that it has been left safe.

8 Lighting, health and safety

Many studies have shown that the frequency and severity of industrial accidents can be significantly reduced by the provision of 'sufficient and suitable lighting', this being confirmed by the extensive experience of users of good lighting. As may be expected, these benefits tend to offset the cost of the lighting, and it is commonly found that the quantifiable benefits due to the improved output or improved quality of work, as well as the reduction in accidents, may be many times greater than the cost of bringing the lighting system up to reasonable modern standards (chapter 21).

8.1 Management responsibility

8.1.1 A report of a five-year study by the Accident Advisory Unit of the Health and Safety Executive (HSE) in the UK[6] states that the most safety-conscious companies are those which tend to be commercially successful. The report also states that the management characteristics needed to achieve a high standard of health and safety for employees are the same as those required for efficient production.

8.1.2 One of the reasons for the failure of some companies to take effective measures is the lack of appreciation and lack of involvement on the part of senior executives. Decisions about health and safety should rank equal in importance to those regarding the efficient operation of the business. It may well be that many managers do not realize that there is a close link between the standard of lighting provided and the standard of safety to be achieved in industrial premises. Thus, the responsibility for lighting is often placed with an executive of limited budgetary and decision-making powers.[5]

8.1.3 The law requires that all business premises shall be provided with sufficient and suitable lighting (appendix A). Lighting installations which have been designed and installed to the recommendations given in this book will undoubtedly contribute significantly to the safety of all persons entering the premises. It is thus prudent for management to provide such lighting in order to reduce their risk of being prosecuted in the event of an accident occurring.

8.1.4 Accidents may cause fatalities or injuries, and often cause damage and loss of production, all of these being potent causes of financial loss. Losses may be minimized by insurance, but (quite apart from humane concern for those who may be injured) it is usually far more economical to take steps to prevent the accident occurring in the first place. The effects of the *Electricity at Work Regulations, 1989*[18] are far-reaching. It should be noted that the HSE inspector has merely to ascertain that an electrical accident involving injury or death has occurred to establish that an offence under these regulations has been committed (appendix A).

8.2 Poor lighting a cause of accidents

Occupiers have a legal duty to provide 'sufficient and suitable' lighting on their premises, to ensure the reasonable safety of employees, visitors or members of the public who may enter (appendix A). It has long been recognized that the state of the lighting in a workplace has important effects on the frequency and severity of accidents, both those which cause personal injury and those that merely result in damage or loss. For every actual accident there are probably several 'near misses'. People who work in places were accidents occur frequently tend to work a little slower as an unconscious precaution, so such premises often have low productivity.

8.2.1 Accidents are costly; they not only affect the injured person, but involve other workers and senior staff who have to deal with the matter[6]. The drama of an accident causing personal injury can affect the output of those nearby for many days, and a fatality in the workplace will depress productivity and profitability of the whole organization for a long period. Such considerations are additional to the humanitarian necessity of protecting people from harm.

It is regretted that the form used by employers to report an accident with personal injury under the Health and Safety at Work Act[25] does not require the state of the lighting at the scene of the incident to be recorded, and described in detail. Were this done, valuable statistics would be available similar to these available from the reporting of road accident injuries, where the police always record the lighting conditions at the time of the incident.

8.2.2 Poor lighting can produce adverse psychological effects on occupants. A workplace that is gloomy is depressing, and tends to make the occupants easily fatigued and irritable, a factor which might predispose them to acts which can lead to mishap. There is plenty of evidence to show that poor lighting does contribute to accident causation by its psychological effects. This is not a new discovery; the author wrote his first paper on the topic in 1949[15], and continues to deplore the injuries and deaths suffered by persons in the accidents which are reported[23], most of which are preventable.

Ill-lit premises soon become dirty and neglected. Untidiness, accumulations of rubbish and floor obstructions set the scene for accidents and fires. Bright premises tend to be kept neat and clean, and such 'good housekeeping' promotes safety and also reduces fire risks.

8.2.3   The most common fault with industrial lighting is simply not having enough of it, and failing to keep the lighting installation in good order. Such practices are false economies and cannot be financially justified (chapter 22).

Over-spaced luminaires will produce patchy illumination and increased dangers, as will failed lamps or dirty luminaires. Lighting that is insufficient in quantity, or poor in quality, i.e. causing serious glare (section 8.3), can lead to employees failing to perceive a danger until it is too late to save themselves or a workmate from injury.

Dense shadows can lead to tripping and falling accidents. In poor light employees may not see a projection upon which they may impale themselves, or they may not see a stationary crane-hook, and walk into it. Shadows and glare may prevent employees noticing a gap between the tailboard of a lorry and the edge of the loading bay.

8.2.4   Poor lighting can lead to persons making mistakes about depths or distances, particularly if the light flow is in an unusual direction (section 2.3). *Discontinuous light*, i.e. from flickering lamps, or light seen intermittently (for example, through the spokes of a rotating wheel) can lead to confusion and danger (section 2.8).

8.2.5   Lights which are unexpectedly extinguished (e.g. by someone at a remote switch position) can leave employees exposed to danger (section 22.4), while lighting which fails because of power failure can leave occupants without means of escape at a time of special danger, e.g. in a fire (chapters 14 and 19).

8.2.6   Despite the provision of machine guards, many accidents occur because subjects put their hands into dangerous places, often while trying to gain information manually that is not available visually (section 9.1). A few years ago, when the UK paper manufacturing industry had a campaign to enhance the lighting in paper mills, there was a significant reduction in personal accidents within twelve months. The improved lighting enabled operatives to *see* the tightness of paper webs without 'patting the reel' and placing their hands in danger. Lighting should enable workers to obtain the information they need without exposing themselves to danger, i.e. to 'use eyes, not fingers'.

Defective vision, as well as defective lighting, can lead to accidents. Short-sighted people may 'peer', bringing their eyes closer to danger. The practice of regular eye examinations or vision screening is recommended (section 1.10).

8.2.7 There are potential dangers of electric shock and burns from improvised lighting. It would be ironic as well as tragic if lighting intended to make the workplace safer was itself a source of danger. If 'temporary lighting' is properly installed and used it can be very safe (section 13.2). If portable lighting must be used, considerable increase in safety is obtained by use of reduced-voltage (chapter 13). The accidents that occur while electrical installation work or maintenance of lighting is being carried out are usually easily preventable (section 23.2).

8.3   Glare and adventitious light

Light which comes to the eye in an uncontrolled fashion and which does not contribute to effective vision is termed *adventitious light* and frequently is a cause of discomforting or disabling glare (section 1.5). Glare may also be caused by ill-designed lighting installations, though following the established design procedures can reduce glare from general lighting systems to an acceptable level (chapter 9).

In the presence of glare, the 'seeing value' of the available light is diminished. The dazzling effect can prevent the subject seeing a danger, while mental confusion may cause him to fail to take avoiding measures in time. Small points of glaring light (e.g. from spotlights of any kind) can have a hypnotic effect, making the subject more accident-prone.

Examples of adventitious light include shafts of direct sunlight entering a building, and so increasing the range of illuminances as to seriously handicap vision. Light from hot processes (e.g. furnaces, drop-forging) can swamp the normal lighting. In a study of lighting conditions in one drop-forge, it was found that the amount of light available for movement depended entirely on the chance of how much light was being produced by the process at that moment – an intolerably dangerous practice. In all locations where luminous processes occur (e.g. glass industry, foundries, drop forging, hot metal extruding etc.), the general lighting should be provided at an illuminance that will ensure (a) that there is adequate general illuminance for safety in the absence of adventitious light, and (b) that, if the adventitious light cannot be controlled, the general illuminance is sufficient to prevent disabling glare from the adventitious light. The provision of light-coloured walls and ceiling will help with this problem, though maintenance of such surfaces may be difficult or costly in dirty atmospheres (suggesting that smoke and dusts should be more efficiently exhausted from the building).

A possible cause of accidents is the effect of moving lights, for example, down-lights fitted on an overhead gantry crane. Persons working below can become confused in their orientation when the shadows about them start moving, and can fall. Moving overhead lights are always potentially dangerous.

8.4   Eye protection

A common and avoidable cause of injury is *welder's flash*. This is a painful and serious injury of the eye (properly termed *retinitis*) caused by exposure of the eyes to the ultraviolet radiation in the welding luminance, and is a possible cause of blindness. Protective goggles to screen the eyes from harmful ultraviolet radiation should be worn at all times in welding shops. The welding area should be provided with fixed or portable screens to protect the eyes of anyone nearby or passing through the department who may not be wearing protective goggles. Goggles and screens also give eye-protection against sparks and metal splashes.

It is noted that in workshops where the general lighting is inadequate, the welders often touch off the electrode to make a flash for the purpose of orientating the rod to the task, for they cannot see the tip through their dark goggles. The provision of one or more focusing spotlights on the task area to give a local illuminance of 2000–3000 lx will aid the task and lead to better quality welding, as well as preventing many cases of welder's flash.

A form of personal injury that is almost entirely preventable is that due to sparks or foreign bodies entering the eye. Every employer has a duty in law to cause persons exposed to this danger to wear eye protection[26]. Where workers are required to wear eye-protection continuously, it is sometimes found they are reluctant to do so if the illuminance in the work area is very low. Good practice is therefore to increase the general illuminance in such areas (both interior and exterior) by at least one 'step' above the design service illuminance as determined by the flow chart in the *CIBSE Code*[1] (as represented by the questionnaires in Appendices B and C).

For many workplaces, the most convenient form of eye-protection is a transparent wrap-around face screen device. Particles of dust adhering to the surfaces of such eye-protection screens tend to handicap vision. It will be found that the brightness of the particles will be reduced if the subject wears an eye-protection device that incorporates a peak above the screen to shade it. The underside of the peak should be of a dark colour. The use of a peak is also effective to shade goggles and the windows of respirator masks. Transparent screens and goggle windows should be cleaned as frequently as is practicable.

8.5   Lighting and health

Throughout this book it is emphasized that good lighting conditions promote health and efficiency by enabling proper use of eyesight, and a number of examples are quoted of persons suffering eyestrain, headaches, fatigue and depression as the result of working in adverse lighting conditions.

8.5.1   Some doubtful claims

From time to time there are attempts by persons and organizations of doubtful motivation to exploit our quite natural fear of eye damage. Often, they offer spurious products having some connection with ultraviolet radiation. For example, after the introduction of the three-colour triphosphor tubes, certain foreign lamp-makers marketed what they described as 'continuous spectrum tubes', claiming that working under them was healthier and, in some unspecified way, better for the eyes. It happens that triphosphor tubes tend to emit less ultraviolet light than do some other tubes, so the vendors of the 'full spectrum' or 'continuous spectrum tubes' made a feature of the fact that their tubes emitted a higher ultraviolet level. These matters were written up in the press in a misleading way, and an organization appeared which had the objective of persuading the general public that 'only continuous spectrum tubes with enhanced u.v. are conducive to good health' – a statement of obvious untruth. It is believed – and hoped – that that organization is now extinct.

In contrast, other unscrupulous traders attempt to exploit public fear and ignorance by offering lamps, or devices to put over lamps, which are claimed to have features that 'reduce ultraviolet radiation to a safe level' – again an entirely unjustified and false claim.

From time to time organizations appear which sell some very strange fluorescent lamps indeed, generally made in the USA, for which almost miraculous claims are made. One, for example, recently claimed that its lamps not only made foodstuffs look fresher, but actually prevented them from going stale while on display! – a claim which they were unwilling to justify in court when legal action for false representation and a claim for damages was threatened by a dissatisfied purchaser.

The reader is advised to examine all such claims with a very critical eye, and not to accept any unproven statements regarding the properties of lamps or other lighting equipment.

The normal physiological phenomenon of night myopia has recently been described in the press as a disease, and an unscrupulous person has offered a 'cure'. Night myopia was first described by Lord Rayleigh about 1899, and appears to be nature's way of enabling us to focus our eyes on the ground ahead of us when walking in very low light conditions (section 1.7.3). There is certainly no need for a cure, though a better understanding of how it works, and how it relates to the bland field effect and the sleeping sentry syndrome would be valuable in reducing some dangers due to visual error. (Night myopia should not be confused with true *night blindness* which results from disease or damage of the retina with loss of peripheral vision.)

8.5.2   Light-related medical conditions

Miner's nystagmus

To remain healthy, our eyes need to be used under acceptable conditions, for gross abuse can cause disease or permanent damage. For example, there used to be a high incidence of a condition called *miner's nystagmus*, in which the eyes of the sufferer tremble in their sockets, making all fine tasks such as reading quite impossible. Miners who worked underground for many years in the minute illuminance which was provided by Davy oil safety-lamps were the only sufferers of this disease. From the mid-1940s onwards, there was considerable modernization of UK pits. Electric cap-lamps with xenon gas-filling became available, better tunnel lighting and the limewashing of tunnel walls was introduced, and better methods of coalface lighting were developed using small flameproof luminaires containing electric generators powered by compressed air. As the illuminances employed below ground were increased, so the incidence of miner's nystagmus declined. The disease has now totally disappeared from the UK mining industry.

Seasonal Affected Disease (SAD)

It is common knowledge that many people become depressed and lethargic during the short days of winter. It is noted that the incidence of depression and suicide in the Scandinavian countries is highest during the months when there is little daylight. One of the worst deprivations prisoners may suffer is not having a window through which they may glimpse the sky. Newly blinded persons may go through a period of deep depression as well as disturbances of sleep rhythm and digestive functions, the more so if they remain constantly indoors.

The depressing effect of lack of exposure to daylight is thought to relate to disorganization of the normal diurnal increase and decrease in the hormone melatonin which regulates our degree of activity and is linked to our sleeping rhythm.

In earlier agricultural societies, people spent far more time out of doors than we do today, and had greater exposure to daylight. It is perhaps for this reason that many people, nearly always town-dwellers, suffer depression and malaise during the winter months. The condition is now recognized by the medical profession as a real illness, and is called seasonal affected disease or 'SAD' (also known as 'Personal Affected Disorder').

The treatment of patients suffering from SAD is simply to expose them to more light. They benefit from spending more time out of doors, as well as being in well-lit environments. Sight of the sun helps them greatly, but even glancing for a few seconds at a bright lamp several times a day seems to help them get their biological rhythms back in step with the daily passage of the sun, and thereby lift them out of their condition of pathological depression.

The prevention of SAD is merely to ensure that one takes time to walk outdoors for a period every day – at all seasons, and in all weathers – and to provide 'sufficient and suitable lighting' in our living and working environments.

Repetitive strain injury (RSI)

This is another disease which seems to be a result of life in our modern environment. It is a painful and disabling condition which may affect virtually any part of the body, but which typically affects the wrists, fingers or neck of the subject. The sufferer is often a typist or keyboard operator, or a person on a production line performing a repetitive task (which may not involve the application of much physical effort). Professional writers seem to be particularly vulnerable to RSI[43].

The condition appears to come about from maintaining an unnatural posture while performing a repetitive movement. It results in painful inflammation of tendons and muscles. Treatment consists of a long period of rest – possibly of months – perhaps with pain-relieving medication and physiotherapy.

Prevention is a matter of ensuring that persons performing repetitive tasks (such as working at keyboards and using VDT screens) are provided with such lighting conditions that they do not take up unnaturally strained postures in trying to do their work while distracted by reflections from data screens or other objects, and perhaps wearing special long-focus spectacles while working (section 9.3.2). Some booklets on RSI (for employers and employees) are available from the Health and Safety Executive (appendix G).

Part 4

Interior lighting

9 Interior lighting design

9.1 Objectives

The ease with which mathematical design of lighting can be performed with the aid of calculators or computers may lead lighting designers into the false belief that lighting design is mainly a matter of number crunching. It should be recognized that designing the lighting for any interior may require far more than merely blanketing the space with uniform general lighting. In any situation where visually demanding work is carried out, in addition to providing the operatives with sufficient general illumination of suitable quality, there are a number of practical factors which may be considered which will further improve the performance of visual tasks, and which may also contribute to comfort, productivity and safety. These factors are illustrated by the examples in the following sections.

Enhancement of contrast between the object and its background

In a factory assembling small relays, the workers often cut their fingers when picking sharp components out of fibre tote-boxes. The action taken was to replace the tote-boxes with shallow trays into which the workers could see easily; this not only enabled the workers to avoid cutting their fingers or running sharp objects under their nails, it also enabled them to pick up the components in correct orientation for immediate assembly, thereby gaining in work-rate.

Enabling the operators to see the object with better contrast or in silhouette.

In the case quoted above, the components were matt black, so the insides of the new trays were painted white to aid contrast between the component and the inside of the trays. Also, the relays had to be finally adjusted by hanging them up at the workstation and adjusting the springs and contacts while a weight simulated the pull of the magnetic coil. The action taken was to provide a well-illuminated white board at the back of each workstation, so the mechanisms could be seen in silhouette. This simplified the task, and improved both output and accuracy.

Improvement of the colour-scheme in the workplace

In a drawing office, neutral light tones on all visible surfaces produced an uncomfortable 'bland field' effect in which the eye found no resting place (section 2.3). The illuminance at drawing-board level was 1200 lx, and the calculated glare index was 18.8, fully satisfying the requirements for illuminance and glare limitation, yet there were continuing complaints about the lighting. The action taken was to introduce some small areas of vivid colour into the décor by application of paint to convenient surfaces, and the addition of some strongly-coloured paintings on selected walls. The paintings and one or two of the new colour patches were spotlit with a few low-power spotlights on track, and – purely for psychological reasons – the tubes in the overhead general lighting were changed to others of a slightly different colour-appearance (but having the same light output and colour-rendering). The staff expressed complete satisfaction with the improvements.

Introduction of directional lighting

Example: A department where hides and reptile skins were graded and sorted for cutting, had been moved to an area with rooflights and no wall windows, i.e. the flow of light was substantially vertical. The staff complained that the lighting was inadequate, even when the measured mix of daylighting and general lighting was over 1500 lx. The action taken was to provide blinds to be drawn over selected rooflights on bright days, and some asymmetrical-distribution wall-mounted luminaires were placed at one end of the worktable. This introduced a near-horizontal flow of light which revealed the surface texture of the hides and aided the work. It was noted that output and quality improved in the winter when the daylight was limited. Similar effects have been noted in 'salles' (inspection rooms) in the paper industry, where light flowing at a flat angle to the paper sheets aids their inspection.

It cannot be too strongly emphasized that the design of any lighting installation should satisfy the eyes of the occupants as well as meeting the theoretical desiderata. The degree of satisfaction that occupants feel with their visual environment affects them emotionally, and this in turn affects their work performance. It would be quite wrong to neglect the aesthetic aspects of the lighting in even the most functional of industrial locations.

9.2 Lumen method calculations

This section summarizes the essentials of the basic calculations used in designing interior general lighting installations. It is intended to familiarize the reader with the concepts and terms used in calculations by the lumen method, but it is not intended to supplant the book *Interior Lighting Design*[8] upon which much of what is written here is based. The reader should also refer to the *CIBSE Code*[1] for further information.

9.2.1   Lumen method

There are now available sophisticated lighting calculation programs which enable the results of any proposed lighting layout to be computed rapidly and with great accuracy. Such a program can determine how light will be reflected and re-reflected throughout the cavity of the room, even to the extent of being able to calculate accurately what will be the illuminance on the underside of the tables! Such programs are but an extension of the simple mathematics of the lumen method.

The lumen method of lighting design is still extensively used, and is of great utility. However, it is by no means the whole technique of how to plan interior lighting. It is merely an arithmetical procedure for designing luminaire layouts for the general lighting of interiors.

In respect of an interior of stated dimensions which it is desired to light to a specific average illuminance (lux) on the working plane, it enables the designer to determine the following:

- the required number of luminaires of a selected type (taking account of the available and practically possible mounting-height, and the recommended spacing-to-mounting-height ratio for the chosen luminaire type);
- the required total lumen output, and hence the power and number of lamps required per luminaire.

A further calculation enables the designer to confirm that the glare index of the resulting proposed layout will not exceed the limiting glare index specified for that application in the *CIBSE Code*[1] (section 9.3).

9.2.2   Basic equation

The basic equation is:

$$E_h = \frac{l \times n \times N \times \text{UF} \times \text{MF} \times \text{Abs}}{A}$$

where E_h is the illuminance (lux) on a plane, e.g. the floor or a working plane (section 9.2.3); l is the number of lumens per lamp. Unless otherwise specified, this will be the lighting design lumens (section 9.2.4); n is the number of lamps per luminaire (section 9.2.5); N is the number of luminaires (section 9.2.6); UF is the utilization factor (section 9.2.7); MF is the maintenance factor (section 9.2.8); Abs is the absorption factor (section 9.2.9); and A is the area (m^2). The equation can be transposed to discover any unknown.

9.2.3   Horizontal illuminance (E_h)

If not otherwise specified, the working plane is assumed to be 0.85 m above the floor. The general illuminance on the working plane is usually required to be provided at a uniformity of 0.8, i.e. the ratio of the minimum illuminance to the average illuminance should not exceed 0.8. If task lighting is provided (section 9.5), the general illuminance in areas adjacent to the task areas should be not less than one-third of the task illuminance.

9.2.4   Lumens per lamp (l)

The lumen value employed is normally the *service lumen output* per lamp, i.e. the average through life. If it is desired to know what the illuminance will be at the end of lamp life, this figure may be replaced with the *end-of-life lumen output* per lamp.

9.2.5   Number of lamps per luminaire (n)

The number of lamps per luminaire will usually be one for HID lamps; for fluorescent tubular lamps there may be some choice.

9.2.6   Number of luminaires (N)

The objective of the calculation is often to determine this factor, all the other factors having been inserted and the equation transposed. The minimum number of luminaires which may be used to achieve satisfactory uniformity is determined by the maximum *spacing-to-mounting-height ratio (S:H_m)* specified by the luminaire maker. The spacing between luminaires is governed by how widely the light from them is spread about the vertical axis. For typical luminaires (fluorescent tubes and HID lamps) used at mounting heights up to around 6 m, the maximum ratio is usually 1.5:1. Thus, if the mounting height above the working plane is 5 m, then the maximum spacing between luminaires will be 7.5 m. For *high-bay luminaires*, typically for use above 5 m mounting height, the ratio is usually 1:1.

As one cannot have a fraction of a row or a fraction of a luminaire, calculated numbers of rows and luminaires may need to be rounded up. There is generally no objection to spacing closer than the maximum permitted, provided it is economical. Indeed, with fluorescent luminaires it may be economical and very practical to mount them in continuous lines end-to-end (e.g. linked, or mounted on trunking).

9.2.7 Utilization factor (UF)

In general lighting installations, not all the lumens emitted from the luminaires reach the working plane; some are trapped in the luminaires and some will be emitted towards the walls or ceiling, and only a proportion of these will be reflected back to the working plane. For any luminaire, the 'direct ratio', i.e. the proportion of its downward light that goes directly to the working plane, varies with the size of the room and the mounting height employed. In a low-ceilinged large room, a particular luminaire might project all its downward flux on to the working plane. The same luminaire in a small high-ceilinged room might project part of its downward light output on to the walls, we say that the utilization factor (UF) in this application is lower. The UF for any interior lighting layout is a decimal fraction of the downward flux from the overhead luminaires that is delivered to the working plane (including those reflected from the walls and ceiling), a factor that varies between between about 0.26 up to about 0.54.

The UF is defined thus:

$$UF = \frac{\text{Lumens reaching the working plane}}{\text{Total lamp lumens}}$$

The UF for a proposed installation may be found by referring to tables published by the luminaire manufacturers for their products, or by reference to *Interior Lighting Design*[8].

In order to find the appropriate UF from the tables, it is necessary first to calculate the *room index* (RI), thus:

$$RI = \frac{LW}{H_m(L + W)}$$

Where L is the length of the room (m), W is the width of the room (m), H_m is the mounting height of the luminaires above the working plane (m).

To derive the UF from the tables it is also necessary to enter the reflection factors of the walls, ceiling cavity and floor cavity (Figure 9.1).

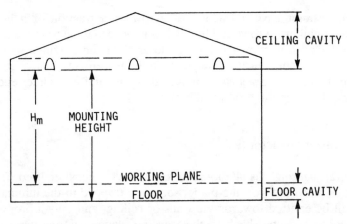

Figure 9.1 *Vertical dimension parameters in lighting calculations.*

9.2.8 Maintenance factor (MF)

The MF is defined as:

$$\text{MF} = \frac{\text{Lumens emitted by the luminaires when soiled}}{\text{Lumens emitted by the luminaires when clean}}$$

In heavy industrial locations, cleaning may need to be carried out at intervals not greater than once every six months, but exceptionally dirty locations may justify quarterly cleaning. In typical 'normal clean dry interiors', cleaning is carried out annually, and it is only at exceptionally clean locations that the intervals between cleanings may be extended to two years (section 23.4).

Past practice was for the lighting engineer to be concerned only with the light lost due to dirt on the lamps and luminaires; modern practice is to take account of light that is absorbed by soiling of all the room surfaces by the calculation of a *light loss factor*.

9.2.9 Light loss factor

The MF normally employed as defined in section 9.2.8 is the ratio of the lumen outputs of the luminaires in clean and soiled condition, and normally assumes that the calculation will be made with the lamps emitting their lighting design lumens. This is anomalous, and has a small inaccuracy because, by the time the luminaires have become soiled, the lamps will have aged to some extent. Further, the room surfaces also will have become soiled and be less reflective. These inaccuracies can be overcome by introducing the concept of the *light loss factor*.

The light loss factor is the product of the *lamp light loss factor*, the *luminaire MF* and the *room surface MF*. Its calculation and use is described in the *CIBSE Code*[1]. Some typical factors are demonstrated in Table 7.

Table 7   *Typical light loss factors*[a]

Recommended illuminance basis	Lamp lumens	Lamp light loss factor	Room category	Luminaire and room surface light loss factor	Total light loss factor
Maximum illuminance	Initial lumens (100 h)	0.8	Clean	0.85	0.68
			Average	0.75	0.60
			Dirty	0.60	0.48
Minimum illuminance	End of life lumens[b]	1.0	Clean	0.85	0.85
			Average	0.75	0.75
			Dirty	0.60	0.60
Service illuminance	Initial lumens (100 h)	0.9	Clean	0.90	0.72
			Average	0.80	0.72
			Dirty	0.70	0.63
Service illuminance	Lighting Design Lumens[c]	1.0	Clean	0.90	0.90
			Average	0.80	0.80
			Dirty	0.70	0.70

[a] Based on the table in the 1984 edition of the *CIBSE Code*[1]
[b] At 70% of rated life
[c] At 2000 h

9.2.10   Absorption factor (Abs)

The absorption factor is a factor used in Lumen Method calculations to allow for the light loss due to absorption and scattering of light during its passage from the luminaire to the working plane. This factor is often completely omitted from design calculations, or if employed is often much underestimated.

From the author's work in studying the absorption effect of atmospheric pollution in interior lighting installations, it has become apparent that even a small degree of atmospheric pollution in a high-roofed building can not only cause significant light loss due to absorption, but can have such a scattering or diffusing effect as to change the nett shape of the luminaire distribution. In bad cases this may be equivalent to raising the BZ number by one or two units, resulting in the effective UF being much reduced. In other words, users get significantly less light than they expect.

In high-roofed industrial buildings there may be serious underlighting due to relatively small concentrations of steam, dusts, atmospheric moisture, oil haze or printing-ink haze. As there is no way of forecasting this loss accurately, it is recommended that experiments are conducted at the actual location.

To give an appreciation of the magnitude of light loss that can occur, measurements made in a metal stamping shop on two successive nights showed a difference of 13% in illuminance, the change being accounted for by an almost imperceptible degree of mist that percolated into the building from a nearby canal on one of the nights. In another location (a drop forge department) a difference of 24% was recorded between clean conditions (before work started) and polluted conditions (six hours later). The absorption loss in a food factory 'kettle room' containing autoclaves was estimated at around 12%. In a badly polluted foundry, the absorption loss was estimated at around 25% during pouring; in this case, soiling of the luminaires was visible within one shift after cleaning.

9.3   Glare in interior lighting

9.3.1   Direct glare in interiors

A glare index is a non-quantitative number representing the degree of subjective glare directly received from overhead luminaires in an interior general lighting layout. As explained in CIBSE Technical Memorandum No. 10[17], glare indices range from 10 (low) to 30 (high).

The *CIBSE Code*[1] recommends Limiting Glare Indices in the range 16–22 for fine work, where reduction in glare will materially improve operator performance. In the performance of coarse tasks, a glare index in the range 22–28 may be justified if reduction would be costly, for any task improvement is likely to be marginal and the fatigue reduction notional, especially if the exposure to glare is of short duration. It is necessary to keep a sense of proportion in this matter. Striving for very low glare indices is not usually rewarded by any marked economic improvement, and very glare-free environments tend to be lacklustre and uniform to the point of being boring.

We can tolerate fairly high glare indices for short periods, e.g. driving at night on unlighted roads we cope with the headlights of oncoming vehicles. However, if severe glare is likely to affect performance (e.g. output rate, quality) or safety (e.g. by masking dangers, preventing adequate vision for drivers) then steps must be taken to reduce direct glare. In many industrial situations, the important glare will be that reflected from specular or wet surfaces (section 9.3.2).

The glare index in any particular situation is affected by factors such as:

- the intrinsic brightness of bright parts of lamps and luminaires at angles at which they can be seen by the room occupants;
- the contrast between the brightness of the luminaires to the ceiling or upper walls against which they are seen;
- the angle between the glare source and the subject's object of special regard while working;
- the reflection factors of the décor of the room;
- the mounting height of the luminaires above the working plane.

The method of calculation of glare index is explained in *Interior Lighting Design*[8], and the importance of the control of direct glare is one of the quality factors in lighting which is stressed in another useful reference, *BS 8206*[40].

The British Zonal Classification is a way of classifying interior general lighting luminaires by the shape of their polar curves. Ten standard distributions are defined by indices which range from BZ1 for luminaires having a narrow angle of spread (as used in some high-bay units, and giving a spacing ratio ($S:H_m$) of 1:1 or less), up to BZ 10 having a wide dispersive distribution (as used in some dispersive luminaires, giving a spacing ratio of 1.5:1 or even greater).

Luminaires having low BZ classifications, if correctly applied, have the capability of producing less direct glare to occupants, and the BZ number allows the calculation of a glare index. The BZ classification also affects the UF (section 9.2.7). Utilization tables and glare index tables are provided in CIBSE Technical Memoranda Nos. 5[39] and 10[17].

Generally, an installation with luminaires of low BZ classification will project light strongly downward so that little light goes directly to the walls. Conversely, luminaires with higher BZ numbers will tend to direct more light towards the walls. It will be seen that the UF is determined by the interaction of the room proportions (via the room index) and the luminaire distribution (e.g. by the BZ number).

The economic value of having light-coloured walls and ceilings in a factory is considerable, no matter what type of distribution is given by the luminaires (chapter 23). If the BZ number is high, light coloured walls will reflect some light to the working plane; if the BZ number is low, light walls will help to lessen the sombre effect. Light-coloured ceilings are of value in helping to avoid the 'tunnel effect' (section 7.6.6).

The British glare index system does not take account of time, but in many real situations, the time taken to adapt to changes in field brightness is important, particularly in moving between areas lit to markedly different illuminances (e.g. between indoor and outdoor areas) (section 9.7).

An important factor in reducing glare in interiors is increased brightness of the background against which the luminaires are seen. The provision of light colours in the ceiling cavity will help reduce the glare sensation whatever type of luminaire is used. If the floor and bench-tops etc. are of light colour, they will reflect light upwards which will tend to reduce glare by brightening the ceiling.

Manufacturers sometimes offer luminaires described as 'glare free' or of 'low glare type'. Such terms are misleading, for it is the combination of the luminaire and the interior which determines the degree of glare experienced.

No useful purpose would be served by taking the reader through the mathematical steps to compute a glare index, for this is done excellently in *Interior Lighting Design*[8] to which the reader is referred. A very good method of approximation is given therein, which, even in the absence of proper photometric data from the luminaire manufacturer, will enable the possibility of unacceptable glare conditions to be predicted using some simple steps.

Purists may not agree, but it is the author's opinion that very great accuracy in calculation of glare indices is not justified. It should be noted that the limiting glare indices given in the *CIBSE Code*[1] were not scientifically calculated, but were empirically chosen on the basis of the experience of the compilers.

9.3.2 Reflected glare in interiors

In general, direct glare is not a difficult problem, for it can be minimized by selection of suitable luminaires and by good design of the lighting layout. However, indirect *reflected glare* tends to be more prevalent than may be realized, and has considerable ill effects on worker comfort and productivity. With computers now on virtually every desk, and used throughout industry in many diverse applications, the effect of reflections from VDT screens is presently receiving considerable attention. It should be made clear that this particular nuisance differs to some extent from other experiences of glare, for the reflections on the VDT screen rarely cause any discomfort – they merely mask the luminous images of characters on the screen, making the data more difficult to read.

It is held that the strain of trying to see VDT data against screen reflections can create stress in the subject. The Royal Society for the Prevention of Accidents has issued a booklet relating to this[41]. Regulations under the Health and Safety at Work Act, 1974[42] require employees engaged in such tasks to have regular eye tests (section 1.10) and that VDT operatives are to be allowed breaks from the work or a change in activity.

It is the author's opinion that most of the strong feelings that are generated about the alleged stress of VDT work are due to a failure to understand the ergonomics of working at a screen in the vertical plane instead of at a task on the horizontal plane. If users of VDT screens wear prescribed reading lenses made up to a suitable long focus, say to 600 mm, they would find that they can sit comfortably and relaxed at their work, and not suffer the constant neck-ache and back-ache which many operatives suffer through crouching and peering at the screen. This is hardly a new idea; professional pianists and orchestral musicians have been wearing such glasses (known as 'piano glasses') for fifty years. Unless they are very long-sighted, all workers at VDTs can benefit from wearing suitable long-focus spectacles when they sit before the screen. Persons who normally wear prescription lenses for close work will need lenses of their usually required corrective prescription, but made up with a long focus. (The provision of long-focus spectacles for this use is recognized as a proper business expense for which tax relief in the UK is available.)

Before leaving the topic of VDT reflections, it must be noted that there is not universal agreement that direct lighting with strongly downward directional lighting is always the best method of lighting these rooms. Some excellent installations use uplighting (section 9.8).

In industrial task situations, glare due to the images of overhead bright luminaires being reflected at the surfaces of polished metals, liquids etc, can not only mask what is to be seen, and make the subject uncomfortable, but can also create a degree of dazzle that actually disables vision (sections 1.5, 3.4).

Some modern luminaire distributions have been engineered to produce reduced direct glare and to enhance the illumination of vertical surfaces (e.g. the *batswing* and *trouser-leg* distributions), and these may produce a worse contrast rendering factor. In work locations where reflecting surfaces abound, e.g. in a sheet-metal works, it is impossible to control both direct and indirect glare in all directions, and the designer must make compromises. Very often the best method of reducing glare sensation will be to use large-area, low-brightness luminaires, or an overall luminous ceiling, or perhaps uplighting and a white ceiling. Note that the use of polarized light can enable the brightness of reflections at selected angles of view to be controlled (section 2.9).

As previously stated, glare conditions are not calculated on a time basis. The experience of disabling glare does not cease at the instant that the subject turns away from the glare source. An industrial worker might be placed in danger by the fact that an after-image of the glare source will remain with him for some seconds, during which time the worker will not have full visual ability until he or she adapts to the ambient luminance again.

9.4   Directional lighting

This section briefly summarizes the essentials of the basic calculations used in designing directional lighting installations. It is not intended to supplant the book *Interior Lighting Design*[8] upon which much of what is written here is based. The reader should also refer to the *CIBSE Code*[1] for further information.

9.4.1   Point-by-point method of calculating E

As explained in section 3.5, this type of calculation applies only to a source that is a 'virtual point source', i.e. one having negligible projected area in the direction of measurement in relation to the distance over which the calculation is to be performed. Such calculations ignore effects of reflection of light from nearby surfaces.

The relationship between illuminance, intensity and distance as stated holds good only if the plane of measurement is normal to the incident ray. When the incident ray strikes the plane of measurement at any other angle θ, the correction $\cos\theta$ must be applied as explained in section 3.6.

Some lighting manufacturers provide very helpful calculation aids for use with their products which are used for spot-lighting, local lighting etc., and these may obviate the need for calculation.

9.4.2   Designing for enhanced vertical illuminance

Calculations of general lighting deal only with illuminance on the horizontal plane, but in many practical situations the working plane is not horizontal; it may be vertical (e.g. a chalkboard, or the side of a machine), or angled (e.g. a drawing-board). The special measures of illuminance (vertical, cylindrical, scalar and vector) are discussed in section 2.5.

Valuable as they are to the proper study of illuminating engineering problems, the special measures of illuminance are not likely to be used in practical design work unless a suitably programmed computer is available which also carries a great deal of photometric data on the luminaires under consideration. To bring these ideas into a practical framework, let us state simply that the measurement of the horizontal illuminance is still a good measure of the amount of light in an interior. However, it must be stated that any steps taken to improve the mean cylindrical illuminance will probably enhance visual comfort as well as enhancing the illuminance on the vertical surfaces in the interior.

To achieve enhancement of the E_c/E_h ratio, the following steps could be taken:

- Use the type of luminaire that has the highest possible BZ number (e.g. the widest possible angle of distribution) compatible with the control of glare. Such would not exclude the use of batwing or trouser-leg distributions if such luminaires are appropriate and available.

- Employ a luminaire type that gives some upward light. This will not only help overcome the tunnel effect, but it will help to keep the luminaires clean if of a through-vented pattern (section 7.6.6).
- Use the lightest possible colours for ceiling, walls, bench-tops, machinery and floor, and institute a programme of regular cleaning and redecorating as necessary. A word of caution: by the application of light colours, do not create a featureless bland field (sections 1.7.3 and, 2.3.3).

In addition to the enhancement of vertical illuminance by the measures mentioned, there may be the possibility of employing local lighting to provide preferential illumination in desired directions (section 9.5).

Some control of vertical illumination can also be effected by positioning and orientation of overhead luminaires. For example, the penetration of light into shelf cavities and to the lowest shelves in a library or store is limited because of shadowing and the cosine effect. A practical point to note is that when lighting shelving facing into a narrow aisle, it is better to place fluorescent tubular lamp luminaires transversely to the aisles (Figure 9.2).

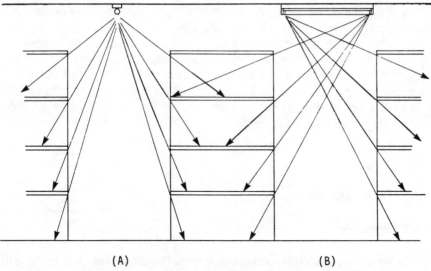

(A)   (B)

Figure 9.2 *Lighting of vertical faces of aisles in stores, libraries etc. (A) longitudinal alignment of luminaires gives poor penetration into cavities; (B) transverse alignment of luminaires allows light to penetrate better into shelf cavities.*

Improvement of conditions will be achieved by painting the shelving white (including the undersides), and painting the floor with a light-coloured floor paint to give upward reflections at the bottom of the stack.

In lighting large interiors, steps which may be taken to improve the vertical component of the illuminance include the use of batwing or trouser-leg distribution luminaires which give restricted output at the near-vertical angles, and distribute their lumens preferentially in a zone roughly between 40° and 70° from the downward vertical axis. This distribution, while giving restricted brightness above the 70° cut-off line (to limit direct glare) preferentially lights vertical surfaces facing towards the luminaires. Other methods include mounting asymmetric distribution luminaires on walls, and perhaps, in larger areas, at the heads of stanchions. If the enhancement of lighting of vertical surfaces is needed mainly in one direction, angled or parabolic reflectors may be used, positioning them so as to restrict discomfort glare being caused to persons who must work facing them.

9.4.3 High-roofed areas

In areas in which high mounting of luminaires is necessitated (by the presence of a gantry crane, for example, or the need to keep clear headroom for operational reasons), there may be problems regarding lighting distribution. If luminaires are mounted at 6 m or higher, 'high-bay' or concentrating luminaires of low BZ Number (section 9.2.11) will be required, resulting in the horizontal illuminance E_h greatly exceeding the vertical illuminance E_v so that vertical surfaces will tend to be underlit.

If luminaires with wider, less-concentrated beams are used, the wastage of light on the walls will make the scheme uneconomic (for it is usually impossible to maintain industrial wall surfaces at a high reflection factor). The problem may be ameliorated by the use of *cross-lighting* to enhance the illuminance on vertical planes by mounting directional luminaires on walls and possibly on stanchions (Figure 9.3).

9.5   Task lighting

9.5.1   Principles

There are some production activities requiring the continuous performance of very difficult visual tasks, e.g. assembly of small components, matching of printing inks, work on microelectronics etc., and for many of these *task lighting* is needed. The terms 'general lighting', 'task lighting' and 'localized lighting' are explained in section 2.7. Some information given in chapter 10 relating to inspection tasks may be applicable to the lighting of visually difficult continuous production tasks.

In all cases where localized or local lighting is employed, it should be to augment the illumination at points where visually demanding work is performed. It should never be used without adequate ambient lighting (either

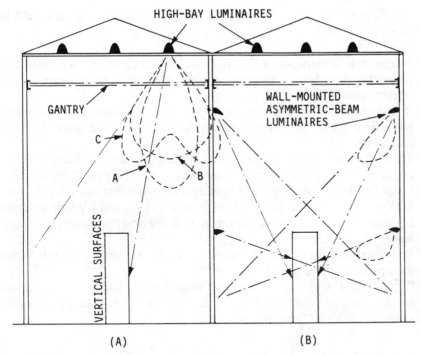

HIGH-BAY LUMINAIRES

GANTRY

WALL-MOUNTED
ASYMMETRIC-BEAM
LUMINAIRES

C

A   B

VERTICAL SURFACES

(A)   (B)

Figure 9.3 *Enhancing vertical illuminance in high rooms. (A) Conventional high-bay luminaires achieve the required horizontal illuminance, but fail to light vertical surfaces adequately. Typical distribution curves shown: A, high-bay; B, wider angle (dispersive); C, batswing or trouser-leg distribution. (B) the addition of wall-mounted cross-lighting luminaires to augment the vertical illuminance.*

from daylight or from the general lighting), and should not be provided merely to make good deficiencies in the general lighting.

The value of localized or task lighting in helping to make significant reductions in energy cost and running costs is simple to demonstrate. Consider the case of a workshop 20 × 10 m, in which visual tasks requiring an illuminance of 750 lx are performed. Assuming a UF of 0.6, and an MF of 0.8, lighting the whole area to that illuminance would require approximately 312 000 lamp lumens. But, if the visual tasks requiring 750 lx are confined to only small areas of the room, a general illumination of say 300 lx would suffice. If, for example, there were four benches for the visually difficult work, these could be lit to at least 750 lx by two local 85 w fluorescent tubes at low mounting level and well shielded. The total lumens required then would be 125 000 lm for the general lighting at 300 lx, plus 8 × 6200 = 49 600 lm for the local lighting, a total of 174 600 lm – making a saving of around 44% of the lumens used, with proportional savings in capital cost and energy consumption.

Local lighting must be used with care, and the following points should be observed:

- The general illuminance of the area should be not less than one-third, and preferably not less than one-half, of the local illuminance at the workstations.
- If task lighting is used, care must be taken to ensure that directable luminaires under the operators control do not cause glare to other occupants.
- The colour-appearance of the general lighting and the local lighting should be the same or as close to each other as possible. It is possible to use HID lamps for the overhead general lighting and fluorescent tubes for the local lighting, or fluorescent tubes for the general lighting and LV filament lamp lighting on tasks. The better the match of colour-appearance, the better will be the system.
- Illuminances at workstations should not create excessive contrasts within the operator's field of view.
- The *CIBSE Code*[1] recommends that, whatever the horizontal illuminance on the working plane, the interior surface illuminances should be in the ratios to the task illuminance shown in Table 8.

Table 8   *Ratios of illuminance in an interior*

Location	Relative illuminance	Relative reflectance
Task	1.0	Immaterial
Ceiling	0.3–0.9	0.6 minimum
Walls	0.5–0.8	0.3–0.8
Floor	1.0	0.2–0.3

9.5.2 Applications

The *CIBSE Code* recommends that the reflectance of the immediate background to the task should preferably be in the range of 0.3–0.5, a level of reflectance that is difficult to achieve in some factories. It also suggest that the night-time average reflectance of windows might be enhanced by the use of blinds, but this is entirely impracticable in many factories because of fire risk and the difficulty of keeping surfaces clean.

Multiplying out the relative illuminances and relative reflectances in Table 8 suggests that, for any task illuminance, the ceiling should have a relative luminance of 0.18–0.9, the walls of 0.15–0.64, and the floor of 0.2–0.3 – all conditions that may be very difficult to attain in factories where dirty processes are performed. Practical experience indicates that the visual conditions under local lighting should be such that the brightness (luminance) of the task should not be more than ten times that of the background formed by the general surroundings. Preferably the immediate background to the task will have a luminance of 30–50% of the task illuminance. This can sometimes be arranged by providing mobile sightscreens faced with easily cleaned plastic surfaces. The brightness of such screens can be controlled by colour, and perhaps by being washed with spill-light from the local lighting or by special lighting provided for the sightscreen surface.

A ratio of local/background/general brightnesses within the field of view of 10:3:1 is about the limit that can be tolerated for long, and a ratio of 10:5:2 will be far more acceptable as it is less tiring for the operator.

A great virtue of task lighting is that the operator can control both the intensity and the direction of the light flowing on to his task. A local-lighting luminaire with a means of adjustment will enable the operator to direct light into cavities or under projections, as well as sending light at a glancing angle to reveal surface characteristics of the object. Such requirements are identical to the lighting needs of persons engaged in inspection (chapter 10).

If the task does not require the light to be directable under the operative's control, then a fixed system of additional lighting at workstations may be employed, using either local or localized luminaires.

A system of localized lighting works best if the workstations are grouped in areas or in lines and the overhead lighting is generally in a regular pattern. The light at the workstations may be enhanced by several means, such as:

- spacing the lines of overhead luminaires closer together over the workstations;
- keeping the lines of luminaires at uniform spacing, but spacing the luminaires closer together over the workstations;
- mounting lines of luminaires over workstations lower;
- make the luminaires over workstations of higher lumen output by having more lamps per luminaire, or by using higher-powered lamps;
- using a combination of any of the above.

If the general lighting is augmented by task-lights at the workstations, the operatives may switch the task-lights at will, for the contribution the task-lights make to the general illuminance is negligible. In the case of localized lighting, the whole installation should preferably be switched as an entity, for extinguishing the localized lighting would reduce the general lighting level impairing the safe movement and amenity of other occupants.

Yet another method is to provide a suitable general lighting system over the whole area, and then to add additional luminaires or groups of luminaires over selected areas. When these additional lamps are in use, there may be a degree of bonus light near the preferentially-lighted area, and this will marginally reduce the overall cost-effectiveness, but the system is suitable and economic when certain operations are only intermittently performed in part of the area.

For departments laid out with lines of benches or machines, it is sometimes possible to localize the rows of lighting over the lines of workstations, and so arrange that the spill-light from these luminaires satisfactorily lights the gangways. This works well when the workstations or machines occupy a large percentage of the floor area. Indeed, at a high density of floor utilization, there is only a notional difference between such a scheme and a general lighting scheme with an unusually low mounting height – perhaps down to 2 m above floor level. If luminaires are mounted as low as this, they must be located only over the benches etc., or people will collide with them. Installations of this kind are extensively used in laboratories and in the clothing industry, with opaque-reflector trough luminaires being mounted with their base openings below the eye-level of a standing person. It is generally best to use luminaires having louvres to restrict the brightness towards the seated operatives. Slotted-top luminaires are recommended for such applications, to throw some light to the ceiling.

In some situations, instead of fitting a local luminaire, it may be safer or more convenient to use a local mirror to redirect light from general or localized luminaires to the object of special regard. Mirrors for this purpose are less vulnerable to damage, and less obstructive to the operator, than local luminaires. Metal mirrors are virtually indestructible, and carry no electrical risks in the presence of water, soluble oil etc. Use may be made of what is termed *remote local lighting*, in which a narrow-beam spotlight is placed high up in a safe position, and its beam directed to the task area. To avoid obstruction of the beam by the operator's head or body, the beam may be deflected to where required by the use of a local mirror which may be adjustable by the operator.

9.6   Integrating electric lighting and daylighting

9.6.1   Daylight factor

The design of interior lighting should take account of the availability of daylight entering the building. Because the illuminance due to daylighting within a building is constantly varying, the amount of daylight arriving at a point is described by a *daylight factor*. This is the ratio between the illuminance due to daylight in an interior due to a 'standard sky', and the illuminance which would be due to such a 'standard sky' at the same place if it were unobstructed by the building and its surroundings.

Daylight factors between 0.5% and 10% may be specified, and are capable of calculation from data about the interior, its fenestration, and the orientation of the window wall. Typical recommended daylight factors in the *CIBSE Code*[1] are: corridors 0.5%, drawing offices 2%, typing and business machine rooms 4%. The last of these recommendation is not supported by current experience in the use of VDTs, where screen reflections from windows – and the varying level of illuminance due to daylight – cause serious problems for operators (section 9.3.2.) Where the Code's recommendations are for daylight factors greater than 4%, they are intended to apply to task areas, and not necessarily to the whole interior.

For working areas with northlight roof windows, the *CIBSE Code* recommends a daylight factor of 5%, but roof windows that do not face north are not recommended because of their great heat gains in summer. In high-stacking stores, use of roof windows cannot be recommended at all because of the glare caused to forklift truck drivers when picking and setting at high level.

9.6.2 Combined electric and natural light

Various methods for combining the use of daylighting and electric lighting have been advocated. The best known are permanent artificial lighting (PAL) and permanent supplementary artificial lighting in interiors (PSALI).

- PAL more or less ignores the effect of the windows, and treats the contribution of daylight as a bonus. PSALI is a method of supplementing daylight in an interior according to conditions hour by hour and through the seasons, and involves reducing the electric lighting in areas receiving adequate light from the windows, instant by instant (Figure 9.4).

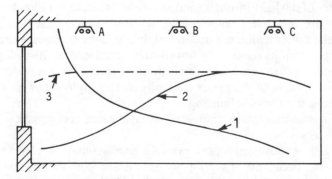

Figure 9.4 *PSALI; daylight supplemented with luminaires 'B' and 'C' during the day, but luminaires 'A', 'B' and 'C' are used when daylight is insufficient. Curve 1, Typical daylight illuminance at desk level; Curve 2, Illuminance due to luminaires 'B' and 'C'; Curve 3, All luminaires in use, no daylight (not to scale).*

Regulation may be effected by switching selected lamps or luminaires, or by dimming, control being either manual or by photoelectric switching (section 22.4).

9.6.3 Windowless buildings

Bigger windows are sometimes advocated as a means of reducing the lighting bill, but practice does not usually support this theory. The hours of availability of daylight in the temperate zones are limited, and do not always coincide with the hours of working. Substantial and continuous savings throughout the life of a building may be achieved by excluding daylight totally and relying only on electric lighting. In deciding whether to construct a conventional fenestrated building or a windowless one, the following factors may be considered:

- The capital cost and the annual cost of a building with windows (including heat losses and gains, cleaning and repairing, and insurance costs etc.) are far higher than the cost of a building which has plain walls.
- Windowless buildings are more secure, and therefore have reduced insurance cost.
- Without windows, it is easier to achieve high standards of cleanliness as there is better control of entry of external air; total filtration becomes practicable if required (section 7.4).
- A windowless building will have lower heat losses in winter, and reduced heat gain in summer. It thus becomes possible to provide all-year-round comfort conditions with smaller capital and operating costs for air-conditioning.
- The entry of daylight through windows – and especially of direct sunlight – causes glare, and the heat-gain from the sun's rays (termed 'insolation') may make the environment uncomfortable, or may occasion extra energy consumption for its removal by forced-draft ventilation or air-conditioning. Systems of automatic blinds and shutters have not generally proved successful because of their poor reliability and the high cost of maintenance throughout the life of a building.
- Cold windows may cause draughts, and may cause condensation in moist atmospheres.
- In very large fenestrated rooms, even if a good standard of electric lighting is provided, occupants at work positions distant from the nearest window may feel a sense of deprivation, particularly if the room has a low ceiling. As no provision has to be made for the penetration of daylight, a building without windows may have lower constructional cost because lower ceiling heights may be employed.

• Mixtures of daylight and electric light vary continuously in colour properties, so that it may be necessary to exclude daylight, at least from areas where fine colour work is performed. Constant lighting conditions without entry of daylight or sunlight may obviate the need for inspection booths in colour-sensitive tasks (section 10.5).

If given the choice, most people would prefer to work in a room with windows for the pleasure of the view out, as well as for the feeling of being in a normal environment. However, for industrial buildings, we see that windows are a mixed blessing. There are some thousands of windowless industrial buildings in the UK, and early fears that occupants of them would suffer psychologically or physically from not being able to see the sky during working hours have not been realized.

If the visual environment within the windowless building is well designed, and the lighting quality and illuminance provided are adequate, occupants are unlikely to complain about their working conditions. Indeed, during a research on employee attitudes relating to working in windowless environments in the food industry, the author interviewed a number of employees who, until questioned, had not realized that they worked in a building without windows!

9.7 Zones of intermediate illuminance

Where people have to pass between areas which are lit to substantially different illuminances, it may be necessary to provide a *zone of intermediate illuminance* to allow their eyes to adapt when passing from the higher to the lower, or from the lower to the higher illuminance. Figure 9.5 shows

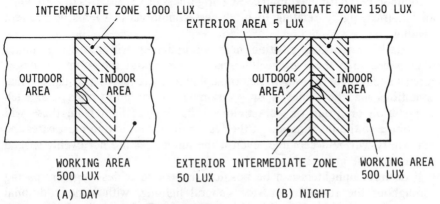

Figure 9.5 *Zones of intermediate illuminance may be provided for the safety and comfort of persons passing between differently lighted areas (not to scale).*

conditions by day and by night in relation to a lighted building, where people have to pass from the well-lit area into an exterior area lit to only a low illuminance. Two conditions have to be satisfied:

- *By day*, an intermediate zone of higher illuminance than that employed in the building may be provided to enable the eyes to adapt as people come in from bright daylight, or leave the 500 lx area and go out to the bright daylight.
- By night, two intermediate zones of lower illuminance may be necessary, one internally, and one externally, to provide a gradient to allow the eyes to adapt as people enter or leave the building.

Switching arrangements to deal with day and night conditions can easily be arranged (section 22.4).

9.8   Uplighting

9.8.1   Introduction

Uplighters produce softly diffused interior lighting, with virtually no shadows, creating a comfortable visual scene. Because the lamps are concealed from view at eye-level, there is no direct glare from uplighters. A room lit by uplighters is bathed in gentle light reflected from the ceiling.

The evenly distributed relatively low brightness of the ceiling in an uplit room produces virtually no reflected glare from tasks, even from objects having glossy surfaces. There is some reflection from VDT screens, but it is evenly diffused without highlights, and therefore the distractive effect is limited. In low-ceiling rooms, the ceiling brightness might exceed the current recommendation for maximum ceiling brightness in VDT rooms of 200 cd/m^2, an empirical figure relating to brightness conditions in rooms with direct lighting.

A desk or working position near an uplighter will receive a higher illuminance than the average illuminance throughout the room. In larger rooms and offices in which several uplighters are used, excellent visual conditions may be achieved by positioning one or more units near each workstation or group of workstations. The flow of light from these will contribute to the illumination of the rest of the room, and in most cases the uniformity will be acceptable, even if the luminaires are not evenly spaced throughout the area.

If desired, uplighters can be positioned at more or less regular spacing throughout the room to provide general lighting, with some additional lighting for tasks being provided by desk-lamps. Such combinations of task lighting and uplighting are economical in operation, for the level of general lighting can be somewhat lower than would otherwise be required (section 9.5).

9.8.2 Practical considerations

If the necessary 13-A 3-pin power points are available, freestanding uplighters are easy to install, without the need for the expense of special wiring. This is convenient for temporary arrangements and when occupying rented accommodation. Once installed, the positions of the uplighters can be adjusted to suit changes in the room layout or repositioning of workstations.

Uplighter luminaires may be free-standing, wall-mounted, suspended or furniture-mounted, with a choice of external colour finishes, so that the lighting equipment can be visually integrated with the décor and furnishings. Standard uplighters are suitable for use in rooms of ceiling heights between about 2.5 m and 4.25 m, but rooms with higher ceilings can be successfully lit with suitable units.

Uplighting can be used in rooms having ceilings of almost any configuration (including textured or coffered ceilings), but the ceiling surface must be matt (non-glossy). The ceiling and the upper walls should preferably be white or of a very light colour. However, uplighting of rooms with coloured ceilings can be very effective as a feature of décor; non-white ceilings will reflect less light, so the efficiency of such installations will be lower. Light reflected from a coloured ceiling would, of course, not be suitable for tasks involving fine colour discrimination (section 4.5.2).

Direct lighting produced by conventional overhead luminaires has to be designed to achieve a fair degree of uniformity, but uplighting produces such soft graduations of illuminance that transitions between different levels of brightness between parts of the room are almost imperceptible. Because considerable departures from uniformity of illuminance are tolerable, some flexibility in the positioning of the uplighters is permissible, and it is not essential to locate them precisely at the theoretical positions indicated by the design procedure.

9.8.3 Designing uplighting installations

When planning layouts of uplighters, it is not always necessary to perform conventional lighting calculations such as are used for designing direct lighting layouts. However, instructions for doing this are given in section 9.8.7.

In most cases the 'Isolux Diagram Method' of design (section 9.8.4) is a satisfactory means of determining the number and wattage of uplighters required to produce the desired illuminance in typical interiors. This easily-applied method uses small charts called *isolux diagrams* (Figure 9.6), these commonly being supplied by luminaire manufacturers. These may be used to select and position a single uplighter, or to design the lighting layout

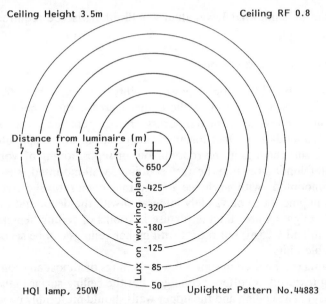

Figure 9.6 *Typical isolux diagram for an uplighter (not to scale; do not use).*

for larger installations with a number of workstations. This procedure enables the designer to see what illuminance will be produced at any point in the room from any proposed arrangement of uplighters.

When designing uplighting layouts by either of the methods discussed in this section, the achieved illuminances may be a little higher or lower than the calculated values. This may be due to the light-reflection characteristics of the particular ceiling, walls and floor not being known accurately. (Methods of measuring the reflection factor of surfaces are given in appendix E.) The illuminance is also affected by the furnishings and other contents of the room. Further, the procedures assume that the floor reflectance will be low. However, the accuracy of these design methods will be found to be satisfactory for most practical purposes. If it is desired to carry out conventional calculations for contract purposes, complete photometric data for the chosen uplighters should be requested from the luminaire manufacturer. In most cases, the manufacturer will supply isolux diagrams for their products.

9.8.4   Isolux diagram method of design

Step 1: Select the required illuminance by reference to the *CIBSE Code*[1]. The recommendations will need to be interpreted. For example, for offices it is found in practice that around 500 lx on desks and tasks is sufficient for most purposes, but the whole room does not need to be lit to this level. Between desks, an illuminance of 200–300 lx will be adequate, and in areas where people move about but where no specific work is done, the illuminance can be as low as 50–100 lx. The illuminance in any part of a room should not be less than 50 lx.

Step 2: Isolux diagrams supplied by manufacturers (Figure 9.6) are commonly drawn at a scale of 100:1 (10 mm = 1 m), and are usually printed on transparent paper or plastic sheet. Obtain or prepare a sketch plan of the room to this scale. If the only plan available is drawn at one-eighth of an inch to a foot (scale 96:1), this can be used without introducing very much error. It will be convenient if the sketch plan is drawn on tracing paper, but this is not essential. It should show the walls, and the positions of desks, equipment, columns etc. It may be helpful if the plan also shows the positions of power points to which freestanding uplighters may be connected.

Step 3: Select the correct isolux diagrams. The diagrams are usually computed for several nominal ceiling heights, e.g. 2.75 m (9 ft); 3.25 m (10 ft 6 ins); 3.75 m (12 ft) and 4.25 m (14 ft), for each of which there may be diagrams showing the lux values which will be produced by specified luminaires, and a choice of types and powers of lamps. Typically, diagrams may give data for uplighters containing one metal-halide HQI lamp of 70 W, 150 W or 250 W rating.

If the actual ceiling height approximates to one of the nominal ceiling heights for which diagrams are drawn, use that set of diagrams. If the actual ceiling height is intermediate between two of the nominal ceiling heights, the illuminance produced will be intermediate between the values shown for the two adjacent nominal ceiling heights.

Step 4: Place the centre of the isolux diagram over the position where it is proposed to locate an uplighter (or, if the plan is drawn on tracing paper, place the plan over the isolux diagram.) The concentric circles of the diagram show the illuminances (lux) that will be produced at desktop height at various distances from the uplighter.

Step 5: Plan the lighting layout. If the installation comprises more than one uplighter, the light arriving at any point in the room will be the sum of the illuminances contributed by all uplighters within about 7 m of that point. By adjusting the number and positions of uplighters, and selecting appropriate lamp wattages, it will be possible to plot the required distribution of light. A mixture of lamp wattages may be employed. Schemes may use combinations of free-standing, suspended, furniture-mounted and wall-mounted uplighters.

9.8.5 Effect of wall reflections

Due to multiple reflections of light, it will generally be found that there is some enhancement of the average illuminance and the uniformity of illuminance produced by uplighters if the walls are light in colour. This effect is most marked in small rooms.

In the case of wall-mounted uplighters and uplighters positioned within about 3 m of a light-coloured wall, a significant amount of light will be reflected from the wall, so that the illuminance reaching adjacent work tops and desks may be somewhat higher than is indicated by the isolux diagrams.

Unless they are fitted with very carefully designed optical systems, wall-mounted uplighters tend to overlight the patch of wall above the point where they are located. This effect can be overcome by mounting a mirror on the wall immediately above the uplighter unit, to intercept its light and reflect it into the room.

9.8.6 Large rooms

Planning of uplighting for large interiors can be simplified if the uplighters can be spaced in a more or less regular square grid pattern. If the outer uplighters are near light-coloured walls, there will be some local increase of illuminance due to reflection, but the positions of the outer luminaires can be adjusted. For any particular room and configuration of workstations, there may be several possible alternative lighting layouts. In some cases it may be worthwhile to cluster work positions around uplighters.

Typical average illuminances in large rooms having ceiling heights between 2.75 m and 4.75 m lit by regular arrays of uplighters are shown in Table 9, calculated for rooms with clean white ceilings (reflection factor (RF) 0.8). If the ceiling has an RF less than 0.8, the resultant illuminance will be reduced in proportion. For example, in a room with a ceiling RF of 0.7, the resultant lux levels will be seven-eighths (0.7/0.8) of those shown in Table 9. In rooms having light-coloured walls, there will be the benefit of reflected light

Table 9 *Uplighting in large rooms. Typical average illuminances*

Uplighter spacing (m)	250-W HQI (lux)	150-W HQI (lux)
8.5 × 8.5	100	55
7.0 × 7.0	150	85
6.0 × 6.0	200	115
5.0 × 5.0	300	170
4.0 × 4.0	500	280

Note: The illuminances shown are typical *average* illuminances. Significantly higher illuminances will be obtained close to the uplighters (which can be calculated by use of isolux diagrams). It is thus possible to provide 300–500 lx on desks, in a room in which the average illuminance is between 150 and 200 lx.

(especially from the upper walls) which will enhance the illuminance. In general, the lighter the colours of all surfaces within the room, the higher will be the illuminance and the better will be the uniformity.

9.8.7   Lumen Method of uplighter installation design

This method may be used to calculate the average illuminance in larger rooms. The procedure is exactly the same as for applying the Lumen Method to direct lighting (section 9.2).

It is the author's opinion that the table of UF for totally indirect luminaires given in *Interior Lighting Design*[8] shows UFs which are far too low in relation to modern uplighters which generally have a much better photometric performance than the types of indirect luminaires which were under consideration when that table was compiled. It would appear that the figures in the *ILD* table take account only of the 'first bounce', i.e. the light reflected directly from the ceiling to the point of measurement, and not the aggregate of light that has been reflected once or more than once from the room surfaces. It is suggested that the UF values given in Table 10 be employed for schemes using uplighters which contain metal halide lamps in efficient specular reflectors giving wide beams with minimum 'hot spot' above the luminaire.

In clean, reasonably dust-free interiors, a maintenance factor of 0.8 may be employed for uplighters which are fitted with heat-resisting glass covers that are cleaned frequently.

Table 10   *Utilization factors for uplighters with luminaires having an upward light ratio of 75% and the top of the reflector mounted at 2 m above floor level*

Room reflectance		Room index							
C	*W*	*1.0*	*1.25*	*1.5*	*2.0*	*2.5*	*3.0*	*4.0*	*5.0*
0.8	0.5	0.35	0.39	0.42	0.44	0.50	0.53	0.56	0.57
0.8	0.3	0.29	0.32	0.32	0.41	0.45	0.47	0.51	0.53
0.7	0.5	0.32	0.36	0.39	0.43	0.46	0.48	0.51	0.53
0.7	0.3	0.28	0.31	0.34	0.39	0.42	0.45	0.48	0.48

C = ceiling reflectance; W = wall reflectance

The author thanks Mr Eric Chapman (Belvoir Lighting Consultancy, Nottingham) for permission to reproduce this table, which was calculated by Mr Chapman and subsequently used by him in practical trials in which the general accuracy of the data was confirmed.

10 Lighting for inspection

10.1 Principles

An analysis of methods for revealing faults in an inspected product or object is given in Figure 10.1 (after Bellchambers and Phillipson[22]). Amongst production engineers, the term 'inspection' is strongly associated with mensuration and the use of test instruments. However, a review of common inspection tasks in many manufacturing industries shows that most 'inspection' is performed by unaided use of the inspector's eyes, i.e. it is inspection by 'direct vision'. The term direct vision means that no optical aids are used other than the wearing of ordinary corrective spectacles. It is important that the inspector's eyesight is of satisfactory standard for the

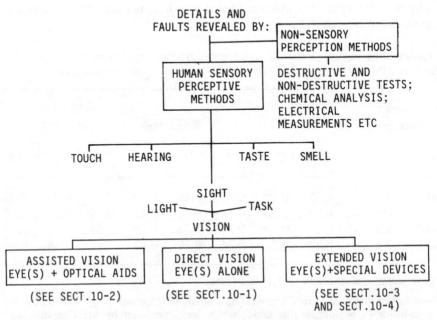

Figure 10.1 *An analysis of methods for revealing detail and faults in an inspected product or object (after Bellchambers and Phillipson[22]).*

work. Vision screening (section 1.10) or examination by a qualified optician or opthalmologist is recommended, particularly for the engagement of new inspectors.

With training and practice, a person having a satisfactory standard of vision may achieve remarkable performance in accuracy, consistency and speed of inspection tasks. Although they appear to be 'just looking', inspectors may be exercising a high degree of skill in operations that are vital to the achievement of the required quality standards. Inspectors cannot perform well without the correct lighting conditions. Of all external factors, lighting is the environmental condition that has the greatest effect on the efficiency of inspectors.

The reputation and prosperity of companies depend on the quality of their products. The satisfaction of customers, and indeed their safety, depend on the goods being 'of merchantable quality and fit for their intended purpose' as required by the UK *Sale of Goods Act*. The Act embodies the principle that, should the user suffer loss, damage or injury, the seller, to avoid paying heavy penalties, must be able to show that the goods were of 'merchantable quality' etc, and that steps were taken to prevent the buyer from receiving faulty goods by the use of quality control, testing and inspection. According to their type, goods may be subject to 100% inspection, or to inspection of a statistically representative sample.

The philosophy of specialization and 'division of labour' is foolishly practised in some organizations to the extent that production workers may fail entirely to inspect their own output, with the attitude that 'inspection is the work of the inspection department'. In studying inspection problems and their lighting needs in factories, the author has often found that lighting in production areas was insufficient to enable production workers to inspect their own work. In some cases, simply bringing the lighting up to *CIBSE Code*[1] standards enabled this to be done, and enabled simple on-line inspection to be performed. When defects were found, the operators could cease production until machine setting adjustments had been made, thereby reducing the scrap rate.

In some factories it has proved economic to disband the inspection department and to distribute the inspectors throughout the production departments, some as 'roving inspectors' and some manning 'inter-process inspection stations'. This, of course, requires that the lighting in the production areas (or at least at the inspection stations) is of a high standard, but the economic benefits can be enormous. Savings come about because goods do not have to be transported to the store after processing, then issued to the inspection department, returned to store, and then issued to the next production station (Figure 10.2).

A review of the lighting aspects of inspection by direct vision is given in section 10.2, by assisted vision is given in section 10.3, and by extended vision in section 10.4, while a review of considerations in colour inspection is given in section 10.5.

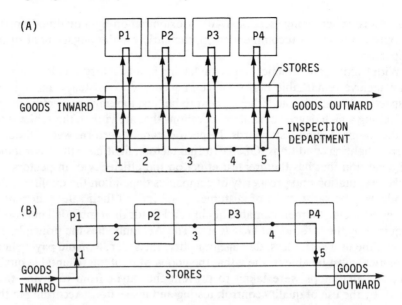

Figure 10.2 *Central versus dispersed factory inspection. (A) Conventional flow, with central stores and inspection department. The four manufacturing processes, (P1, P2 etc) and five inspection processes (1, 2 etc) require 20 stores transactions per batch (shown by arrowheads). (B) Linear flow with dispersed inspection stations. For the same production, only four stores transactions per batch are required (shown by arrowheads). This system is only possible if the lighting in the production area is good enough to permit inspection processes. It saves time and cost in moving goods about, and is space saving.*

10.2 Inspection by direct vision

10.2.1 As an illustration of the effect of improvement of lighting on rates of fault creation and fault discovery in a situation where all inspection was by direct vision simply using good eyesight, consider the case of a company producing domestic appliances which included some zinc-alloy gravity die castings. At the beginning of the consultancy (which had been instituted because of the high frequency of customer complaints about failures in use of the die-cast components) the general lighting in the production and inspection areas was found to be 180 lx and 270 lx respectively. The fault rate (as determined by the inspection department) was 7.5%, but it was apparent from the level of customer complaints that a significant proportion of faulty castings was getting through the inspection process.

As a first step, the inspection department was re-lighted to 500 lx, using fluorescent tubes of good colour-rendering, and local lighting producing

1000 lx was installed at each inspection station. It was quickly established that the fault rate was far higher than had been previously thought; it was determined at about 17.8%, a figure far too high for economic operation. The company then introduced better lighting in the production areas, putting in a suitable general lighting scheme to produce 500 lx. Within a few weeks the true fault rate fell from 17.8% to around 3.8%, due largely to the ability of the operators now to detect faults in the products they were producing, and to take corrective action. As a result, the reject rate in inspection fell to around 3.2%, indicating that at least 96% of all faults were being discovered. This still meant that 4% of the goods sold were faulty, but the management decided to live with this on the grounds that (a) this was such a good improvement over the previous 17.8% fault rate, and (b) experience soon showed that the undiscovered 4% of faults were mainly minor blemishes which did not presage failure of the casting in normal use. The management were able to revise their criteria for 'pass' and 'not pass' defects, on the basis that more accurate assessment of faults was possible with the improved lighting, this leading to further minor savings. The nett cost of the improvements to the lighting (which involved replacing outworn fluorescent tubular lamp luminaires with new high-pressure sodium-lamp luminaires) was estimated to have been fully recovered in a period of 37 weeks by (a) a reduction in faulty production, (b) an improvement in output, (c) a reduction in labour turnover and lost time, and (d) energy savings, and thereafter there was continued enhancement of profitability.

10.2.2 In production and inspection tasks involving critical vision, any method of emphasizing and clarifying the 'object of special regard' will speed up the process and add to accuracy. Every industry has its special visual difficulties, and many kinds of optical, mechanical and electrical devices are used to reveal particular features of the product under inspection. The lighting may need to be modified, perhaps not only increasing the illuminance on the task but also by change or control of one or more of its attributes, e.g. to provide parallel, divergent or convergent beams of light. Light may need to be from virtual point sources or from large-area diffuse sources. The angle of incidence of the light may need to be controlled (section 9.3). Light of high colour quality may be needed for colour discrimination (sections 4.5 and 10.5), monochromatic or non-full-spectrum light may be needed to reveal one particular colour which would otherwise be difficult to see, or UV or polarized light may be needed.

These measures of special lighting may be difficult to apply to a whole department, and it therefore becomes a practical necessity to create a miniature environment in which the desired lighting conditions can be produced, i.e. to use an inspection booth of some kind. There are difficulties relating to adaptation to illuminance and colour on entering (and sometimes on leaving) a booth, necessitating allowance of time for adaptation. Some subjects experience claustrophobia on entering small spaces, and there may be problems in providing adequate ventilation for operators' comfort and for cooling of lamps. In these situations a 'head and shoulders booth' may be suitable (Figure 10.3). An inspection booth may be a moveable item which can be placed on a bench of convenient height. Where *dispersed inspection* (section 10.1) is adopted, such booths can be placed within the manufacturing area for inter-process inspection of products, and can be moved to new locations as required to match production activities.

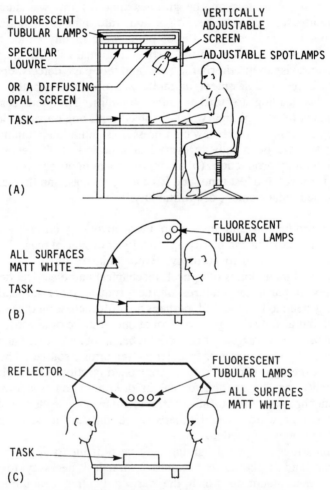

Figure 10.3 *Types of 'head-and-shoulders' inspection booths. (A) using direct lighting through a specular louvre grid or diffuser; (B) indirect lighting; (c) indirect lighting with workstation accessible from both sides. (Examples (A) and (B) are based on Bellchambers and Phillipson).*

10.2.3   A valuable inspection method is to place a sample of a translucent or transparent material over a light-box or similar device. It is a minor doubt whether, strictly speaking, this constitutes assisted or unassisted vision. This method is extensively used for such diverse operations as: checking the positions of stiffeners in shirt collars; inspecting sheets of dried wood pulp for inclusions (such as dead insects) before the material is repulped to make into medical filters; examining artists' and photographic transparencies (section 10.5); routine in-line inspection of glass products, e.g. bottles, and products in glass containers, e.g. pots of jam, ampoules of drugs etc; and inspecting the watermarks in banknotes and securities (section 11.3).

Light-boxes containing fluorescent tubes or tubular architectural tungsten-filament lamps have long been used in drawing-offices for tracing drawings on to linen or plastic, and are widely used for lithographic platemaking. Care has to be taken to prevent the lamps or the diffusing surface of the light-box overheating. Overheating of fluorescent lamps will cause colour-shift, while general overheating may damage the equipment. The art of designing light-boxes to obtain highly uniform luminance on the diffusing surface lies outside the scope of this book. The fitting of dimming control is usually advantageous. It is usually necessary to shield the surface of a light box from the ambient light so that the luminance due to transmitted light through the sample dominates. To reduce glare, an opaque mask (with an aperture of similar shape to the object being examined) may be applied to the bright surface. A footswitch enables the light-box to be switched off between uses to prevent operator eye-strain.

10.3   Inspection by assisted vision

10.3.1   Requirements

Assisted vision consists of employing an optical device which uses visible radiation to enable the eye to see something that cannot be seen by direct vision. It must be emphasized that such devices are not substitutes for sufficient and suitable lighting which is still needed. If, for example, because of poor lighting we cannot generate sufficient acuity to see a particular detail, use of a magnifier will produce a larger but greyer image which will also be unreadable. This is due to the optical system gathering light over a narrow collection angle and distributing it over a wider angle, further reducing the brightness of the image. Also, lenses absorb a proportion of the light passing through them. While it is desirable to increase the illuminance on a task which is to be viewed through an optical system, such increase must be accompanied by an increase of the background illuminance, or unbearable glare may be generated. Thus the problem is not solved by simply spotlighting the task.

Some examples of assisted vision are given in this section, extending the information given by Bellchambers and Phillipson[22].

10.3.2 Assisted vision devices

Watchmaker's glass

A watchmaker's glass (also known as an 'optic' or 'loupe') is worn by the operator and, of course, gives only monocular magnification. The device may be worn as a monocle or mounted on a spectacle frame, in which case the optic may be hinged to enable it to be swung up out of the line of vision. Working with a monocular device for long periods is tiring, particularly if the same eye is constantly used, and always results in the eyes being differently adapted. Continuous monocular work can lead to a pathological condition called 'lazy eye', in which the the under-used eye becomes defective in alignment and focus.

Binocular magnifiers

Binocular magnifiers consist of two watchmaker's glasses mounted on a spectacle frame. These are less likely to cause eyestrain than monocular devices. In use, the optics effectively 'blind' the operator for distance viewing, and for safety should therefore only be used when seated.

Hand-held magnifiers

Hand-held magnifiers tend to become scratched in use, plastic lenses being worse in this respect but lighter in weight and not subject to breakage. Because the hand is not steady, a magnification of about X5 is the practical limit. There may be marked chromatic aberration from magnifiers of X3 or more.

Magnifiers on stands

These magnifiers may have a local lamp or lamps built into the bezel and screened from the operator's view, e.g. miniature fluorescent lamps. It is difficult to obtain such lamps in high colour quality versions. The device may incorporate a small fan for blowing away smoke and fumes from tasks such as soldering.

Prismatic magnifiers

These magnifiers are used when magnification is required in one dimension only, e.g. for observing a thin column of mercury in a thermometer, when a prismatic magnifier consisting of a half cylinder of glass or clear plastic may be used.

Microscopes

Microscopes are magnifiers having compound lenses with a fairly narrow collection angle and magnification from about X5 upward.

Bench microscopes may be monocular but are commonly fitted with a prismatic image-splitter and two eyepieces. Some provide binocular viewing plus a position for a camera, and some provide double binocular viewing (for two operators). Colour filters may be used. The object may be back-lit or front-lit via a mirror which reflects light from the local or general lighting (avoiding the colour difference which can occur if a microscope light is used).

Machine-mounted microscopes and magnifiers are built into some machine tools, and are used to aid reading of vernier scales or for examining a workpiece.

Telescopes

Telescopes are occasionally used in industrial plant so that an operator can observe distant equipment or read a distant gauge without leaving the operating position, though closed circuit television is now extensively used for this duty.

Mirrors

Mirrors may be used to direct light into locations where the local or general lighting does not penetrate. Mirrors may direct light into hostile environments (e.g. high-temperature or explosive) which prevent the introduction of electrical apparatus. In some applications the mirror may consist of a plate of metal, either polished or matt according to requirements. Combinations of mirrors may be used to direct light to difficult locations (e.g. inside a machine), but there is a light loss of 8–14% at each reflection from rear-silvered mirrors. Front silvered mirrors (which are easily damaged) may have a light loss as low as 4% per reflection. Note that when light has been reflected twice in the same plane it becomes substantially polarized (section 2.9).

Mirrors are used to reflect a virtual image to the observer, and may be placed where it would be impossible or dangerous for the operator to place his eyes, e.g. where physical space is confined, or there is heat or danger from moving machinery, high voltage etc. Combinations of mirrors are used to view inaccessible objects (periscopy), and there are methods of using a system of mirrors to both transmit light to the object and to bring back a virtual image to the observer (duplex periscopy) (Figure 10.4). There are practical limitations to the length of the light path and the number of reflections that can be effectively used because of the light loss at each reflection.

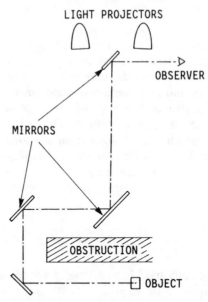

Figure 10.4 *Principle of one configuration of self-illuminating (duplex) periscope.*

For short light-transmission paths, and where the dimensional constraints are severe, mono or duplex fibre optic devices (discussed below) have much to recommend them. But for situations where there is a long light path and a need for high resolution of the image, or if there is a need for two observers to see the image and possibly also to take photographs of the object, a plane-mirror periscope will have many practical advantages. A periscope can produce a bright high-resolution virtual image, which can be magnified by adding telescope optics.

Mirrors may be used on inspection lines to see the tops or backs of objects on a conveyor belt. In the paper industry they are used to get a good view of a continuous web of paper in production, looking for 'thins' and inclusions, and also for colour checking (section 11.2).

Another valuable application of mirrors in inspection is to take advantage of the fact that when a mirror rotates through an angle θ, the reflected ray rotates through an angle of 2θ (Figure 10.5). Thus components may be tested for angular accuracy by fixing a mirror thereto, and observing the displacement of the incident ray, e.g. by a collimated narrow beam of projected light, or by movement of the virtual image.

Profile projectors

Profile projectors ('shadowgraphs') exploit the fact that the shadow thrown by an object by the light of a small source is larger than the object (Figure

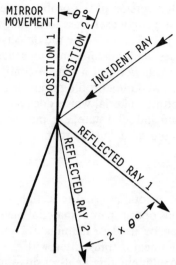

Figure 10.5 *Angular displacement of a reflected ray. When the mirror rotates through an angle of θ, the reflected ray rotates through 2 × θ.*

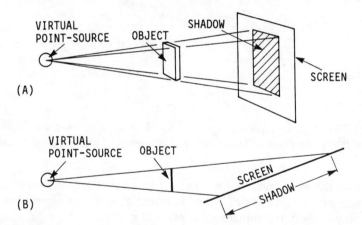

Figure 10.6 *Principle of the profile-projector. (A) shadow projected on to white opaque or translucent diffusing screen; (B) tilting the screen enlarges the shadow in one direction.*

10.6). The shadow may be cast on to a white opaque screen, or may be viewed from the reverse side of a translucent screen of, for example, ground glass, flashed-opal glass or a diffusing plastic material. The magnification effect due to divergence of the rays may be enhanced by tilting the screen to produce an elongated shadow. If the positions of the light source, the object and the screen are controlled, the shadow profile will be of a known magnification and its outline can be compared with a standard outline.

In a more sophisticated version of the profile projector, the light beam is generated by a mirror and condenser glass to produce a near-parallel beam. This beam, after being partly occluded by the object, is then magnified by a further system of lenses and mirrors and presented in silhouette on a translucent screen. The silhouette of a female configuration (e.g. a female helical thread) can be viewed on a profile projector by first making a cast of the cavity in a quick-setting material which does not have much variance in volume between its liquid and solid state, e.g. pure sulphur which liquefies on heating, and returns to the solid state at room temperature.

Light pipes and fibre optics

These use the principle of total internal reflection which enables light to be 'piped' through solid rods of transparent material such as glass or clear plastic (Figure 10.7). They can be used to illuminate a small object without any scattering of light. As no heat is transmitted with the light to the object, the method can be used for illuminating heat-sensitive material, e.g. biological samples.

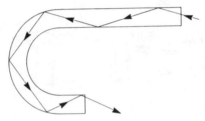

Figure 10.7 *Principle of 'piping of light'. An incident ray, entering the end of a polished transparent rod, will travel along it by a series of total internal reflections.*

Glass filaments of comparable diameter to a hair (0.05–0.25 mm) exhibit the piping effect. Bundles of small-diameter glass fibres (termed a 'fibre optic') may be used to conduct light in place of a solid rod, with the advantage that such a bundle is flexible. Fibre optics may be used to direct light into deep narrow apertures, around corners or into enclosed spaces with small entries. If it is arranged that each fibre at the distal end is in the same position in relation to the other fibres as at the proximal end, the bundle is said to be 'coherent' and is capable of transmitting an image. The resolution of the image will depend upon the density (number of fibres per unit area) and diameter of the fibres.

A fibre optic may be used to illuminate an object in a hazardous area, the lightsource being located in an adjacent area of normal atmosphere. Random bundle fibre optics can be used to transmit light to the distal end; coherent bundles can either bring back an image to the proximal end or, if of sophisticated design, can operate in duplex mode, transmitting light to the distal end and bringing back an image through the same fibres in a similar manner to a duplex periscope. As an example of use, a duplex coherent fibre optic could be used to inspect the interior of a vessel having only a small aperture and containing flammable material.

By the use of a suitable optical system to collect light from a source and enter it into a fibre optic channel, it is possible to deliver very high illumination intensities to a small point of use without any heat being transmitted (section 5.13.4). Illuminances of the order of tens of thousands of lux can be so delivered. These illuminances are high enough to cause light to be transmitted through many solid materials which are generally regarded as opaque and thus it is possible to examine the internal structure of objects made of such materials. The technique has possibilities for use in such diverse processes as checking the integrity of moulded plastic components, and perhaps in medical examination, e.g. for detecting cryptocaries within apparently healthy teeth.

Illuminated grids

Such grids marked on background boards, or painted on back-lit diffusing panels have several valuable uses in inspection for checking for flatness or examining transparent materials. To examine the flatness of specular surfaces, an illuminated grid (typically comprising black lines 2 mm wide in a square grid pattern of lines at 15 mm centres) may be used to inspect a specular plane surface by observing the reflection of the grid from the surface. Because of the 'doubling of the angle effect' (Figure 10.5), small deviations from flatness will be revealed by quite marked deviations in the image of the grid (Figure 10.8A).

An illuminated grid may be used for the inspection of transparent sheet materials. Because of the varying refractive effects due to quite small variations in the thickness of the sheet, or lack of parallelism between its surfaces, the grid appears distorted when viewed through the sheet, enabling the detection of faults (Figure 10.8B).

10.4 Inspection by extended vision

The term *extended vision* is applied to inspection by means other than the ordinary use of the eyes alone (direct vision) (section 10.2), or with some form of optical device (assisted vision) (section 10.3) plus white light. Although extended vision techniques lie outside the scope of this book, some examples are given here for reference.

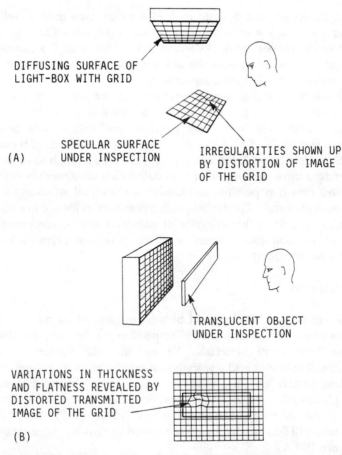

DIFFUSING SURFACE OF
LIGHT-BOX WITH GRID

(A)   SPECULAR SURFACE
       UNDER INSPECTION

IRREGULARITIES SHOWN UP
BY DISTORTION OF IMAGE
OF THE GRID

TRANSLUCENT OBJECT
UNDER INSPECTION

VARIATIONS IN THICKNESS
AND FLATNESS REVEALED BY
DISTORTED TRANSMITTED
IMAGE OF THE GRID

(B)

Figure 10.8 *Inspection of specular and translucent materials by the use of illuminated grids. (A) observing the reflected image on a specular surface; (B) observing the transmitted image through translucent material.*

Ultraviolet irradiation

This is used to create fluorescence in substances as a means of analysis or detection. A spectrometer may be used to analyse the fluorescent radiation from substances irradiated with UV. For crack detection, an object may be dusted with a fine powder of fluorescent material, or painted with solution of such a material. When the surface has been cleaned, irradiation with UV will reveal minute traces of the fluorescent material in cracks and crizzles (surface defects) on the surface of the object, revealing faults which might be impossible to see with the naked eye in normal light.

X-ray irradiation

Objects may be irradiated with X-rays, and a visible image formed on a fluoroscopic screen by the rays which have passed through the object. Internal faults may be detected by this means. Detailed physical analysis of materials may be made by means of an X-ray spectrometer.

Low light level irradiation

For substances and objects which would be adversely affected on exposure to strong visible light (for example, biological samples, or unstable chemical mixtures), examination may be carried out under controlled illuminances as low as 0.0001 lx, the object then being scanned with a scintillator or photo-multiplier optical system. Photographs may be taken with sensitive visible-light film at very long exposures.

Infra-red irradiation

Objects which would be adversely affected by other forms of radiation may be irradiated with infrared radiation. The objects may be viewed with an infrared closed circuit television camera, or an infrared-sensitive film used to photograph the object and produce instant prints for examination.

Photography

Photography may be performed under the various types of radiation mentioned (visible light, u.v., X-ray, IR). Photography by a single very short flash can 'freeze' a movement and reveal information about an object that is moving too fast to be seen by normal vision. Ciné film or video recordings may be taken by pulsed light from a strobe-flash at very short time intervals to allow a movement to be analysed. Video has the advantage of instant playback. Slow-changing phenomena may be studied by time-lapse photography.

Interference bands

Flatness of glass surfaces in contact, or a glass surface in contact with a metal surface may be assessed by observation of the coloured interference bands (Newton's Rings) created by diffraction occurring between two nearly optically-flat surfaces.

Photoelectric detection

A method of measuring flatness by the use of gratings has been devised by the National Physical Laboratory. This detects changes in current through a photoelectric cell when light passes through two diffraction gratings, one of which is held stationary and the other moved across the surface to be measured.

Light as a straightedge

The alignment of gaps, holes etc can be checked by projecting a collimated beam of light through the apertures and detecting it with a photocell. Alignment of even very large structures or those with components separated by distances of kilometres, can be checked by means of laser beams, with visual or instrumental supervision.

Coloured light

Visible light of a colour other than white may be used to aid the performance of some visually difficult tasks. For example, in print-works it is necessary to align accurately three or more colour impressions. The yellow alignment marks are the most difficult to see, and become more clearly visible under blue light, this being obtained by a cyan filter held before the eye or placed over a tungsten-filament lamp local luminaire, the general lighting being shielded from the test point.

Polarized light

Studies of stress patterns in objects can be made by creating an exact copy of the object (full size or to a reduced or enlarged scale) in clear plastic material. This is then observed under polarized light while it is subjected to stresses in simulation of those which will be applied to the real object. Areas of opacity and clarity appear in the transparent copy, and these can be related to stress concentrations.

Polarized light may be used as a means of standardizing the perceived colour of certain objects (sections 2.9, 10.5).

Temperature analysis

There are test methods that enable a photograph to be taken that reveals differences in temperatures of objects and gases (including convected air currents), using a heat-sensitive film which registers distinctive colours for ranges of temperatures (Schlieren technique). A recent development is a temperature-discriminating closed circuit television camera, which shows a picture not in actual colours, but in colours which are representative of the temperatures of the objects viewed.

10.5 Colour inspection

Colour inspection during production may be necessary to monitor such processes as the mixing of paints, inks and pigments (where the standard formulation may need to be adjusted to take account of tolerances in the chromatic characteristics of the constituents). Differences in colour arise in production processes due to many possible causes such as: variation of paint thickness over a substrate; differences in browning of baked foodstuffs; differences in freshness of meat; tolerances in formulation of various products which consist of mixed constituents; variations in processes such as plating, buffing, anodizing and dyeing of metals; differences in production or dyeing of fabrics, etc.

In past times when the cost of installing lighting with good colour characteristics was high, it was almost invariably necessary to limit the use of such lighting to inspection booths or small inspection areas. Difficulties for inspectors under such conditions included passing from the general factory area to the booth, for time must be allowed for colour adaptation, and inclusion of uncontrolled areas of lighting and colour in their field of vision which could degrade their colour judgement (section 4.5). It is now sometimes economic to light a production area to colour-matching standards and overcome these difficulties. Of course, to ensure constancy of lighting conditions, it is necessary to exclude daylight from all areas where colour inspection is carried out. Most industrial colour-matching tasks involve either *side-by-side matching* (section 10.5.1) or *differential matching* (section 10.5.2).

10.5.1 Side-by-side colour matching

In side-by-side matching, the sample under test is compared with a control sample. Good matching cannot be done if the inspector has to move his or her point of vision very far between the samples, particularly if there is something of any other colour or of different brightness between the samples. Ideally, the two samples to be compared should be contiguous and capable of being seen simultaneously under identical lighting conditions, distances and angles. Where it is physically impossible to achieve this, a small control sample may be laid on a large test sample, or vice versa. It is essential to ensure that both samples are of the same degree of specularity, and that the control sample does not become damaged or dirty in use. The control sample must be standardized, e.g. inspected by at least two inspectors independently, and the matching must be carried out under exactly the same lighting conditions under which the control sample was standardized and approved.

An example of a particular difficulty in control of colour concerns the browning of meat pies. An important feature of bakery products is colour of pastry and breads; a pale pie is unappetizing, a dark pie is suspected by the shopper of having been 'freshened up' in the oven, while just the right colour of pastry makes the product look tasty. Because of variations in materials, oven temperature and speed through the oven, one bakery concern checked every batch of pies against a 'perfect sample'. Unfortunately, in the heat of the bakery, the 'perfect sample' soon deteriorated and changed colour. Therefore, a firm of model-makers was employed to fashion a replica of the 'perfect sample' pie, and the colour of this was checked under the same kind of illumination as that used for monitoring the real pies (in this case, Artificial Daylight tubes).

An ingenious method of getting a conjoined image of two samples of fabrics so they could be closely compared was devised by the Furniture Industry Research Association, in which mirrors were used to bring virtual images of the two materials side by side. In another application, the control sample was placed above the test sample and facing downward; a small mirror placed on the test sample produced an image of a small area of the control sample visually right in the middle of the test sample. Such mirror techniques will work if a high quality mirror is used, preferably a front-silvered one (section 10.2).

10.5.2   Differential colour matching

This is a sophisticated method of matching that lends itself to accurate colour control, and with the aid of a computer, variations in the performance of individual inspectors or groups of inspectors can be monitored and compensated for. The method involves presenting the inspector with four samples: (a) the control sample, (b) the sample under test, (c) a 'dummy' sample having a deliberate slight variation of hue, chroma or greyness from the control sample, and (d) another dummy sample with a different slight deliberate variation. Only the controlling inspectors know which sample is which. One or more inspectors examine the samples under excellent lighting conditions, and attempt to identify them. If they can identify two identical samples, this is recorded; failing that, they have to try to decide which sample is which. The results recorded over a period of time for different groups of four samples as inspected by different inspectors can be analysed to yield such useful information as the minimum perceptible colour difference that can be detected by each inspector, the percentage variance of various colour components, the percentage variance with the time of day or the day of the week etc., the constancy of the performance of individual inspectors and the effect of slight environmental changes, including changes to the lighting. With a suitable program, the data being fed to the computer by a carefully designed form or screen menu, the software can 'learn' about the inspectors, and be programmed either to weight their reports according to their past record, or to directly vary the colour constituents in the product to bring it closer to perfection.

10.5.3   Performance of colour inspectors

Before appointment, all inspectors should be asked to submit to vision screening (section 1.10), and this should include appropriate testing of their colour vision. Testing should be repeated at 2-year intervals, with annual testing for inspectors over the age of 50. Colour discrimination ability tends to decline with age due to the yellowing of the cornea and lens of the eye, but this is to some extent compensated for by experience. The experience factor is important, and should not be dismissed. As an illustration of this, surgeons at a hospital were achieving high standards of accuracy and colour judgement in normal surgery and microsurgery under filament-lamp lighting, which has a relatively poor colour-rendering property. On changing to modern light-sources of better colour rendering, the surgeons had to undergo a period of relearning before they regained their former ability, and then progressed to even more skilled and meticulous work. Similar effects have been noted in industry. For example, there may be complaints that 'the new lighting is not as good as the old', but after a period of relearning the inspectors find that the new lighting actually makes the job easier, and the quality of their work may improve.

The performance of an inspector on fine colour work will not be constant. Accuracy will decline a little through the working day due to fatigue, and may decline a little more each day through the week, the onset of fatigue occurring earlier each day. Thus, the accuracy of inspection work will be highest on the first day of the week after the weekend break, and lowest in the latter part of the last day of the week when the accumulation of fatigue is greatest. Therefore, a short working day, with frequent breaks, is recommended for best performance.

Various patterns of working hours have been employed in factories, including switching inspectors between colour and non-colour inspection work on alternate days or half days. It is probably simple body rather than eye fatigue, that causes colour judgement to deteriorate.

Testing of individual inspectors' performance by the differential colour matching technique (section 10.5.2) has given rise to the belief that performance is worse the day after imbibing alcohol, and that deterioration of performance after taking even modest quantities of alcoholic drink persists for many hours. A heavy meal, particularly with meat (which stimulates the bile) will cause temporary deterioration of colour judgement. Older inspectors, whose colour judgement appears to be gradually deteriorating, may find their skill restored by being provided with higher illuminance – say, 50–100% more than is needed by younger inspectors.

10.5.4   Practical arrangements for colour inspection

The arrangements for special lighting for accurate colour inspection will vary according to whether the work is performed in an inspection department, or at dispersed inspection points along the production line (section 10.1). Because lamps of better colour rendering generally have lower efficacies than 'white' lamps, there will be additional costs if it is decided to light the whole production area with the former rather than the latter. Some ways in which this problem can be tackled include:

- Exclude daylight, and provide general lighting of the whole area from better colour rendering tubes but at a lower illuminance than that required at the inspection points, with inspection booths or additional lighting at such points.
- Exclude daylight, and provide general lighting at a generous level by use of high-efficacy tubes of colour appearance similar to that of the better colour-rendering tubes used for inspection, and provide inspection lighting only at the inspection points or use inspection booths.
- Without necessarily excluding daylight (but preventing the entry of sunlight), provide general lighting as above plus inspection booths.

10.5.5   Inspection booths

An inspection booth fitted internally with better colour rendering tubes can theoretically provide ideal viewing conditions for colour judgement, but the following points should be borne in mind:

- The lamps in the booth should be switched on 30 minutes before use so they are at the correct operating temperature.
- The booth must be ventilated for the comfort of the inspector and to ensure that the lamps are not overheated (which will cause change of colour).
- Before commencing work, inspectors must allow sufficient time for their eyes to become colour-adapted and adapted to the illuminance within the booth.
- 'Telephone kiosk' booths can be rather claustrophobic, and many inspectors prefer to work in a 'head and shoulders' booth (section 10.2).

11 Interior lighting practice

Environmental conditions in various industries differ so greatly one from another that it is impossible to devise a standard lighting method that will suit all industrial interiors. Much that is written in this book is of general application, but in applying the principles outlined to particular premises, one must take account of the actual visual tasks performed and the particular environmental conditions, and select appropriate equipment and lighting methods. The brief survey of lighting requirements in a number of industries given in this chapter will indicate the range of problems likely to be met, and some solutions.

It is suggested that ideas that have been successful in one industry may perhaps be utilized in another having some common visual problem or environmental condition. Such cross-pollination may take place between industries which are grouped in the following four sections of this chapter. Alternatively, an idea may be taken from an entirely different sort of industry and successfully transplanted.

This chapter also contains a section on lighting for the non-manufacturing areas of industrial premises (section 11.5).

11.1 Food, drink and pharmaceutical industries

11.1.1 Food industries

Lighting requirements for these industries are dominated by the need for lighting of good colour-rendering (chapter 4) and meeting the needs of hygiene. Lighting in food premises must comply with the *Food Hygiene (General) Regulations*[34] and other Regulations relevant to the food industries.

Many foodstuffs have unexpected optical properties. For example, some exhibit a degree of fluorescence under daylight and UV-rich sources and others exhibit dichromaticity (section 4.5.3). A constant problem is that foodstuffs in manufacture or preparation may be inspected under a 'cool' high colour-quality lamp type (e.g. Northlight lamps of 6500 K), yet may be exposed for sale under virtually any type of lamp, including DeLuxe Natural lamps, triphosphor lamps or 'warm' lamps down to 2800 K. It can only be advised that food manufacturers should ensure the quality, freshness and attractiveness of their products by inspecting them under high-quality lamps (such as Colour 96, Northlight, or Artificial Daylight) and under 'warm' lamps (such as Warm White, Colour 93, DeLuxe Natural etc.).

In particular, butcher's meat, fish and poultry should be inspected for quality under a good colour-rendering 'cool' source. Meat that is almost at the point of putrefaction will appear to be passably good if displayed in a retail butcher's shop under DeLuxe Natural tubes, or Warm White tubes sleeved in clear pink plastic – or even under Grolux lamps! (Meat, fish and poultry at or below freezing point have virtually no smell, whatever their condition.)

For lighting to comply with the *Food Regulations*, there shall be no likelihood of contamination or any part of a luminaire or lamp falling into foodstuffs. In meeting these requirements the following points should be considered:

- All lamps shall be enclosed so that no part of a broken lamp could fall into the food area, the lamps either being housed within enclosed luminaires, or fitted with transparent sleeving or other enclosure that will retain all parts of a broken lamp.
- Openable parts shall be hinged or retained by chains etc., and all fixings shall be captive.
- Luminaires that are enclosed, but not dust-tight/dust-proof, present the danger of dust etc. entering a luminaire and later being carried by air-currents and falling into foodstuffs. Only luminaires which are dust-tight, dust-proof or water-jet proof to BS 4533 will satisfy the requirement, and even these are not free of the dust problem if they are suspended so that dust can settle on their upper surfaces.
- Luminaires shall have minimum horizontal surface upon which dust may settle (Figure 11.1).

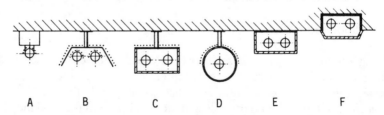

Figure 11.1 *Sketches of the six most commonly used types of fluorescent-lamp luminaires used in food factories, in ascending order of preference from left to right. The dotted surfaces are those on which dust may settle and later fall to contaminate foodstuffs. Types 'A' and 'B' do not comply with regulations unless the lamps are enclosed in transparent tubular enclosures, e.g. as in waterproof batten construction.*

Although high colour-quality lamps are needed for inspection, there is some choice of sources for ordinary production areas. Virtually any fluorescent lamp having a colour temperature of 3800 K or above will suit the baking and confectionary trades, while for large establishments with high roofs, mercury-vapour lamps (MB and MBF) can be used, though metal-halide (MBI, MBIF and HQI) lamps are preferred for colour appearance and overall economy.

It has been observed that when food premises have been relit to modern standards, the management are motivated to have the premises redecorated – probably because they can now see how grubby and dingy the place really is! Staff personal hygiene may also improve when there is better lighting and it is easy to see if finger-nails and whites are clean.

In canning lines there may be overhangs of equipment and conveyors which obstruct light, often necessitating local lighting, sometimes actually within machines. Such local luminaires must be especially robust and protected against damage, and each workstation should be checked for excessive glare. When locating local luminaires, consider the effect of dripping moisture, steam from autoclaves etc, and the proximity of steam-pipes.

11.1.2 Drinks industries

Milk bottling plants

These have the same visual problem of inspecting glass bottles in single-line progression past an inspection station as in glass container manufacture (section 11.4). Apart from the inspection stations and the local laboratory (which should be lit as for any other laboratory; see section 11.1.3), all areas of milk-bottling plants, interior and exterior, may be efficiently lit using high-pressure sodium (SON) lamps. All interior luminaires should be water-jet resistant to *BS4533*.

Soft drinks industry

The general requirements are as for milk bottling plants, noting that canning lines require special care (section 11.1.1).

Brewing, distilling

It could be a sound move to standardize on one pattern of corrosion-resistant Zone 2 luminaire because many locations in distilleries are wet or dusty, and sometimes flammable conditions exist. The bottling areas will have the same requirements as for milk bottling, and the canning lines have the same requirements as for food canning.

Some breweries have all overhead luminaires suspended on hooks, and connected by waterproof/sparkproof plugs and sockets. If a few spare luminaires are kept, this enables a defective luminaire to be safely removed and replaced without affecting production, and periodic maintenance can be carried out simply. Luminaires may be affected by fungus and yeasts, and may need to be cleaned/swapped frequently. In distilling and brewing there is no conventional 'working plane', but tasks are performed at various levels and few workers remain at particular workstations for long. Lighting should be adapted to these conditions, ensuring that there is adequate light for reading gauges, and that walkways and ladderways are safe to traverse, day and night.

For periodic maintenance at high level in breweries and distilleries, some tungsten–halogen floodlight luminaires may be positioned in the upper parts of the plant, with beams directed upward. These luminaires will be used only during maintenance work.

Alcohol fumes are flammable and explosive. Consultation between the client and lighting provider, perhaps with professional advice, will enable dangerous zones where protected equipment must be used to be identified (section 7.10). Do not overlook the fact that fine particulate dusts, e.g. grain dusts, can burn and explode with violence if suspended in the air at certain concentrations (section 7.6.8).

11.1.3   Pharmaceutical industries

The hygiene requirements in the pharmaceutical industries (including the manufacture of non-medical chemical products and cosmetics etc.) are generally similar to, but rather more stringent than those for the food industries. There may be special requirements for sterile rooms and clean-rooms (section 7.4). Some laboratories use fume cupboards or bench-top environmental enclosures, the lighting of which is similar to that for inspection booths (section 10.5), while others may have flame-hazard conditions (section 7.10). For all these environments, the advantages of using pressurized protected luminaires should be considered (section 7.10.3).

11.2   Clothing, textiles, paper and leather industries

The industries discussed in this section produce products consisting of natural or synthetic materials many of which have been treated with dyes.

Natural light, and light from some fluorescent tubes, contains an ultraviolet component which can cause organic or inorganic dyestuffs or base materials to fluoresce. Products may undergo significant changes in colour appearance according to the spectral composition of the light under which they are viewed. The lighting user may need to inspect both raw materials and finished products under several different test illuminants to ensure that the product will be satisfactory during exposure for sale and during use.

To quote just one example of this apparent colour change effect, a clothing manufacturer had a large quantity of skirts returned because some panels of the skirts appeared to be of markedly different colour from the others when seen in daylight, though all panels appeared to be of identical colour under normal electric lighting. It was only when the cloths were inspected under Northlight tubes that the colour differences (due to variations in the dyeing of bolts from the same supplier) became apparent. More information on this problem is given in chapter 4.

11.2.1 Clothing industry

A study by the author of lighting conditions in the clothing industry indicated that the industry is highly sensitive to improvements in the working environment. It is apparent that lighting is a factor of high importance in helping create conditions fostering productivity and staff contentment.

The study indicated that some 60% of all faults found in garments at final inspection were fabric faults, i.e. had there been better inspection of the cloth initially, the considerable costs of manufacture of unsaleable or substandard garments could have been avoided. In particular, it was noted that lighting at cutting tables rarely came up to the illuminance and colour standards recommended in the *CIBSE Code*[1] as necessary for the cutter to spot fabric faults. It should be asked if the cutting table is the best place for this inspection. Would it not invariably be better for all fabric to be inspected properly before cutting?

Weave faults can best be detected by running the fabric over a back- or front-lit winding frame (Figures 11.2 and 11.3), and/or arranging light to come at a glancing angle.

The beneficial effect on morale from providing a pleasant visual environment is of importance, because in machine rooms, the operatives cannot converse, being isolated by noise and the nature of their task. Good lighting aids the work by helping concentration as well as vision. Lighting is mainly responsible for the appearance of the interior as seen from the working positions, so it is important that the general lighting illuminance should be at least a third of that at the needle.

In some clothing factories, the uncontrolled entry of daylight and shafts of sunlight from windows (and especially from roof windows) is often a serious visual handicap to workers. Anecdotal evidence suggests that productivity in such premises may be significantly higher in winter when the daylight contribution is less. Most probably the same improvement in quality and output could be achieved all year round by fitting window blinds and relying on well-designed electric lighting.

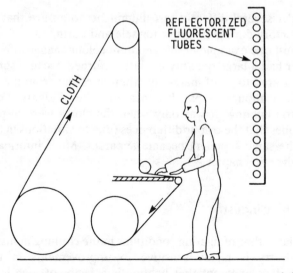

Figure 11.2 *Use of an 'artificial window' for perching (inspection) of cloth.*

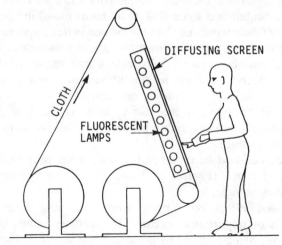

Figure 11.3 *Use of a rear-lighted perching frame. One such apparatus could serve this purpose and as an 'artificial window'.*

11.2.2 Textiles industry

In factories concerned with both natural and man-made fibres, the problem of dusts and linters affecting luminaires must be dealt with. It is not uncommon for suspended luminaires to be so thickly covered with dust as to overheat and fail in service. Recessed and flush ceiling-mounted luminaires present no upper surfaces on which dust can settle.

In weaving, if the lighting is such as to enable the smallest faults to be seen, there will be more stoppages for correction, but the overall downtime will be less. Looms and weaving machines may need light flow from unusual directions according to the visual needs of the weaver, especially when dealing with weave faults. The use of pastel colours for machine surfaces is a proven aid to seeing threads. Localized lighting, with additional local luminaires if necessary mounted within the machines, may be needed.

For final grey perching, an 'artificial window' may be used. This consists of a wall-mounted or free-standing array of perhaps as many as 12 or 15 closely-spaced 65-W fluorescent tubes, either bare, or mounted behind a diffusing cover. The artificial window is placed parallel to the vertical cloth web and about 1.5 m to 2 m from it. The inspector stands with his or her back to the light array which is large enough to throw only an insignificant shadow on the cloth (Figure 11.2).

Some fabrics are better inspected by back-lighting, the fluorescent tubes being placed in a glazed frame which can be tilted and varied in height (Figure 11.3). It is desirable that the lamps used in either of these applications are equipped with dimming. A single array equipment might serve the purposes of an artificial window and provide backlighting, and will do so better if the unit is fitted with castors for ease of movement, and brakes. Such lighting arrangements may use reflectorized fluorescent tubular lamps, though these may be difficult to obtain with high colour-quality phosphors.

11.2.3 Paper industry

The UK paper industry traditionally has set a fine example of conscientious application of good safety practices, there being many dangers to personnel in paper making, corrugating, cardboard making and box making etc. The provision of adequate lighting encourages operators and setters to 'use eyes, not fingers'.

Because this industry is highly mechanized, it employs only small numbers of workers in relation to its turnover. General lighting is required for supervisory duties, but for setting-up, lacing machines etc., the use of light projected into the machines from luminaires mounted at the sides, under or within the machines, is vital. More light is usually needed to deal with setting-up and clearing jams than for normal operation, since control of downtime is essential. Backlighting may be used to enable the operator to watch for holes and thins in paper webs. The atmospheric conditions in paper mills are generally humid, so enclosed and drip-proof luminaires are usually necessary.

A difficulty met in paper making is lighting for the judgement of colour of the wet pulp with the view of achieving an objective dry paper colour (section 4.5).

In paper mills where waste paper and materials are stored out of doors, drivers of trucks and clasp-trucks etc may be frequently passing between outdoor and indoor areas, day and night, so that some zones of intermediate illuminance may be needed (section 9.7).

In salles (inspection rooms), it is generally found that the best method of lighting is to project light onto the paper sheets at a glancing angle (almost parallel to the top of the stack) so as to reveal surface defects.

11.2.4   Leather industries

The environmental conditions surrounding the processes of tanning usually include a humid and corrosive atmosphere, usually necessitating the use of water-jet proof luminaires of corrosion-resisting construction.

Although there is today a generally declining demand for goods made of natural leather, the demand for finely-made leather goods of superlative quality is greater than ever. In the selection, matching, treatment, grading and cutting of tanned hides and finished leather, the provision of excellent standards of lighting is a prime requirement.

By the very nature of shoemaking machinery, particularly in clicking and closing, it is impossible to guard the machines completely. Adequate illuminance is a valuable safeguard for fingers under these circumstances. In some factories, optical guidelines are projected on to the work, requiring that any local lighting be carefully arranged to ensure that there is sufficient light for the tasks without swamping the luminous guidelines and location marks.

The final inspection of finished footwear justifies having a separate inspection department, usually with the packing area adjoining. For the final inspection of glossy finished leather goods and footwear made of natural leathers, synthetics and patent leathers, the use of large-area low-brightness diffusing luminaires localized at the inspection stations will probably bring best results. Northlight or Colour Matching fluorescent lamps should be used in such areas.

11.3   Engineering, plastics, printing and furniture industries

In many production processes, although there is a high degree of reliance on machine processing, the man–machine cybernetic relationship is dominated by the visual sense (section 1.9.2). In many factories, better use could be made of human and material resources by the provision of an environment of cheerful appearance.

The following is an extract from a report on lighting in a machine shop, written by the author for a major industrial organization in 1980:

(a)

(b)

(c)

(d)

Photo 1 *Examples of access equipment.*
(a) Ladder-tower erected;
(b) Ladder-tower lowered for movement;
(c) Man-lift tower;
(d) Mobile platform with clip-on ladder
(Photos: Access Equipment Ltd)

Photo 2 *External apron floodlighting and high-bay lighting in the Civil Aviation Authority's hangar at Stanstead which uses 400 W high-pressure sodium lamps. (Photo: Thorn Lighting)*

Photo 3 *A storage and unloading area floodlit to provide 30 lx, using 250 W high-pressure sodium lamps. (Photo: F. W. Thorpe plc)*

Photo 4 *Exterior lighting at British Coal's Daw Mill Colliery, Warwickshire. (Photo: Abacus Municipal Ltd)*

Photo 5 *An energy-efficient scheme using twinlamp flourescent luminaires with high-frequency controls at Ford's Belfast factory (Photo Crompton Parkinson)*

Photo 6 *Fluorescent lighting using high-frequency control gear with photocell switching for energy efficiency at PDP Pumps, Eastbourne. (Photo: Thorn Lighting)*

Photo 7 *'Prismalume' prismatic high-bay units at Aston Martin Lagonda. (Photo: Holophane)*

Photo 8 *Metal-halide lighting which is in use 24-hours a day at British Sugar, Kidderminster. (Photo: GTE Sylvania)*

hoto 9 *The library reception area at the London School of Foreign Trade which is t by continuous-trough recessed fluorescent units with low-brightness metalised ouvres. Fluorescent lamps on the tops of bookcases provide some ceiling rightness. (Photo: Poselco Ltd)*

Photo 10 An example from a range of freestanding uplighter units. (Photo: Anglepoise)

Photo 11 *A low-brightness fluorescent unit with metal louvres. It can be provided with a 3-hour duration battery pack for duty as an emergency lighting luminaire. (Photo: Tenby Electrical Accessories – Emess plc)*

'In large parts of the factory, the general decor of the building and machinery is green, and the floors are near-black. The result of this combination is to produce a very depressing effect indeed, which cannot be good for morale. In some departments the lanes between machines are very narrow, and machinery extends above eye-level, which, with the dark colours and poor lighting must produce a depressing, even hysterical, response in sensitive workers. One must ask if such conditions are conducive to good mental health and the welfare of workers, let alone their productivity.'

At the time of writing that report, such conditions were not exceptional in UK factories. It has to be recorded that, although the general standard of environmental conditions in UK industry has improved enormously since then (largely because of the introduction of new production methods and the general adoption of numerically-controlled machines), there are still some 'dark Satanic mills' which could be transformed with quite minimal expenditure and with good cost justification.

For older premises where upgrading of the environment has been neglected over the years, the following essential steps should be taken:

- Relight the premises to the standards of the *CIBSE Code*[1], adapting the lighting installation to the guidelines given in this book.
- Redecorate the premises to provide walls and ceilings of a reasonably high reflection factor; remember that machinery and plant can also be painted.
- If the floor is of a very dark colour, consider covering it, or painting it with a suitable floor paint to raise its reflection factor to at least 0.20, preferably higher.

11.3.1 Engineering industries

In microtechnological processes, it is noted that too great a reliance on small-area task-lighting and built-in lights within microscopes, micromanipulators etc., may be counterproductive, as the following case-history demonstrates.

In a department assembling minute components, the operators were initially provided with general lighting at 350 lx, and had built-in lighting within the optical devices that they operated, giving about 1500 lx at the point of work as viewed. The operators complained of fatigue and headaches, and the management found it was necessary to institute a routine of 30 minutes work followed by 10 minutes of eye relaxation. This limited the productivity of the unit, but it was found that longer work periods or shorter rest periods reduced the quality of the output.

After the tasks had been studied, 1750 lx of general lighting was installed, and some workstations were experimentally fitted with small focusing spotlamps controlled by individual dimmers. (The lamps used for this were 35-W car headlamp bulbs on a 12-V supply, being the only lamps available at that time that would physically fit inside the equipment. Currently, a wide range of LV miniature filament lamps is available.) When the improved general lighting was installed, the white plastic benchtops were covered with grey matt linoleum (reflection factor 0.4) to prevent reflected glare, and the room walls were painted a pastel colour with a reflection factor of 0.7.

After an experimental period, it was found that work periods of one hour were possible without loss of quality or complaints of headaches or eyestrain. The operators said that under the new lighting conditions they did not feel so isolated, as they could look away from the microscope for a few seconds without any difficulty of re-adaptation on returning to the task. It seems that the few seconds glancing away from the task without a major change in the luminance viewed rested the eyes and helped concentration.

11.3.2   Plastics industries

The manufacture of plastic base materials is carried on in environments similar to those met in the pharmaceutical industry (section 11.1.3) and the petrochemical industries (section 11.4.4). In plastic moulding shops and extrusion plants the presence of solvents in the atmosphere can create a flame hazard (section 7.10). Cases are known of solvent materials (not necessarily those with a high fire risk) attacking plastic luminaire enclosures, and deteriorating the insulation of electric equipment and cables.

A common problem is how to direct light into the moulds to check that they are clean of flash and trash; the machine-head and the top half-mould mask the lower half-mould. The use of 'remote local lighting' or a metal mirror (section 9.5.2) to deflect light into the lower half-mould may be the solution.

In moulding, pressing and extrusion shops, the air temperature above machines may be too high to allow lighting equipment to be located there. In such hot conditions, luminaires should not be mounted higher than the highest window opening, and forced-draft ventilation may be needed.

11.3.3   Printing industry

Because of the very high visual content in all printing operations, this industry stands to benefit greatly from the application of modern lighting techniques. The technological transformation of this industry continues apace, with the general adoption of text scanning, typesetting from disk, screen graphics and screen page composition etc., which have enabled many printing concerns to operate in much smaller areas than hitherto. For this reason many printers have moved to smaller modern premises, wherein the clean and visually correct conditions can be created for computer-based tasks.

For colour work, illuminances considerably higher than those recommended in the current *CIBSE Code*[1] may be justified, with illuminances of 2000–3000 lx from better colour rendering lamps being adopted.

Many print works employ ultraviolet light for accelerated 'drying' of print (actually a process of almost instant polymerization of the ink molecules). Two cautions: never place an ultraviolet lamp in an ordinary luminaire – the rays will be harmful to eyes and skin; and do not attempt to dismantle or open ultraviolet drying conveyor-tunnels, nor attempt to bypass the interlocks on the covers of such equipment.

Note the use of cyan filters to enable yellow print registration marks to be seen clearly (section 10.4).

11.3.4 Furniture industries

In woodworking areas, the problem of dust must be dealt with. Fine wood flour will penetrate into non-dust-tight luminaires, reducing their light output. Polishing and varnishing shops using flammable solvents will be hazardous zones (section 7.10.1).

In departments using routing and moulding spindles, circular saws, bandsaws etc, there is the possible danger of stroboscopic effect which must be suitably countered (section 2.8). In the saw-doctor's workshop, use may be made of a back-lit screen for checking saw tooth configuration.

Throughout the works, there should be consideration of alternative lightsources. In general machine shops and sawmills, furniture assembly and

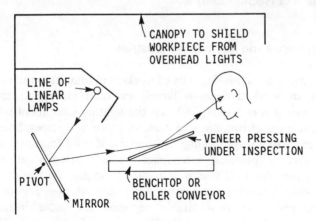

Figure 11.4 *Directional light apparatus for examining veneer pressings for flatness. The lamps may be reflectorized fluorescent lamps or architectural filament lamps. The mirror is pivoted to enable it to be set at a convenient angle for the inspector.*

stores areas, high-pressure sodium (SON) lamps should be the first choice on grounds of efficacy, while fluorescent tubes of high colour quality should be used in veneer preparation and matching, finishing of all upholstered or polished goods. Upholstery materials will be carefully inspected and checked for colour before being cut (section 10.5). Local lighting, perhaps built into the machines, will help quality control; for example, placing a low-mounted dust-proof luminaire behind a belt-sander.

For veneer sorting, a louvred ceiling or louvred luminaires over the workstations will enable the provision of high illuminance without excessive direct glare, but diffusing luminaires may reveal surface quality better. Checking of flatness and surface quality of veneers after pressing may be facilitated by use of an illuminated grid (section 10.3.2), or by an inspection apparatus to project light in a controlled directional manner (Figure 11.4), an arrangement that seems peculiar to the furniture industry.

11.4 Metals, foundries, glass and petrochemical industries

Although these industries use vast amounts of fuels, they cannot afford to waste energy any more than any other kind of business, and the cost of installing energy-efficient lighting is always justified.

The high-roofed buildings common in heavy industry make lamp replacements costly in terms of labour. In many situations, if *raising-and-lowering gear* (section 23.3) is not fitted, it may only be practicable to replace lamps during a periodic shut-down. Because of danger from hot plant, and many floor obstructions, relamping may be more conveniently performed through the roof (section 23.3).

11.4.1 Foundries and hot metals industries

The safety, morale and productivity of workers in foundries and places where hot metals are worked depends largely on their managers ensuring that adequate lighting is provided, and that the lighting equipment is kept clean and serviceable under conditions of heat, vibration and atmospheric pollution – the last of these always tending to be worse if solid fuels are used. Many premises have a polluted atmosphere which actually attenuates the illumination (section 9.2.10) as well as rapidly soiling the luminaires.

High pressure sodium (SON) lamps have been adopted generally in these workplaces, and have now largely replaced the older installations of fluorescent lamps and mercury (MBF) lamps. However, if colour-coded moulding sands are used, the use of metal-halide lamps could be justified. Tests should confirm that the colours of moulding sands, and the colour-code marks on metal stock, can be readily recognized under the chosen lightsource (section 4.2).

In drop-hammer shops, where opportunities for personal injury abound, the provision of very good lighting is of great importance. The drop-forging industry had a poor record for the provision of lighting in past decades, and environmental conditions were very poor, but with the more general use of controlled-atmosphere gas heating of billets and the increasing adoption of electric billet-heating, the environmental conditions in the modern drop forge are no worse than in many other sections of heavy industry. However, special attention must be given to providing access for cleaning and relamping the luminaires, and lighting layouts have to be carefully designed to avoid positioning luminaires over hot plant, and to provide good means of access.

In all workplaces where hot metal is worked, the general lighting illuminance must be sufficient to 'swamp' the brightness of hot metals, or the 'adventitious light' from them will be a hazard due to uncontrolled glare. The drop-forging industry has long accepted that the old concept of 'keeping the general lighting level low so the temperature of the billet may be judged by its brightness' is technically unsound, and unjustified on grounds of safety; billet temperature is readily determined by instrumentation.

11.4.2   Glass industry

In this industry sector there are a number of operations having different specialized lighting needs. For example, special attention has to be given to lighting for the inspection of pressed or blown glass containers (bottles, jars etc.) at inspection stations where the products pass through on single-line conveyors.

The essence of efficient inspection of such containers (which pass the inspector at 50 to 90 pieces per minute) is to enable the inspector to use his peripheral vision. This is done by providing inspection stations in which the approaching and receding containers can be seen by the inspectors 'out of the corner of their eye' while they are ostensibly viewing the container that is passing the nominal inspection point straight in front of them (Figure 11.5). This type of inspection station utilizes an illuminated white backboard ruled with a grid to reveal inconsistencies in wall thickness (section 10.3.2). The grid device enables an inspector to spot containers having thicks or thins in their walls etc.

After a period of training and practice, inspectors seem to carry in their mind's eye an image of the perfect bottle or jar, and, as though by instinct, will seize and cast into the cullet box any container with 'seed' (tiny bubbles), 'airlines' (long thin bubbles), 'crizzles' (surface defects due to air between the glass 'gob' and the mould, 'hammocks' (strands of glass across the interior of the container) etc, as well as those which are obviously defective at the finish

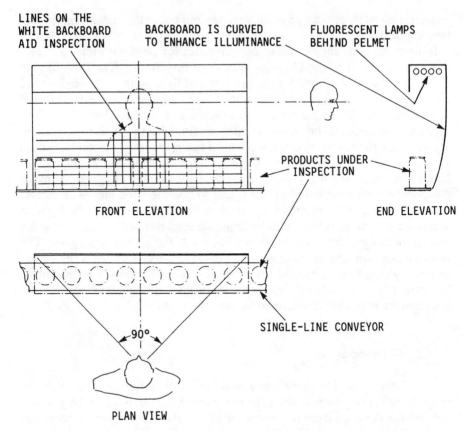

LINES ON THE
WHITE BACKBOARD
AID INSPECTION

BACKBOARD IS CURVED
TO ENHANCE ILLUMINANCE

FLUORESCENT LAMPS
BEHIND PELMET

FRONT ELEVATION

END ELEVATION

PRODUCTS UNDER
INSPECTION

SINGLE-LINE CONVEYOR

90°

PLAN VIEW

Figure 11.5 *Inspection station for inspecting glass containers on a single-line conveyor.*

or are asymmetrical. Often an inspector will seize and discard a container as it enters his zone of peripheral vision, apparently without ever looking at the defective item!

Tests have shown that the area of the inspection station must have good general lighting, and be reasonably free of direct glare; the approaching and receding conveyor should have an illuminance of at least 500 lx, and the backboard needs an illuminance of around 750–1000 lx from reflectorized fluorescent tubes concealed behind its canopy. The illuminated backboard should subtend an angle of 90° to the inspector's eyes in the horizontal plane (Figure 11.5).

Some glass hollow-ware is better inspected by bottom lighting (Figure 11.6), with the product seen against a fairly dark background in subdued and well diffused general lighting.

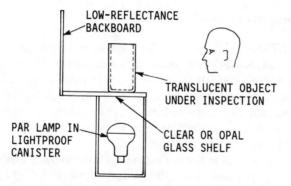

Figure 11.6 *Bottom light to aid inspection of translucent object.*

Various means of introducing light into the edges of sheets of glass are used, so that light is 'piped' through the sheet to reveal faults and surface scratches. Glass tubes and cylinders are similarly inspected, either by piping the light in from one end, or by placing the cylinder in contact with a prism of clear plastic with a light source behind it.

11.4.3 Petrochemicals industries

In order to minimize hazards from fumes and toxic gases, petrochemical plants these days are usually built as open-air structures, and the lighting techniques needed are largely those applicable to other exterior industrial lighting (section 16.4). Various sections of chapter 7 relate to the environments met in this group of industries. There is need for especially good colour discernment in chemical testing carried out in the laboratory (section 10.5) which may be a hazardous zone (section 7.10).

11.5 Offices and non-manufacturing areas

This section deals with lighting for the non-manufacturing areas of industrial premises which, like every other part of the premises, should make their contribution to the efficiency of the user's organization. Lighting of entrance areas, offices and reception areas is important to creating a pleasant and prestigious appearance. Lighting in the offices and design department greatly facilitates efficient work, while that in the staff rooms and amenity areas can help shape the environment of the staff in a favourable way.

11.5.1   Lighting of offices

Refer to the *CIBSE Code*[1] for specific recommendations for the lighting of offices. Offices associated with typical factories are usually of limited extent, and high-quality treatment of the décor and lighting may not be justified, but, even on a limited budget, there is no difficulty in providing lighting of good quality and appearance using standard equipment. Most modern offices have suspended ceilings which permit the easy installation of recessed fluorescent luminaires and recessed downlighters.

Typical ceiling heights in factory offices do not generally enable the use of HID lamps except perhaps in uplighter installations (section 9.8). Uplighting is becoming very popular for offices, particularly those in which VDTs are in use. General lighting plus individual workstation lighting using well-shielded small fluorescent-lamp luminaires, is often adopted.

For larger office areas, particularly those designed on open-plan lines, air-conditioning may be employed, and the use of air-handling luminaires may be considered. These have the following advantages:

- The air in the room is extracted through the luminaires, tending to keep the tubes cooler and thus giving enhanced light output.
- Because it may be possible to dispense with ceiling grilles, the appearance of the ceiling may be less cluttered.
- In large installations, in conjunction with suitable air-conditioning equipment, it may be possible to extract the heat from the luminaires and reuse it.

For rooms used for meetings and training, special lighting arrangements may be justified. A very flexible method of lighting conference rooms is to install a

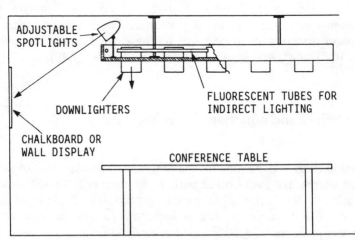

Figure 11.7 *Construction of a 'lighting raft', a lighting fitment appropriate to conference rooms and boardrooms.*

lighting raft (Figure 11.7). This is a simple structure which can be constructed in-house, and carries several systems of lamps. These may include downlighters to illuminate the conference table, indirect lighting for a soft lighting effect across the room (suitable for social occasions, or when showing slides or audiovisuals), plus some directional lighting for decorative effect or to illuminate exhibits or wall displays. If correctly dimensioned, the lighting raft will appear to be 'floating', i.e. the down-rods by which it is suspended will not be visible from normal angles of view.

11.5.2 Lighting for design offices

The *CIBSE Code*[1] recommendation of 750 lx of general lighting for drawing-offices can be interpreted in various ways. For smaller offices, the rooms may have general lighting of 500 lx, with adjustable tasklights at each workstation. This is a traditional method, and often preferred by the staff. The adjustable tasklights may house two or more miniature fluorescent lamps, but some draughtsmen prefer to have a filament-lamp luminaire. If VDTs are in use, uplighting may be preferred, or a lower illuminance of general direct lighting using sharp cut-off luminaires may be employed.

For large drawing offices having 20 or more boards or having boards of A0 size or larger, a good lighting method is to install a *coffered ceiling* designed to give adequate illuminance on drawing boards inclined at steep angles (Figure 11.8). In the example shown, the cut-off angle is 70° from the downward vertical.

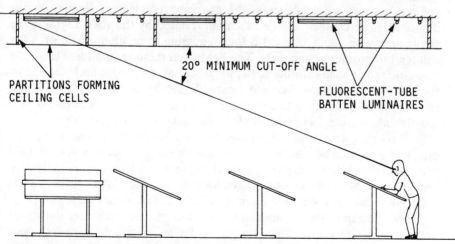

Figure 11.8 *Coffered ceiling for drawing offices and areas where visually demanding work is done. If the coffers are square in plan, operatives may face in any direction with equal visual comfort.*

A typical coffered ceiling system consists of a pattern of ceiling cells comprising vertical laminae attached to the structural ceiling, so that the whole ceiling is covered with open-bottom cells, each containing one or more fluorescent batten luminaires. Because of the large area of the cells in such a ceiling, it is necessary to construct them of fire-resistant materials to the satisfaction of the insurers and the Fire Prevention Officer. For example they may need to be of sheet metal, asbestos board, or fire-resistant building board.

If fine colour judgement is to be exercised in the room, the ceiling and the walls of the cells must be coloured white or a pale shade of grey devised from a mixture of black and white paints only; in other cases, the ceiling surfaces and walls of the cells may be painted a pale pastel tint with a reflection factor of 0.7 to 0.85.

A coffered ceiling as described will give high cylindrical illuminance (section 9.4.2), but the light may be so diffused that it may be difficult to see the minute holes in drawing surfaces made by compasses and dividers, necessitating the use of drawing-board lights.

The size of the ceiling cells will depend on the length of the fluorescent lamps to be used and the height of the ceiling. For 1200 mm and 1050 mm tubes, cells of 1500 mm and 1350 mm should be considered.

11.5.3   Lighting for canteen, staff rooms and medical room

Luminaires in kitchens and areas where food is stored or prepared should be of the types discussed for food factories (section 11.1.1).

Even in small establishments, the provision of pleasant lighting in the dining room or canteen will be much appreciated by the staff. Lighting the whole area with lamps of good colour-rendering to an illuminance of 300 lx will tend to promote cleanliness. The lighting in these areas should tend to a domestic rather than institutional appearance. Many people prefer to eat by the light from ordinary filament lamps. Some lighting track, with a few adjustable dichroic lamps to emphasize a picture or some indoor plants, will greatly enhance the feeling of comfort and relaxation in these areas.

Staff rooms sometimes have to double as the dining room, the works club etc., and may even be used for lectures and meetings. Thus it is wise to build some flexibility into the lighting by providing several different ways of lighting such rooms. The provision of dimming for at least some of the lights is a great asset for social occasions and when showing slides or videos etc.

The writer can reflect on the hundreds of factories he has visited during his career in lighting, recalling that an almost invariable sign of poor management has been dirty toilet areas. The cause and effect: poor lighting engenders lack of respect for the surroundings; the cleaners will not take a pride in their work, and users will leave mess and dirty basins etc. Well lighted lavatory and washing areas are conducive to better hygiene and to better respect for the premises. The cost of lighting these areas, including lighting for the mirrors, is so small that economy is pointless. If there is a problem with damage to luminaires or stealing of lamps, vandal-resistant equipment should be fitted (section 7.8.2).

Medical and first-aid rooms need suitable lighting for hygienic purposes, and to facilitate diagnosis and treatment. The use of better colour-rendering lamps is necessary, as people's complexions are an important diagnostic aid. Enclosed, dust-tight luminaires are appropriate, for all the room surfaces will need to be thoroughly cleaned periodically. A wheeled pedestal lamp as used in doctors' surgeries will be a considerable asset when dealing with foreign bodies in eyes or extracting small splinters from the skin. In a treatment room, a general lighting level of at least 300 lx is needed, with 500 lx where patients are examined. These levels are readily obtained by localizing suitable luminaires to the examination area. The ability to dim the lighting so that a sick person may sleep is desirable. Dimmed lighting would be appropriate when vision-screening (section 1.10) is being performed. Eye-test wall charts (Snellen Charts – section 1.6.3) require an illuminance on the chart of 300 lx; near-vision tests require 500 lx on the test.

11.5.4   Stairs, corridors, circulation areas and entrances

Where a staircase forms part of the reception area for visitors, it is tempting to provide lighting that is rather more decorative than functional. Lighting of stairs is important, for falls on stairs account for a significant number of industrial injuries. If the staircase has open balustrades and/or the steps have no risers, position the luminaires so as not to cause glare to people ascending or descending. If the risers and treads are of uniform colour, try to locate the luminaires so as to leave the risers in shadow by positioning the luminaires on the landings and half-landings rather than over the flights (Figure 11.9). In particular, check the visibility on the stairs under emergency-lighting conditions. The use of 'pilot lighting' on stairs is recommended, i.e. strategically placed lights or emergency lights ('sustained' or 'maintained' – chapter 14) positioned to give guidance to a person seeking the exit under emergency conditions.

In the *CIBSE Code*[1], the only areas for which scalar illuminance (section 2.5.3) is specified are circulation areas. The objective is to give the appearance of a 'well-lit space' (though of course, it is only the planes defining the space that are lit), in which persons will feel confident and comfortable. But note that such lighting is virtually shadowless, and should be relieved by some decorative spotlights, table lamps, wall lamps etc, or other suitable features of lighting and décor. In waiting rooms and interview areas,

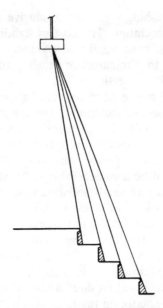

Figure 11.9 *A method of lighting stairs. Luminaires are positioned to throw shadows on the risers and thus make the treads more easily seen.*

if armchairs and a coffee table are provided, with one or two modern floor standards or table lamps (of kinds which might well be used in the home), an appropriate informal and relaxed atmosphere will be created.

Outdoors, luminous signs and building floodlighting help to make the premises readily recognizable, while exterior lighting of roadways, and entrances from the main road are important for safety and security. Visitors should be provided with a well-lit car-park and the exterior lighting should help them find their way to the visitors' entrance. At entrances, there may be the need to provide gradations of illuminance to allow people to adapt to day and night conditions on entering or leaving (section 9.7).

A well-lit entrance and foyer, with good signing and a pleasant reception area, is a welcoming and pleasing approach to any organization. The relatively small cost of providing it may well be repaid by the benefit to prestige. First impressions of organizations, like those of people, are often long-lasting and tend to shape the subsequent relationship.

12   Interior security lighting

The principles of exterior security lighting (chapter 17) are well known and are widely applied for the protection of business premises. This short chapter is to introduce the concept of *interior security lighting* which can be a valuable addition to the night security precautions at industrial and commercial premises.

12.1   Principles

12.1.1   When business premises are closed, their night defence may be effected by a security system including detection devices and alarms, and exterior security lighting. The premises may be guarded by on-site guards, or visited by mobile security patrols at intervals during the quiet hours.

It is a general objective of all security lighting systems to enable the guarding and inspection of the premises to be carried out without the security personnel having to use handlamps. The use of a torch or handlamp has several disadvantages:

- Security guards tend to see only those things to which they happen to direct the torch, and may not see vital matters outside the moving beam of light.
- The eyes of security guards tend to become adapted to the brightness of the bright patch from the torch, so that they may fail to see vital matters which they cannot pick out of the shadows because of their adaptation level.
- The bright torch would signal the guard's position and his movements to an intruder concealed in the shadows inside or outside the premises.

12.1.2   Interior security lighting is lighting within the premises that is kept on during all the hours that the premises are unoccupied, and serves several purposes:

- To obviate the need for security personnel to use a torch while patrolling or searching the premises.
- To facilitate safe and thorough patrolling or searching of the interior of the premises.
- To enable the interior of the premises to be viewed through the windows by security personnel or the police.

- 'Pilot lighting' to enable persons to move about the premises safely for legitimate purposes during the quiet hours, e.g. cleaners, maintenance staff.

As with all security lighting, interior security lighting has a strong deterrent effect on those contemplating surreptitious entry for illegal purposes.

12.2   Applications

12.2.1   Electrical supply

The design of a system of interior security lighting has some factors in common with the design of interior emergency lighting (chapter 14). However, in most cases the interior security luminaires will not be fed from a maintained supply. Although it certainly would be advantageous if these luminaires could be fed from a battery-backed supply, the most that can usually be done in practical situations is to make the supply arrangements to the internal security lighting luminaires as secure as possible by the following means:

- The supply to the security luminaires to be on a separate fuseboard from other lighting.
- The fuseboard and switches for the security luminaires to be in a locked room or cabinet to prevent unauthorized interference.
- If a standby generator system is in place, the interior security lighting luminaires should be arranged to be automatically switched to that supply on failure of the mains supply.
- For simplicity and reliability, the security luminaires may be constantly energized day and night provided no glare to occupants is caused; some energy economy may be achieved by switching these circuits using a solar-dial control contactor. Manual daily switching is not recommended.

12.2.2 Design of internal security lighting systems

In the absence of recommendations from the professional lighting institutions, some suggestions are made in the following paragraphs.

The enclosure of the luminaires used must match the environmental conditions of each location (chapter 7). Practical considerations may dictate that robust well-glass or bulkhead luminaires fitted with compact fluorescent lamps be employed for this duty in all areas having 'normal' atmospheres. With regard to illuminance, bearing in mind the limited objectives of this lighting (section 12.1), providing a minimum anywhere in the interior of 1 lx at floor level is likely to satisfy the requirements. No specific recommendation is made on uniformity. Any of the following luminaire positions may apply to a particular interior:

- For ground floor rooms, luminaires may be positioned above windows so that vision inward from the exterior is not handicapped.
- Luminaires may be positioned above doorways so that vision inward from the doorway into the room is not handicapped.
- If supervision hatches are cut in interior walls (to permit the security guard to see in from an adjacent area), the luminaires should be positioned above or beside such hatches so that vision inward from the hatch is not handicapped.
- Luminaires may be positioned at vulnerable points, e.g. over safes and filing cabinets, or over or near valuable plant.
- For the safety and convenience of the security guard, luminaires may be positioned over walkways and routes through the plant, and used to illuminate all ramps and staircases.

12.2.3 Integration with other security systems

Interior security lighting luminaires may be positioned to illuminate firefighting equipment. If closed circuit television security monitoring is installed, the positions of luminaires should take account of the lighting needs of the cctv cameras.

13   Interior portable and temporary lighting

13.1   Interior portable lighting

13.1.1   Applications

The disadvantages of using torches and handlamps for security patrolling have been noted (section 12.1). However, there are tasks carried out in industry which really necessitate the use of portable lighting. Some examples are as follows:

- Work on plant which cannot be adequately illuminated by general lighting, e.g. dealing with jams inside paper-making machines (section 11.2.3).
- Maintenance operations in places which are not normally lit, e.g. cleaning the inside of large boilers, de-scaling tanks and vats, replacing refractories in furnaces, relining glass-melting kilns.
- Carrying out electrical installation work and refurbishment in unlit premises, including installing lighting and emergency lighting.

For work to be done that is of short duration, e.g. carring out a periodic inspection of normally unattended plant, then *battery-powered handlamps* may be the most convenient means of providing a small amount of light where and when it is needed (section 13.1.2). If the light is to be used for a longer duration, or a larger working space must be lit, then *mains-powered portable lighting* is preferred (section 13.1.3). If the lighting requirement is temporary but of fairly long duration, then a system of *temporary lighting* could be used (section 13.2).

13.1.2   Battery-operated handlamps

Handlamps with separate rechargeable batteries are not recommended for industrial use, for unless new batteries are fitted each time the handlamp is taken out of store (a costly procedure), inevitably some handlamps will be issued with partially-discharged batteries which will give short duration of use. It is generally better to use handlamps having sealed rechargeable batteries which can be recharged in situ by simply placing the handlamp in a storage rack which connects it to the charging circuit.

13.1.3 Mains-powered portable lighting

The use of any portable lighting equipment operated on mains voltage is strongly deprecated. All portable lighting equipment should preferentially be operated at reduced voltage. Robust portable lighting equipment operating on 110 V as described and recommended for use on construction sites (section 16.1) is very suitable for indoor and outdoor industrial use, subject to the equipment satisfying the requirements for safety in the particular environment (section 13.1.5).

13.1.4 Reduced-voltage distribution systems

Where the need for portable lighting in a factory is re-occurring, it is preferable to install a permanent distribution system at reduced voltage, with socket outlets of appropriate design at all positions where the need for portable lighting is likely to occur. Such a system should be fed from a double-wound step-down transformer at 110 V (in the future 100 V) having an earthed screen between the windings and its low-voltage winding centre-tapped to earth. With such a system, under fault conditions, the highest voltage to earth that can appear on the system is 55 V (future systems 50 V). A distribution system similar to that described in *BS 4643*[49] could be employed (Figure 13.1). To prevent voltage mismatch, socket outlets complying with *BS 4343*[50] could be used if these are suitable for the particular environment.

13.1.5 Cautions

● Personnel should not enter an otherwise dark area which is lit with only one handlamp or portable light, in case it fails or is damaged leaving the personnel exposed to danger.
● Handlamps and portable lights containing miniature tubular fluorescent lamps or compact fluorescent lamps which are operated on high-frequency control gear may not be fully screened for radiated radiofrequency interference (section 7.5), and such handlamps should not be taken into sensitive areas or near sensitive plant.
● It is recommended that people entering potentially dangerous unlit locations and areas into which they take portable lighting equipment,

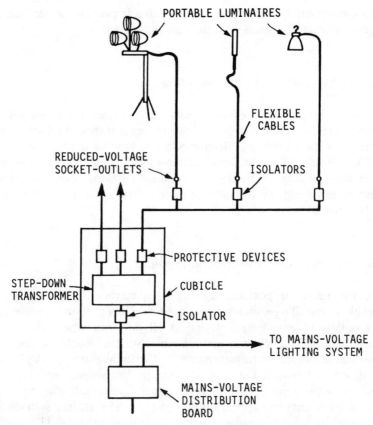

Figure 13.1 *Reduced-voltage distribution system for a permanent factory installation.*

should also carry one or two chemiluminescent lightsticks (section 5.13.1) as a routine precaution.

- No electrical apparatus of any kind (including battery-operated hand-lamps) may be taken into any designated hazardous zone (chapter 7) unless there is full compliance with the safety requirements for that zone in terms of construction and enclosure of the equipment, and its use. Under no circumstances should the enclosure of any electrical apparatus be opened in such a zone.

13.2   Interior temporary lighting

13.2.1   Temporary lighting installations, often improvised hastily to cope with some unexpected difficulty, are a potent cause of accidents, electrocutions and fires. The word 'temporary' must never mean 'of a lower standard of safety'. All temporary electrical installations should be tested before being put into service, and should comply with the *IEE Wiring Regulations*[27] as regards insulation resistance, correctness of polarity and earth continuity.

13.2.2 The possible need of temporary lighting for special occasions, e.g. during the installation of new plant, for descaling large boilers and tanks, or during major periodic maintenance operations, should be anticipated by the premises manager, and suitable lighting equipment purchased beforehand and held in readiness. Such lighting may be powered from the mains (if available), or self-powered by generators or batteries (chapter 18).

13.2.3 It is strongly recommended that reduced-voltage systems be employed for interior temporary lighting. An example of a 110 V adaptable fluorescent-lighting system which would be appropriate for indoor temporary lighting in a factory or warehouse is the 'Flori-stoon' system manufactured by Blakley Electrics Ltd (appendix G). This uses luminaires which each have a short pre-wired lead terminating in a socket connector by which the unit may be connected to the distribution cables. The important feature of this equipment is that a temporary installation can be erected speedily, without any improvized connections, taped joints or other irregular and potentially dangerous installation methods. For safety, such a system would best be fed from a double-wound step-down 110 V transformer (section 13.1.4)

13.2.4 Under conditions of protracted failure of normal lighting (chapter 14), engine-driven self-contained mobile generator units may be brought to the vicinity of the building, and temporary lighting fed from them. Engine-powered self-contained lighting units (chapter 18) can be used indoors if no other lighting means is available, but electrical equipment and prime movers may only be introduced into an interior area if they comply with environmental regulations (chapter 7), and if the exhaust fumes from the engine are evacuated safely.

14 Interior emergency lighting

14.1 Introduction

Emergency lighting is a vital facility to enable people to escape from buildings when the normal lighting has failed. Failure of the normal lighting may occur during a fire or other emergency. The risk of death or injury in a fire is high – the annual death toll in the UK is around 900, and the number injured around 10 000. Emergency lighting, together with proper means of fire prevention, escape, and fire management, can help to reduce this loss of life and degree of suffering.

14.1.1 Function of emergency lighting

The function of emergency lighting is to enable people to escape from premises during a failure of the normal lighting. It also enables essential things to be done during such a failure, e.g. to bring dangerous plant to a state of safety, or to continue the safe operation of processes that cannot be quickly stopped. If the duration of such activities is likely to be longer than the feasible duration of the emergency lighting batteries, additional lighting termed *stand-by lighting* (section 22.5) having an independent power supply may be employed to provide illumination for a longer period.

Emergency lighting is either in continuous operation while the premises are occupied, or it comes into operation automatically on failure of the normal lighting, either immediately or after a delay of a few seconds. The term 'emergency lighting' includes luminous exit signs and luminous devices that indicate the exit route (e.g. luminous arrow signs, illuminated handrails etc.) as will be discussed.

14.1.2 Current requirements of the law

Many industrial premises in the UK were equipped with emergency lighting before the UK *Fire Precautions Act (FPA)*, 1977[51] made the provision of such equipment in business premises mandatory. The life-saving value of such lighting is clearly established. Management's task is to determine the nature of the emergency lighting system that will best serve the needs of their organization, and to seek out those methods of emergency lighting which, while complying with the law, are economical, easy to maintain, and will give long technical life.

The FPA states in chapter 40 that 'the means of egress shall be capable of use at all material times', and this is construed to mean that lighting shall be provided along escape routes to enable persons to use them to reach a designated place of safety. The *Health and Safety at Work Etc Act (HASAWA)*, 1974[25] extends the general duty of care that occupiers of business premises must exercise to protect persons on their premises from harm, and imposes a further duty on occupiers to demonstrate that they have complied with the requirements of the FPA. Inspectors appointed under HASAWA are empowered to issue Fire Certificates indicating that proper fire precautions have been instituted, and that facilities for the speedy and safe evacuation of the premises (including the provision of suitable emergency lighting) are in place.

The occupier is not excused his duty to provide lighting for escape because of failure of another party to perform any act, for example, if the mains electricity failed. So the need for independently powered emergency lighting is clearly established as a legal obligation. With certain exceptions, without a valid Fire Certificate the premises may not be occupied. Fire Certificates are not required for factories or offices etc. in which:

- not more than 20 persons are at work at any one time; and
- not more than 10 persons are at work at any one time elsewhere than on the ground floor.

The above exemptions do not apply to premises in multiple occupancy where the aggregate number of persons at work in the premises exceeds the above numbers, nor does it apply to premises where highly dangerous materials are stored or used. The fire authority has power to decide if materials are 'highly dangerous', and if they are present in sufficient quantity to justify the requirement for a Fire Certificate even if less than the stipulated numbers are employed. Conversely, for certain low-risk premises, the fire authority has the discretion to grant exemption from the need for a Fire Certificate, even though the actual numbers of persons working on the premises exceed the normal limits for exemption. These rules may shortly be modified by new legislation (section 14.2).

The inspector who authorizes the issue of Fire Certificates for workplaces may be the Fire Prevention Officer of the local fire brigade, or an officer appointed by the local authority. Such officers are valuable sources of information and practical advice regarding the prevention of fire, the management of fire situations and the evacuation of premises. They may not, however, be able to advise on the technology of emergency lighting systems and the selection of equipment.

Because it is costly to provide illumination from battery supplies, escape lighting employs very small illuminances which are just adequate to enable dark-adapted persons having normal vision to see well enough to find their way out of the premises.

Few people outside the lighting profession have detailed knowledge about visibility at low lighting levels. It might be imagined that each enforcing officer who is responsible for issuing Fire Certificates under the FPA would have his or her own ideas about how escape lighting should be engineered, and to some extent this is true. The enforcing officers work generally within the guidelines of *BS 5266*[47] which has acquired something of the force of law, since under the FPA the enforcing authorities may take such a document as a Code of Practice to decide how the requirements of the FPA should be met. In the same way, the *CIBSE Code*[1] serves the purpose of a Code of Practice in relation to enforcement of HASAWA. Although the *Code* is not legislation its recommendations are enforceable by HSE Inspectors. Thus a good defence against a charge of failing to provide 'sufficient and suitable lighting' would be to demonstrate that the lighting complied with the *Code*.

Careful study of the Regulations is necessary to ensure that all obligations under the law are properly discharged. A convenient reference is the publication *Guide to Fire Precautions Act*[57].

14.2   Emergency lighting standards

14.2.1   Current UK technical standards

Reference should be made to appendix A. The important current references to the subject of emergency lighting are:

- *BS 5266*[47] This serves the purpose of a code of practice and contains much wise guidance, but it does not provide sufficient information to enable the detailed design of emergency lighting systems to be carried out.
- *CIBSE Technical Memorandum TM-12*[53] This gives technical guidance on applications and design of emergency lighting systems.
- *Lighting Industry Federation Application Guide*[14]. This covers much the same ground as the foregoing but is concerned in more detail with the construction and performance of specific types of emergency lighting luminaires.

14.2.2   Technical standards, 1992 onward

Technical and legal requirements for emergency lighting and safety signs in the UK are about to undergo significant change as a result of integration and harmonization of standards within the European Community (appendix A).

It is believed that the most significant changes will be in respect of safety of emergency lighting systems and in control of electromagnetic contamination (radiofrequency interference).

At the time of writing (Autumn 1991), it is understood that the following Directives have been issued by the European Commission:

- *Low Voltage Directive*, concerned with electrical safety. (The present Directive may be extended.)
- *Work Place Directive*, concerned with safety in the workplace, and including exit signs, safety signs and emergency lighting in workplaces (anticipated introduction date 31 December 1992).
- *Construction Products Directive* which lays down methods of making buildings safer, specifying the design of escape routes, lighting and other matters (anticipated introduction date 27 December 1991).
- *Safety Signs Directive* (anticipated introduction date 1 January 1994).

These will be interpreted by CENELEC as European Norms and converted to national standards. Each EC government must then introduce legislation for their enforcement.

An important change in practice is the increasing use of illuminated signs with symbols (pictograms) to overcome the confusion of languages. It is recognized that safety signs should be understood by people speaking any language, and even by the illiterate. Consequently, the requirements of

Figure 14.1 *Exit route pictogram to indicate escape route (BS 5499).*

BS 5499:1980 and *BS 5260:1978* as regards externally and internally illuminated signs are replaced by those of *BS 5499*[54]. Exit signs are now required to show a graphic symbol or 'exit pictogram' (Figure 14.1). For a time, the word 'exit' will continue to be used, but will probably be dropped in the future.

14.2.3 Exit route guidance

Important developments are afoot regarding the development of illuminated means of guiding persons along escape routes in conditions of smoke when visibility is very limited. Two areas of new technology may be noted:

- low-mounted illuminated signs and luminous devices recessed into the floor or stair nosings to give guidance in darkness and smoke to a person crawling towards safety (section 14.4.6);
- illuminated handrails and balustrades to give guidance on stairs, inclines etc (section 14.4.7).

14.3 Emergency lighting equipment

The terminology used in emergency lighting practice is confusing to those coming new to the subject, and there is no universal agreement on some of the terms. This is recognized in the *CIBSE Technical Memorandum on Emergency Lighting*[53] which states that some of the terms used therein may be defined differently in other documents. (Note: In this section, references to 'lamp' may be taken to mean 'lamp or lamps'.)

14.3.1 Types of emergency lighting luminaires

The following categories of emergency lighting luminaire have been described:

- *Single-point luminaire.* An emergency lighting luminaire containing a lamp, a secondary-cell battery, and a charging circuit to charge the battery from the mains. During failure of the mains supply, the supply to the lamp is derived from the integral battery.
- *Slave luminaire.* An emergency lighting luminaire containing a lamp but no batteries. During failure of the mains supply, the supply to the lamp is derived from a remote source.
- *Combined luminaire.* A single-point or slave emergency lighting luminaire which also functions as a normal luminaire (see 'Sustained mode' in section 14.3.2).

14.3.2 Modes of operation

An emergency lighting luminaire may be designed to operate in one of three modes, discussed in the following sections.

Maintained mode

The lamp is lit during normal use, deriving its power from the mains supply, and is switched automatically to the emergency supply (derived from an internal or remote battery) on failure of the mains supply.

In a variant design (known as 'switched maintained'), the lamp may be switched on and off while the mains are healthy, but, on failure of the mains, the lamp will light automatically from the emergency supply, irrespective of the switch being 'ON' or 'OFF'.

The emergency supply may be applied via a solid-state converter within the luminaire to provide an a.c. or d.c. supply to the lamp at the same or at a different voltage and/or frequency to that of the normal supply. Under emergency conditions the lamp may give a lower lumen output than when operating normally.

'Maintained' emergency luminaires employing fluorescent tubular lamps are known. The use of high-intensity discharge (HID) lamps for emergency lighting has hitherto been regarded as impracticable because of their long restrike and run-up times. However, if such lamps were fed from an 'uninterruptable' electrical supply, or fitted with ignitors, their use might well be considered.

Sustained mode

A sustained luminaire has two or more lamps (or groups of lamps), at least one of which is energized from the emergency supply, and the remainder from the normal supply, thus ensuring that illumination is sustained at all times.

Non-maintained mode

In non-maintained emergency lighting luminaires the lamp is not lit when the mains are healthy. This has the disadvantage that, even with the recommended standard and frequency of routine testing, the condition of the lamp and luminaire cannot be known until they are called on to perform.

14.3.3 Enclosures

To provide emergency lighting in a factory or offices where there is a 'normal dry atmosphere' is a simple operation, for there are plenty of suppliers offering suitable luminaires, but it may be difficult to find suitable equipment for food premises (section 11.1) or for hazardous areas and other aggressive environments (chapter 7). There may be applications of emergency lighting in areas requiring, for example, flameproof, corrosion-resistant, or hose-proof luminaires, and for these the choice of products on the market is limited. The best solution may be to locate a central battery in a 'normal atmosphere' zone nearby, and to use bulkhead luminaires having appropriate enclosures (section 7.1) as slave luminaires in the areas where special protection is required.

14.4 Emergency lighting installations

14.4.1 Central and zonal batteries

Batteries for emergency lighting are usually secondary-cell batteries which may be *integral* (located within the luminaire, together with a charging circuit) or *remote*.

A remote *central battery* may serve more than one building. On large installations the battery may be located in a plant room, and there may be associated generators for stand-by supply (section 22.5).

Instead of having a single central battery, a number of *zonal batteries* may be employed. These are self-contained battery cubicles each incorporating their own chargers, and each providing power for the emergency luminaires in an area. This arrangement has several advantages, not the least of which is that it is often the cheapest method. The zonal battery units are metal cabinets which can be located in any well-ventilated room that has a 'normal atmosphere' (i.e. no special environmental hazard). The reliability of such an arrangement should be as good as for any other conventional system, but an extra degree of security can be achieved by connecting a few selected key lights in each area to the zonal battery cubicle in another area (Figure 14.2).

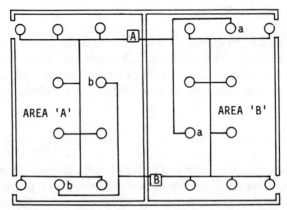

Figure 14.2 *Zonal battery system. The two areas have individual battery cubicles 'A' and 'B'. Each cubicle supplies a small number of selected luminaires (shown as 'a' and 'b') in the adjacent area to give greater security of lighting.*

This concept can be utilized in a horizontal single-floor system, or to provide interconnections between floors in a multistorey building.

14.4.2 Battery duration

BS 5266[47] recommends that the battery duration of escape lighting for even the smallest premises should be at least one hour (plus any period of occupation permitted by the inspector for essential duties to be performed), and that for larger premises durations two or three hours should be specified.

When determining the required battery life, consider the time it would take to search a large building – especially a multistorey one – to ensure that everyone is out by the subdued illumination of the emergency lighting, possibly with the presence of smoke. Consider what would happen when, during the search, an ill or injured person is found on an upper floor, consider the time it would take to descend by the stairs (the lifts being inoperable), to fetch medical aid, to carry the stretcher to the upper floor, render first aid, and then to transport the patient on the stretcher down the stairs to safety. It seems unlikely that the above scenario could be enacted in just one hour; and if there were more than one patient, the time taken would be even longer.

The specification of battery life is thus seen to be an important factor in the choice of emergency lighting equipment. It will always be better to err on the generous side and specify two or three hours life or even longer, unless there are provisions for stand-by lighting or mobile lighting which can be quickly brought into use (section 22.5).

14.4.3 Routine tests of battery capacity in continuously-occupied premises

BS 5266[47] calls for operational testing of emergency lighting. The Standard does not explain how such partial discharges of emergency lighting battery capacity can be arranged in continuously occupied buildings. In order to follow the law scrupulously, if the battery should be partially discharged and thus be unable to deliver the duration of lighting specified by the Fire Certificate, then logically the building should remain unoccupied until the batteries have been adequately recharged.

Any partial discharge of the required battery capacity would make it illegal to occupy a building. For example, if the emergency lighting in some business premises was inadvertently switched on, say, early in the morning, then the proper course would seem to be for the occupier not to admit the staff or public to the premises until the correct minimum battery duration had been restored – an operation that might take six or eight hours.

One way out of this difficulty is to provide a larger battery capacity than is needed for compliance with the Fire Certificate. Then, a partial discharge would still leave at least the mandatory minimum duration available from the batteries when the premises are occupied, though full tests would still be required.

Another method might be to arrange for a contractor to bring a mobile power supply to the building, and to connect it into the circuits to take over the emergency supply duty while the batteries in the installation were being test discharged and then recharged. The mobile power unit could consist of a bank of batteries, or an engine and generator.

In the case of very large plants with hundreds of emergency lighting luminaires, it could be economic to invest in a mobile power unit which could be brought to buildings in the complex in turn when it was time to carry out the testing of emergency lighting luminaires.

14.4.4   Response time

BS 5266[47] stipulates that, on failure of the normal lighting, the emergency lighting shall be operating within five seconds, though this period may be extended to 15 seconds (at the discretion of the enforcing authority) if the premises are likely to be occupied for the most part by persons who are familiar with them. This proviso is of little practical value, except for systems where the emergency lighting is provided from generators which take an appreciable number of seconds to come into operation. The idea is a bad one, and it would be far preferable for installations powered by generators to have a bridging battery to operate the lighting between the time of mains failure and the availability of power from the generator (section 22.5). Experience suggests that, in locations such as machine shops, those first seconds immediately following lighting failure are a time of especial danger, especially if the machines continue to function. Clearly, the best arrangement will always be for the emergency lighting to come on within a fraction of a second of failure of the normal lighting.

14.4.5   Emergency illuminance

BS 5266[47] lays down the illuminance that shall be provided on escape routes. This is very small indeed, just about the same as full moonlight, and is defined as 0.2 lx minimum along the centre-lines of escape routes, with a maximum diversity of 40:1. For 'undefined escape routes', i.e. open areas, without a defined escape path, the Standard requires that the whole area should be illuminated to not less than 1 lx on average.

These requirements can readily be achieved in small rooms, perhaps by a single luminaire which doubles as an illuminated sign. Applied to office blocks and cellular offices and corridors, the Standard works well, but applied to larger industrial premises which may present many hazards to the escaper, it is doubtful if the recommendations are adequate.

Doubts about the adequacy of the 0.2 lx figure stem from experience of the great diminution of illuminance that occurs in the presence of even a little smoke, or because of dust suspended in the air following an explosion. Further, our eyes take an appreciable time to adapt to such low illuminances following exposure to typical illuminances in the ranges 200–700 lx of general lighting, and far higher under local lighting.

In an emergency, the occupants may be frightened, and their adrenalin reaction would result in their pupils dilating, making their eyes more susceptible to glare, and possibly delaying their adjustment to a very much lower illuminance. Table 11 shows the author's proposals for illuminances for use in interior emergency lighting systems, employing the same values as for Table 19 (illuminances for outdoor emergency lighting).

Under emergency conditions, the illuminance along the centreline of a theoretical escape route (which may now be cluttered with unfamiliar

Table 11 *Illuminances for indoor emergency lighting*

Activity	Standard design illuminance (lux)	Minimum measured illuminance[a] (lux)
Escape along safe and known routes: emergency exit lanes, or walkways and paths where the users are familiar with them or the route is level and not dangerous to traverse	1 or 1% of the normal illuminance (whichever is the greater)	0.2
Escape along dangerous or unknown routes: emergency exit lanes, walkways and paths where the users are not familiar with them or possibly dangerous to traverse, and involves risk of falls, contact with hot or sharp objects etc.	5, or 5% of the normal illuminance (whichever is the greater)	1

[a] Minimum measured illuminance is the actual minimum illuminance anywhere in the lighted area as confirmed by inspection with a lightmeter.

obstructions) is an unreal concept. What may really be needed is sufficient general lighting for orientation and avoidance of dangers. Also needed is some degree of 'pilot lighting', i.e. some light ahead to indicate the direction to move.

The author can report his experience of being in a burning building, when dense choking smoke filled the room to within 300 mm of the floor, and seeing, breathing and escaping were only possible by crawling. This leaves him with doubts about the practice of mounting emergency lighting units at conventional positions at door-head height. Proposals to have low-mounted emergency lighting luminaires are always countered with the objection that such would be obscured by the bodies of other escapees. This leads to the obvious suggestion that an efficient emergency lighting system to aid escape should have conventional high-mounted luminaires plus some low-mounted luminous guidance devices.

14.4.6   Low-mounted luminous guidance devices

A number of products are available, including luminous devices to be recessed into floors and skirting-boards to give guidance to persons crawling under dense smoke to escape, some of these employing fibre-optic light guides[48]. A product that deserves mention is the 'DirExit' luminous arrow sign (Figure 14.3) manufactured by Ring Electronics Ltd (appendix H).

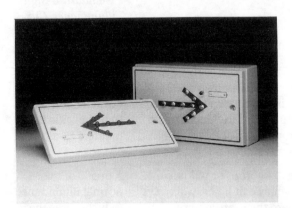

Figure 14.3 *The 'DireExit' luminous exit route guidance device. (Photo: Ring Electronics Ltd)*

14.4.7   Luminous handrails and balustrades

There is considerable interest by public authorities and the transport industry in means of guiding large numbers of people to safety in conditions of smoke.

There are luminous handrail and luminous balustrade systems available to give guidance to people escaping on the level, on inclines, along curved passages, and on stairways. A product worthy of mention is the handrail/emergency escape guidance system manufactured by Industrolite Ltd (appendix H).

Part 5

Exterior lighting

15 Exterior lighting design

15.1 Introduction

15.1.1 Some industrial activities by their nature must be performed outdoors, e.g. building construction, roadmaking, civil engineering, quarrying, opencast mining. It is also logical to store materials such as building materials and steel stocks outdoors. For such activities appropriate outdoor lighting is indispensable for safety and for efficient operation.

Changed lighting needs have come about because tasks and working environments have changed. For example, shipbuilding and ship repairing are in decline, but modern shipbuilding methods enable large sections of vessels to be made in covered factories. Some ports now deal mainly with containers, and so require to have greater outdoor areas lighted. It used to be economic to store large components such as castings, boilers and metals stocks outdoors. Coarse work such as fettling, welding and large assembly work used to be performed outdoors, and many factories kept palleted goods outdoors under shrink-wrapped plastic covers. Industry is moving out of its yards and open-air depots, and making more use of buildings. The factors tending to bring this about include:

- Workers are no longer willing to work outdoors in all weathers. Even if offered good lighting and warm clothes, they prefer to work in reasonable comfort indoors.
- Tasks formerly performed outdoors because they created fumes and smoke can now be done under better conditions indoors with the aid of mechanical ventilation.
- Goods stored outdoors are difficult to protect from vandals and thieves, and attract higher insurance premiums than if stored in a secure building.

Traditional industrial area lighting methods using masts or towers of 25 m or higher are now mainly met in the heavy metals industries, in mining and quarrying, and for bulk storage in a few industries (chapter 16). Typical modern industrial outdoor area lighting schemes tend to cover relatively small spaces, with mounting heights as low as 8 m. The demand is for better quality lighting, better colour appearance of the light used, with good glare limitation and uniformity.

The trends to miniaturization and specialization result in many businesses operating in smaller spaces; firms no longer make every component in their products, but buy in from specialist suppliers. Consequently, there are more firms, but they tend typically to employ less than 30 workers in small factory areas. The new small firms (together with many older concerns which are reducing in size and leaving their out-of-date factories) are moving into industrial estates and science parks.

Most industrial outdoor lighting installations these days are to aid the safe movement of people and vehicles, or for security, and often include decorative floodlighting of buildings, and the lighting of gardens and open spaces. There is a new demand for lighting for estate roadways and the common areas for access and parking which are a feature of modern industrial parks.

15.2   Lighting specifications

15.2.1   In the provision of outdoor lighting, the user (or his consultant or architect) will need to specify the requirements so that the project can be costed and quoted for. Without a clear specification, the lighting designer is faced with so many choices that he may well produce abortive work. He will need to know if the area will regularly be used for work at night, whether high-mast or low-mast designs are preferred, and whether the lighting can be integrated with a security lighting system and the lighting for adjacent areas. The logical approach is to provide the designer or quoting parties with site plans plus an *outline lighting specification* (section 24.2). Such a specification should not be needlessly detailed and specific, for the designer should be given a degree of freedom to evaluate possible alternatives where choices are permissible.

15.2.2   Additional headings for an outline lighting specification for an outdoor industrial scheme may include the following:

Contract parties

- Identify the site owner, developer, consultants, architect, main contractor if appointed, any nominated suppliers.
- Identify the party to whom the lighting proposals must be submitted for approval.

The site

- Identify the site boundaries.
- Identify adjacent properties and the nature of their occupancy if known, so that nuisance to neighbours, road-users etc., can be avoided.

- Ensure that the topography is known to the designer, for even slight variations in ground level can greatly affect how the lighting is to be designed.

Site utilization

- Describe any phasing of work or occupancy, with dates if possible.
- State precisely the use of the site. State if mechanical handling of any kind will be used and the height to which goods will be stacked.
- State the height of loading bays and the dimensions of any canopies.
- Indicate positions of paths, roadways, stacking areas, checkpoints, weighbridges etc. on the plan.

Tasks

- Specify visual tasks to be performed in the area, e.g. if colour coding is to be used on materials etc. or if documents have to be read.
- Describe any critical or dangerous tasks (e.g. the use of circular saws).
- Identify visibility problems, e.g. visual needs of crane drivers, clasp-truck drivers etc.

Environmental data

- Indicate the nature of the soil for supporting foundations of towers and masts.
- Indicate the seasonal and maximum windspeeds expected.
- Describe the climate, particularly with respect to temperature ranges (day/night/summer/winter), rainfall and the possibility of flooding, stating maximum possible height of floodwater above datum.

Environmental hazards (chapter 7)

- Indicate risks due to the presence of corrosive substances.
- Indicate risks due to the presence of flammable liquids, gases and substances, defining areas as Zone 1, or Zone 2.
- Indicate risks due to the presence of conductive or flammable dusts.

Legal requirements

State any local requirements, e.g. UK Health and Safety at Work etc. Act, recommendations of the International Labour Organization for port lighting, or Local Authority planning directives.

Exterior emergency lighting (chapter 19)

- State requirements for standby electrical supplies for lighting, referring to a suitable Standard or specifying the minimum service illuminance required on the centreline of escape routes, the delay before emergency lighting comes into operation on the failure of normal lighting, the period of operation from batteries, and the recharge period of the batteries.
- Specify use of local battery-powered lights or a central battery.
- State if mobile self-powered lighting will be available for emergency use.

Constraints

- Indicate the availability of power supply on site, the proposed date of connection, the voltage and frequency, and the location of substation/s or intake/s.
- Indicate the locations of any underground or overhead cable routes present or proposed.
- Indicate the cabling method to the lighting system (overhead, underground).

Client preferences

- Indicate preferences regarding choice of lightsources, luminaires, lighting methods etc.
- Indicate restrictions to types and powers of lamps to simplify stocking of spares.
- Indicate preferred or nominated suppliers or contractors.

Economic and other comparisons (section 21.3)

- The designer may be asked to prepare economic comparisons between an existing scheme and a proposal, or between two or more alternative schemes. The designer may also prepare technical comparisons between alternative schemes, taking account of illuminance, uniformity, glare, colour rendering, colour appearance, capital cost, maintenance cost, daytime appearance, and ease of control.

Outline lighting specifications for security lighting may cover all the factors given in sections 15.2.2 and 24.2, with additional information relating to the special functions of an installation designed to aid night security (section 17.9). The better is the briefing to the lighting designer, the better the quality and economics of the resultant lighting scheme are likely to be.

Where the lighting designer is employed by the client or retained on a fee basis, correct specification has a direct economic benefit; where the lighting designer is employed by a contracting or tendering party, the ultimate cost may still be affected, even if there is no stated charge for the designer's services. It will be appreciated that the speculative preparation of schemes is a costly operation, and tenderers always have to include the cost of scheme preparation in their overhead costs. Where a sound outline lighting specification is provided, the preparation of the initial scheme for tendering purposes is much simplified and speeded, and the savings may be reflected in lower tender prices.

Even though the cost of scheme preparation may be much reduced by the use of computer methods, it seems unlikely that lighting suppliers will for much longer continue to provide free schemes for exterior lighting. The work is skilled and responsible, and one would not expect any other qualified expert (say, an architect) to provide even preliminary proposals without a fee.

15.3   Calculation methods

15.3.1   Design methods

It may be thought that high-quality lighting design can only arise from the use of sophisticated design methods and the employment of advanced computer programs. This is not so. The computer can save time and money by obviating repetitive design calculations, but the computer performs only within the limitations of the instructions in its programs and the data entered by the operator. Computers do not make judgements, but only make evaluations against stated criteria. Computation enables approximate estimates of probable costs to be evaluated quickly, and for feasibility of designs to be examined. Computers do not produce creative thinking; they cannot come up with the inspired 'hunch' that sometimes takes a good designer straight to the heart of the problem. Whatever method of design is used, it is always a wise precaution to do a round-figure calculation by the Lumen Method, or carry out one or two point-by-point calculations to check spacing of luminaires etc.

Calculation methods include the *zonal flux diagram method* which is a method of calculating the average horizontal illuminance in an area lit by luminaires of asymmetrical distribution (e.g. type 'Z' distribution luminaires – section 6.4). The method requires the luminaire photometric data to be set out in a special diagram divided to represent angular zones. From this, an overlay is compiled, so that the effect of varying the mounting height or angle of tilt can be investigated – this is conveniently performed by a suitable computer program.

Another variation is the *zenithal web diagram method* which uses isocandela diagrams of a special form. This is claimed to provide simplified geometry and can be used to construct illuminance diagrams, or to sum the illuminance values at points over a regular grid representing the area to be lighted. This method is also used as the basis for computer programs to prepare the simpler isolux diagrams.

Yet another mathematical approach to floodlighting and area lighting installation design is the *spherical chart method* which uses special laboratory equipment which is not available to ordinary users and specifiers, and is thus a tool for photometrists and designers of luminaires rather than for those concerned with lighting applications.

Ingenious methods of computed aiming enable the specialized lighting systems for television lighting at major sports stadia to be trimmed to achieve a high degree of uniformity, where each of some hundreds of individual luminaires have their beams tailored by the addition of optical front glasses and then individually aimed – a method unlikely to be applicable to exterior lighting for industry or security.

15.3.2   Point-by-point calculations

Point-by-point calculations for large schemes are tedious to perform manually. Quite accurate results can be produced by the *nomogram method* described in *Interior Lighting Design*[8]. The method assumes that the lightsource is a 'point source', i.e. not of significant physical dimensions compared with the distances involved – as is invariably the case in exterior lighting installations. It is an objective of good area lighting to ensure that light comes to every point on the ground or working plane from at least two luminaires. Thus, the illuminance from every luminaire that contributes to the illuminance at a point must be added together to discover the total illuminance at that point.

15.3.3   Isolux diagram method

An isolux diagram is a plot of a succession of point-by-point calculations, usually prepared as a standard document for future use. Such are readily prepared by standard computer programs. An isolux diagram for a particular floodlight luminaire may display the data for one lamp power, at one mounting height, and at one angle of tilt. Such a diagram may be prepared in an adaptable form by being plotted for a theoretical lamp of 1000 lm, so that the results must be multiplied by the actual lamp lumens/1000 to find the actual illuminances in the squares or along the isolux lines.

In one form, the isolux diagram is drawn for equal areas on a scale proportional to a nominal mounting height h. This arrangement enables the designer to experiment with the effect of various mounting heights – a process that can be rapidly performed by computer.

It is helpful if *isolux envelopes* are superimposed on an isolux diagram.

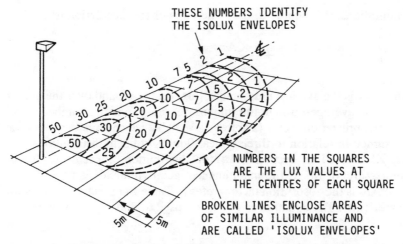

THESE NUMBERS IDENTIFY
THE ISOLUX ENVELOPES

NUMBERS IN THE SQUARES
ARE THE LUX VALUES AT
THE CENTRES OF EACH SQUARE

BROKEN LINES ENCLOSE AREAS
OF SIMILAR ILLUMINANCE AND
ARE CALLED 'ISOLUX ENVELOPES'

Figure 15.1 *Typical isolux diagram with isolux envelopes. The pattern is symmetrical so only half is shown. The diagram is for explanation only; do not use for design.*

These are areas enclosed by isolux lines, in which the illuminance will be at least the value stated (Figure 15.1). Such diagrams save time, for the profiles can be conveniently traced off, drawn out or photocopied at any convenient scale, and superimposed on the site plan for planning purposes. The usual scale for site plans is 1:500, in which each 10×10 mm square in a grid represents an area of 5×5 m. Luminaire manufacturers can usually provide isolux diagrams at this scale. Such a diagram, drawn out on a scale of h and with curves only is known as an *equilux diagram*.

It is easy to be misled by trying to compare distribution curves of alternative luminaires, for the power in a polar curve is not proportional to the area enclosed by the curve. Further, the cosine effect dramatically changes the performance compared with the abstract data of the polar curve.

Isolux diagrams show the illuminance (lux) on the horizontal plane (E_h), or on the vertical plane (E_v), the latter being of value in assessing the revealing power of low illuminances at long throws, as in security lighting applications (chapter 17). Alternatively, the data may be presented as E_h and a ratio of E_v/E_h. It is important to define the orientation of vertical plane measurements, for vertical illuminances in different planes cannot be summated.

15.3.4 Area lighting lumen method calculations

The lumen method is traditionally associated with interior lighting design (chapter 9), but is adaptable to area lighting applications and, although it has its limitations, it is simple to use. The basis of the calculation is:

$$E_h = \frac{l \times n \times \text{UF} \times \text{MF} \times \text{Abs}}{A}$$

Where: E_h is the average illuminance (lux) on the ground over the area to be lit; l is the average-through-life lumens per lamp; n is the number of lamps; UF is the utilization factor (Table 11 gives UFs which have been empirically determined in relation to three typical floodlight beam distributions 'X', 'Y' and 'Z' defined in section 6.4.; for distribution 'F' refer to luminaire manufacturers); MF is the maintenance factor (refer to Table 13 in section 15.4); Abs is the absorption factor (refer to Table 14 in section 15.4); and A is the area (m^2). The equation can be transposed to find any unknown variable.

Table 12   *Utilization factors for exterior area lighting*

Mounting height of luminaires expressed as percentage of the throw	UFs for conical beam floodlights Type 'X'*	UFs for fan-shaped beam floodlights Type 'Y'*	UFs for fan-shaped beam asymmetrical floodlights Type 'Z'*
10	0.15	0.20	0.25
20	0.20	0.25	0.30
30	0.25	0.30	0.35

* See section 6.4

The calculation can be performed in similar manner to that for interior lighting design by the lumen method, thus:

Step 1: decide the illuminance required (appendix C).
Step 2: select the most suitable type of lamp (chapter 5).
Step 3: select the most appropriate type of luminaire for the selected type of lamp and the environmental conditions (section 6.4, chapter 7).
Step 4: carry out a preliminary calculation by the Lumen Method, assuming a UF of 0.25 and selecting an absorption factor from Table 15 (section 15.4) and an MF from Table 14 (section 15.4). This will indicate the possible number of lamps, thus:

$$\text{Total lamp lumens} = \frac{A \times E_h}{\text{UF} \times \text{Abs} \times \text{MF}}$$

Step 5: dividing the total lamp lumens by the lumen outputs of several powers of lamps will give guidance on the greatest and least number of luminaires that will be needed, so that some rough layouts can be drawn up.

Step 6: By reference to isolux diagrams and manufacturer's data, take account of how far a particular luminaire will cast light with a range of possible mounting heights. Consider how far apart laterally luminaires in an array may be spaced, remembering that luminaires may be clustered on the supports. Plan out possible layouts, referring to Figure 15.4 and Table 12, noting the limitation on end-of-row spacing. Note the limitations of weight and windage on towers and masts.

Table 13   *Spacing and throw of luminaires in area lighting*

*Luminaire type**	*Maximum spacing (s)*	*Limit of throw (t)*	*Max.end spacing (x)*	*Mounting height (h)*
XN	0.5 h	9 h	0.5 h	12–30 m
XM	1 h	6 h	0.5 h	12–25 m
XN	1.5 h	3 h	0.66 h	10–25 m
Y	3 h	3 h	1 h	10–25 m
Y/Z	2 h	4 h	0.66 h	8–20 m
Z	4 h	12 h	2 h	9–20 m

[a] Luminaire classifications as in section 6.4. Read this table in conjunction with Figure 15.3.
h = mounting height

Luminaires vary considerably in design and performance from one manufacturer to another, and the performance of available luminaires may not match exactly with the three basic distributions described in section 6.4. Therefore it is necessary to plan the scheme using the Lumen Method as described above, and then to check it in detail using isolux diagrams and other data supplied by the luminaire provider.

In particular, note that the pattern of light emitted may not fit the area to be lit, so that a considerable amount of light may be lost outside the designated area and will reduce the net UF. The UF for conical beam floodlights (e.g. paraboloid conical reflectors) does not vary very much with the beam angle. Narrow-beam floodlights have greater internal light losses which are offset by a greater part of the beam falling on to the target area; conversely, wider beam angle floodlights are more efficient at collecting the lamp lumens, but a greater proportion are lost outside the site boundaries, especially on small sites.

15.4   Maintenance factors and absorption factors for exterior lighting installations

15.4.1   Maintenance factors

Airborne dust and pollutants will in time settle on the reflecting or transmitting surfaces of a luminaire. When rain or damp wets these surfaces, entrained dust or dissolved impurities will be deposited. Such soiling tends to act as a diffuser, and changes the luminaire performance generally by widening the beam. This loss of performance of luminaires is allowed for in the maintenance factor.

It is not easy to obtain reliable data on maintenance factors for exterior lighting installations. Some published tables give factors which do not appear to be consistent with practical experience. The accretion of dirt seems to have a most marked effect on light output and control soon after cleaning the luminaires, and thereafter the rate of deterioration appears to slow down. Some users specify half-yearly or quarterly cleaning to gain the saving in energy from being able to install a smaller quantity of lamp lumens, but this is surely cancelled by the cost of labour. The light loss during the first six months after cleaning leads some to believe that six-monthly or annual cleaning is always necessary, but many users clean their luminaires at two-yearly intervals, perhaps carrying out lamp replacement at the same time.

It is theoretically possible to calculate the 'economic cleaning period', but such calculations have in the past been based on interior lighting practice where there is still air (which permits layers of dust to settle) and where the cost of access is relatively low (no towers or masts to be scaled). The empirical data given in Table 14 is believed to give reasonably reliable guidance and is based on field experience.

15.4.2   Absorption factors

Even a slight degree of mist or fog can have a marked effect on the illuminance produced by exterior lighting, both because of light lost by absorption and by the scattering effect which tends to modify the polar distribution of the luminaire and make it perform as if it were a source of larger area and more diffuse. The light distribution of a luminaire will also be temporarily modified due to beads of condensation, droplets of rain etc. on reflectors or cover glasses. In locations where it is vital to provide a guaranteed minimum illuminance (e.g. at sensitive security lighting installations), allowance should be made for atmospheric absorption by the application of an absorption factor. The data given in Table 15 is based upon field experience.

Table 14   *Maintenance factors for exterior lighting installations*

Location and conditions	Frequency of conditions	Maintenance factor
Exceptionally dirty or dusty industrial locations	Half-yearly	0.70
	Annually	0.40
	Two-yearly	0.30
Typical industrial locations	Half-yearly	0.85
	Annually	0.65
	Two-yearly	0.60
Exceptionally clean locations	Half-yearly	0.95
	Annually	0.80
	Two-yearly	0.75

Table 15   *Absorption factors for exterior lighting installations*

Mounting height expressed as a percentage of throw	Conditions	Absorption factor
10	Clear air	0.95
	Slight mist	0.60
	Heavy rain	0.45
	Mist/fog	0.40–0.20
20	Clear air	0.98
	Slight mist	0.70
	Heavy rain	0.70
	Mist/fog	0.50–0.30
30	Clear air	1.00
	Slight mist	0.80
	Heavy rain	0.70
	Mist/fog	0.60–0.40

15.5   Lighting surveys

15.5.1   The information in this section is applicable to site surveys in relation to exterior industrial lighting and also for security lighting, though in the latter case the engineer will make some additional observations relating to the functions of the lighting (chapter 17). There are generally three types of exterior lighting survey:

- surveys of existing installations to determine what, if anything, should be done to improve them or adapt them to changed demands;
- surveys of built or partly built environments as an aid to design; and
- surveys of newly completed lighting installations to verify that the desired performance is being achieved.

15.5.2   The following notes will be of help to those who have to carry out such surveys:

- The site should be visited in daylight and at night. This is particularly important when discussing daytime appearance as well as night performance, and is essential for surveys of, or for, security lighting.
- Obtain site plans before the visit if possible for familiarization with the terrain. Do not start measuring up a site before the availability of drawings has been checked! If there is a choice, ask for plans at a scale of 1:500, i.e. 10 mm = 5 m.
- Go properly dressed for the weather and equipped for the job. Take a hard hat and industrial gumboots as routine. Carry a 2-m wooden measuring stick, preferably the kind that folds in the middle into two 1-m arms. (With a wooden stick the surveyor is less likely to get an electric shock by its accidental contact with live plant or cables.) Such a measuring stick (known in the trade as a 'jumping stick') can be used by laying it end-over-end on the ground very rapidly for measuring distances with fair accuracy; it is also convenient for vertical measurements, as will be explained. Take a note-board with a clip for papers, and a transparent plastic bag to keep it dry, a pencil, a pencil sharpener, possibly red and blue ballpoint pens or crayons, and a stick of chalk.
- Check for dangers. Do not approach machinery or excavations etc. without permission. Do not smoke without first checking that this is permissible. Do not eat or drink anything while working on chemical and waste-disposal sites.
- On dangerous sites a permit-to-work must be obtained, signed by a responsible engineer or manager, as a safeguard against physical accidents or electric shock etc.
- Do not climb if you have no head for heights, and never be alone in any elevated or hazardous situation in case you should need help. Use one of the simple methods given in section 15.5.3 for estimating heights.
- During the survey, check that all visible structures are shown on the plan. Using the jumping stick, perhaps with help from another person, measure any rise or fall of the land.

- While surveying, have in mind the topics covered by the outline lighting specification (section 15.2.2), for all the data to satisfy this are necessary. For example, look outside the boundaries of the site, and take note of what is there. There are a number of additional guidelines in relation to surveys of, or for, security lighting installations (section 17.9).

15.5.3 High structures can be quite accurately measured by simple methods. Height can be estimated accurately enough for ordinary survey purposes by counting the courses of bricks or building blocks up the face of the structure.

A practical and simple method of estimating heights is by triangulation. On a fine day you can employ the simplest method, which uses the shadows cast by the sun. To do this, place your jumping-stick vertically, and note where the shadow of the tip falls – mark this with a stone etc; now, measure the length of the shadow. To find the structure height h, the ratio of the length of the jumping-stick to its shadow will be the same as the ratio of the height of the structure to its shadow. For example, if the shadow of the 2 m stick was 6 m long, and the length of the shadow from the structure at about the same time is 27 m, then,

$$\frac{2}{6} = \frac{h}{27}$$

Thus

$$h = \frac{2 \times 27}{6} = 9\,m$$

An alternative method called *judging-up* (Figure 15.2) is almost as accurate, and can be performed without sunlight being available. Make a

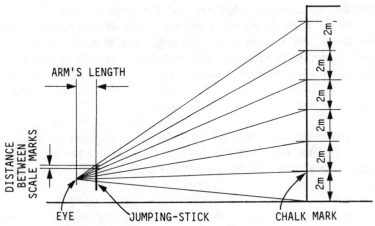

Figure 15.2 *Judging-up a height.*

clear horizontal chalk line on the structure, 2 m above its base. Then get back a convenient distance, say about the same height as the structure being measured if possible. Hold the folded jumping-stick at full arm's length in front of you, holding the stick near its top so that it hangs vertically. Align the top of the stick with the chalk mark, and keeping still, observe where the base of the structure aligns with the stick, and mark that point by grasping the stick and placing your thumb at that point. You have now defined a distance on the stick which represents 2 m vertically up the building. Keeping it quite vertical, raise the stick until the point you are marking with your thumb aligns with the chalk mark, and observe where your line of sight at the top of the stick relates to the building surface, and remember that point well (4 m above the ground). Repeat similar sightings at 2 m or 4 m intervals until you have estimated the complete height of the building.

15.6 Aiming of floodlights

15.6.1 Many exterior lighting installations fail to produce the calculated performance because the floodlighting luminaires are not accurately aimed. Two angles have to be considered: the *tilt angle* (the angle of depression below the horizontal); and the *azimuth angle* (the orientation angle in plan relative to a datum).

Some luminaires can be aimed accurately at a calculated point with an aiming aid fashioned like a gun-sight or a small clip-on telescope, and some are provided with a protractor which 'clicks' at 2.5° or 5° steps, enabling setting to be done on the ground before installing, or even be done in the dark at the tower head.

If the tilt angle can be calculated and the luminaire correctly aimed, it leaves only the azimuth direction to be set. The designer may instruct the installer to set the luminaire at a tilt angle, but may give only a general guidance 'to aim the beam at the calculated aiming-point' – usually a far more difficult task for, because of the cosine effect, the central axis of the beam may not be the brightest point on the ground, and therefore it is common for quite serious errors to be made in this operation.

15.6.2 A convenient method is for the designer to specify an aiming point, instructing that the luminaire is to be aimed to produce the highest possible illuminance at that point. Referring to Figure 15.3, the procedure is as follows:

Step 1. Mount the luminaire on its support at the desired position and specified mounting height. Measure out from the base of the column in the

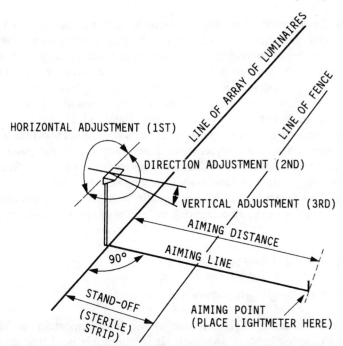

Figure 15.3 *Aiming of floodlight luminaires. (Stand-off etc. refers to security perimeter lights – chapter 17).*

direction in which it is desired to aim. At the specified aiming distance mark the aiming point with a peg.

Step 2. Place the cell of a suitable lightmeter over the peg at a height of 300 mm with the cell vertical and facing back towards the luminaire.

Step 3. Set the luminaire horizontal in the transverse plane, using a spirit level. This is most important for linear tungsten-halogen lamps which must be set within 4° of the horizontal (section 5.4).

Step 4. With only one luminaire lit, slowly swing the lighted luminaire around the support from side to side until the lightmeter reads a maximum (the actual reading is immaterial).

Step 5. If the luminaire has not previously been set for the correct angle of tilt, now tilt the luminaire up and down until the lightmeter again reads a maximum. Lock all the adjustments on the luminaire, for it is now correctly set.

15.6.3 The procedures just described can be performed for area lighting arrays and for perimeter lighting luminaires in security lighting installations. However, in the absence of complete aiming instructions, it is possible to aim floodlights by reference to an isolux diagram (section 15.3), adapting the above procedure accordingly.

15.6.4   In the absence of aiming instructions, and without even an isolux diagram to give guidance, it is still possible to aim an array of luminaires to obtain a reasonably even uniformity of illuminance over an area. The method is to switch on all the luminaires in the array, and aim them 'by eye' as well as possible. Then, the person in control observes the lighted area through a piece of deeply tinted glass from as high a point as possible (e.g. from the head of a mast or from the roof of an adjacent building). If such a glass is not available, wearing two or three pairs of ordinary sunglasses together will give sufficient obscuration. The objective is to see only the bright patch at the centre of each beam, and to co-ordinate the centres of the beams into an even spacing pattern on the ground and thus produce the desired degree of uniformity. This technique will work equally well for aiming floodlights at vertical surfaces, as in the floodlighting of buildings and other structures.

15.7   Towers and masts

15.7.1 Mounting heights for luminaires

The choice of mounting height is one of the most important decisions in formulating exterior lighting proposals. It is commonly held that increasing the mounting height will reduce the glare. Rules of thumb are included in some guides on the subject, suggesting that mounting heights between 15 m and 60 m be used, on the basis that the greater the lumen output of the luminaire, the higher it should be mounted.

It is usually unsound to increase the mounting height in order to reduce the glare. The glare effect of distant floodlight luminaires is only marginally affected by increasing the mounting height. This is because raising the luminaires a few metres will not substantially change the angle of the luminaires above eye-level when the observer's distance from the luminaire is five, ten or 15 times the mounting height (Figure 15.4). For lighting an undulating site, increasing the mounting height will eliminate shadowing in hollows and behind raised parts (Figure 15.5).

The increasing concern about pollution by light has increased the use of luminaires with a flat horizontal glass and no upward light (described in this

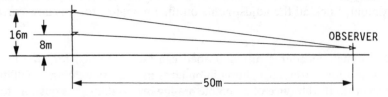

Figure 15.4 *Glare effect of distant floodlight luminaires is only marginally affected by increasing the mounting height.*

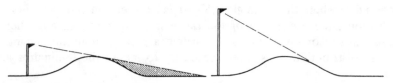

Figure 15.5 *Lighting an undulating site. If luminaires are too low, areas behind raised parts will be shadowed.*

book as type 'F') (section 6.4). Having a fixed projection angle, without the facility to introduce greater or lesser tilt, it may become necessary to increase the mounting height.

The greater the mounting height, the greater may be the spacing between luminaires, leading to a less costly installation, but possibly increasing the cost of maintenance. Each increment of mounting height adds significantly to the cost of the support because of the need for greater stiffness and resistance to uprooting or fracture.

For many industrial applications, lightly constructed tripole towers of 8–14 m have much to recommend them. They are easy to scale, and can have climbing means built in. They can carry large arrays, say up to six 1-kW SON-T luminaires, or up to eight if mounted in two groups back to back. At the other end of the scale, high-masts (which can carry a large group of luminaires) have a very high installation cost, and except with high precision in luminaire construction and aiming, they may give a rather poor economic return in terms of the cost *per lux per m² per annum*.

15.7.2 Mounting height affecting vertical illuminance

For security lighting and other applications where the objective is to reveal an intruder who is within a substantially unobstructed space, the illuminance on the horizontal plane is not the illuminance measure that best relates to a subjective assessment of revealment yielded by the lighting. Experiments show that revealment of distant objects is consistently related to the the vertical illuminance E_v. Now,

$$E_v = \frac{I \cos^2\theta \sin\theta}{H^2}$$

Thus, a reduction in H will increase E_v.

It may be the objective of the designer to find the lowest acceptable mounting height that will satisfy the need for uniformity, and which will give the greatest degree of revealment by enhancement of E_v. For this reason, the preferred range of mounting heights for types 'X', 'Y' and 'Z' distributions (section 6.4) where the tilt angle is 5°–30° is between 8 m and 25 m. For most applications, the lower end of the range is satisfactory, except where an undulating site or obstructions necessitate a greater mounting height. For type 'Z' distribution, it is rarely necessary to mount the luminaires above 14 m.

15.7.3   Typical mounting heights

A summary of current practice is as follows:

- Perimeter glare lighting (security lighting): 1–2 m.
- Perimeter lighting (security lighting): 3–5 m.
- Roadlighting: 5–10 m.
- Industrial area floodlighting – most common range: 8–14 m.
- Industrial area floodlighting – large sites: 14–25 m.
- Special schemes (docks, dry-docks, marshalling yards, large container parks etc): 25–60 m.

15.7.4   Wind resistance

The designer must take into account the probability of a particular maximum wind speed occurring at the place of use. Practice hitherto has been to design to the guidelines of BSCP 3 which sets a windspeed of 46 m/s (106 mph) as the design limit for damage or overturning, but the requirements of EuroNorm EN60958 must now be followed (appendix A). Any reduction in the projected area of the headload (i.e. of the luminaires, supporting metalwork, control gear enclosures etc.) will enable the support to withstand a higher windspeed, and vice versa.

A lighting support has to be designed to an agreed factor of safety, but it should be clear to the contracting parties just what this means. A lighting support could fail by being overturned (failure of foundations), or could fail structurally. A good design is one in which there is little or no risk of structural failure right up to the critical windspeed which could overturn the support. Such considerations hold good for static towers and masts (section 15.7.5), and for portable towers and mobile lighting equipment (chapter 18), though in the latter cases instead of a foundation in the ground, reaction will be provided by guy-lines and stays, or by the mass of the trailer or skid etc.

All structures tend to vibrate in the wind, the frequency of vibration varying with the mass and stiffness of the structure. Some masts vibrate sufficiently to shorten lamp life and in extreme cases metal fatigue leads to structural failure. Concrete columns have also been known to fail in windy conditions. Lattice towers and masts can vibrate to such an extent that the noise causes nuisance.

15.7.5   Static towers and masts

The structural safety of these supports is affected by the nature of the soil and quality of the foundation. The common procedure is to dig a foundation hole of the required dimensions, and to cast into this a concrete foundation in which is implanted the foundation grid or root provided by the tower supplier. In soft soils and sand, or areas prone to flooding, the tower must be provided with a foundation of sufficient mass and dimensions that the structure relies very little on the support of the surrounding soil. For temporary installations, it is usually arranged that the upper surface of the block is well below the finished land level, so that the foundation can be left in the soil after the contract. However, it is sometimes possible to cast a foundation block on the ground surface without excavation, the dimensions and mass of the block being sufficient to resist the overturning moment due to wind of the maximum forecast speed. Surface anchorages of this kind are more easily broken up and disposed of at the end of the contract.

These structures are not well defined, but the following general descriptions apply:

- *Columns* are usually comprised of cylindrical steel sections, hot-swaged or welded together, generally of a tapered form, with heights up to around 15 m. They may be provided with a base chamber for connection; base-hinging types enable the column to be lowered for access to the head equipment.
- *Masts* are generally higher than 15 m, may be parallel-sided or tapered rather than of stepped construction, and usually incorporate means of scaling or a means of lowering the luminaires by winding gear, but the difference between masts and columns is arbitrary.
- *Towers* may be of a tripole or square section lattice construction, fitted with ladder bracing to make for safe scaling, with hooping for the length of the ladder from 2.5 m above ground[52]. Rest platforms may be built in, with handrailing, and a head platform similarly protected. Towers are of particular value in situations where soft, wet or sandy ground makes it impracticable to bring a hydraulic-lift platform alongside (if one is available).

Mobile lighting, including portable towers, skid-mounted towers, and demountable towers, is discussed in chapter 18.

16 Exterior lighting practice

16.1 Construction and civil engineering sites

16.1.1 Introduction

Lighting can make such a positive contribution to the efficiency, economy and safety of a site that it is difficult to understand why, after some 20 years of teaching and exhortation to the construction industry, the subject is so neglected and so little understood by many who are responsible for its provision. The larger developers and contractors now install site lighting as a routine on all contracts but some smaller contractors regard lighting as something that they will install 'later on, if it is needed'.

Although site lighting is of greatest value in the short days of winter, it should not be regarded as only necessary for winter building. The planning and continual updating of site lighting through any contract is a vital function of contract management, the neglect of which cannot be properly made good by the hasty hiring of some self-powered lighting units when the contract starts to run late as the days begin to shorten.

16.1.2 Legal and contractual requirements

Lighting for building operations and engineering construction is subject to the Construction (General Provision) Regulations 1961, Regulation 47 of which states:

> Every working place and approach thereto, every place where raising or lowering operations with the use of a lifting appliance are in progress, and all openings dangerous to persons employed, shall be adequately and suitably lighted.

In addition to the requirements of the common law and legislation regarding safety of workplaces (appendix A), the main contractor may be under contractual obligation to provide lighting to an acceptable standard during the contract period. In the case of failure to do this, the main contractor may be subject to penalties or liable to claims for damages from subcontractors, who might have a sound excuse for poor quality work or late completion if adequate lighting is not provided.

In the programme of work for the site, application to the electricity supply company for a temporary or early connection of power to the site should be made sufficiently early to ensure the supply is available from Day One of site activity. (Lead times of nine months or more are not uncommon on new sites.) If the supply cannot be available when required, then temporary use of mobile lighting may be needed (chapter 18).

16.1.3 The benefits of site lighting

Improved safety on site

The construction industry is sometimes described as 'the deadly industry', having a death toll greater than that of deep-sea fishing, and a horrendous record of personal injuries (many of which are not reported by self-employed one-person subcontractors). Long experience on sites shows positively that good lighting tends to reduce accident frequency. Incidentally, a good safety record will tend to reduce insurance premiums.

Improved quality of work

Electric lighting, if correctly applied, is a valid alternative to daylight for the performance of all crafts and trades. Its use does not imply that lesser standards of work quality compared with daylight working will ensue; indeed, because directional light from task-lights is under the control of the operator, tasks may be performed better than in daylight. This is especially so for work involving flat planes, e.g. laying bricks, screeding, plastering, terrazzo etc.

Flexibility of working hours

The availability of daylight varies seasonally, and is affected by weather conditions. The average annual number of daylight hours is constant everywhere (section 22.10.1), but the ability to work during its availability may be limited by frost, snow, wind, monsoons, floods, sandstorms, and other local climatic conditions. In equatorial regions, day length is more constant, but the twilight periods are shorter. Site lighting is used to augment or replace daylight in open areas, permitting work to continue during all the available hours, extending the working day and permitting night operations when necessary.

Work in areas without daylight

Electric lighting must be used constantly in shafts, tunnels, deep excavations and in enclosed places such as basement floors. Many structures are windowless or have only limited fenestration, so that insufficient daylight enters for the fixings and finishing tradesmen to perform their tasks.

Use of all the available hours of work

Electric lighting is essential for operations that must be performed at hours which are dictated by external factors (such as working between tides, day or night), or which cannot be interrupted (such as pouring massive monolithic structures, or slip-casting tall structures such as silos and chimneys). A benefit of site lighting is that deliveries of materials and collections of spoil can continue throughout a 24-hour period. At city centre sites it becomes possible for vehicle movements to take place outside times of traffic congestion.

Night security

Losses due to theft, vandalism, incendiarism, unauthorized use of plant, sabotage etc. on sites can be crippling to contractors, who usually are unable to insure against all the incidental and consequential losses, for example, the operation of penalty clauses due to late completion of a contract following an incident.

16.1.4   Electrical safety

The need for correct electrical practices cannot be over emphasized (section 22.1). The tradition of so-called 'lash-ups' and improvised wiring on sites is deplored. The following points should be noted:

- The practice of using *festoon lighting* of any form is not recommended and in particular, festoon lighting using lampholders having pointed penetrating contacts which pierce the insulation of the cable should never be employed.
- Lighting equipment under the control of the operatives or mounted within hand-reach height should be supplied from a reduced-voltage system fed from a double-wound step-down transformer having an earthed screen between the windings and the centre-point of the secondary winding connected to earth, with both poles fused and switched, working at a maximum voltage of 110 V. Under fault conditions, the highest voltage to earth that can appear on such a system is 55 V. A distribution system as described in *BS 4643*[49] could be employed (Figure 13.1). To prevent voltage mismatch, socket-outlets complying with *BS 4343*[50] should be used in non-hazardous atmospheres. (Note: The operating voltage is likely to be changed to 100 V (50 V above earth) by future EC legislation.)
- Much of the gain in safety due to use of reduced voltage is lost if the step-down transformer is located near the point of work, and if the mains cables to it are exposed to risk of damage.

- Wiring should be placed as high as possible. Sufficient properly fixed socket outlets should be provided so that cables do not trail on the ground or floor.
- Used cables should be inspected before re-use, and discarded if the insulation is damaged.
- Lampholders, plugs and sockets and line-connectors should be non-interchangeable between the different voltages.

16.1.5 Site lighting techniques

The general objectives of exterior industrial lighting (section 15.1) apply to site lighting. At the time of writing it is not possible to refer the reader to any up-to-date publication on site lighting, though a new publication is in preparation by the CIBSE for proposed issue in 1992[58].

16.1.6 Illuminances required

Appendix C contains guidance on illuminances for site work. The provision of mobile task lighting without provision of suitable general lighting is a bad lighting practice, and is potentially dangerous.

16.1.7 Emergency lighting on construction sites

Chapter 19 contains guidance on applying emergency lighting to sites.

16.1.8 Security lighting on sites

Security lighting offers valuable protection and uses simple technology. The guidelines given in chapter 17 are generally applicable to civil engineering and construction sites, and should be adapted to the needs of the particular site, its size and location. As for any other system of security lighting, protection of the site should involve the three elements of a sound perimeter fence, an efficient patrol or security guarding routine, and effective lighting.

On large sites, the provision of lighting towers or masts, 8–14 m high, or trailer-lights (chapter 18) of about this height, is frequently the most practicable and economic way of providing light for both work and security outside working hours. On strategic and military sites, particularly in isolated areas, the technique of glare lighting may provide a strong deterrent to intrusion. On large unfenced sites, for example, dam-building projects and motorway construction, it may be necessary to create small fenced *citadels*, and provide them with suitable lighting. In such enclosures explosives, valuable tools and materials etc. may be stored.

The use of 'transparent' fences, e.g. wire-mesh, chain-link or palisade fences, is recommended. If temporary fencing of materials such as corrugated-iron sheeting or weatherproof plyboard are used, it is recommended they should have eye-level 'windows' cut in them, these being protected with securely fixed strong wire mesh. Then, with the provision of some area lighting, the windows will permit random supervision at night by police and visiting patrols – and by members of the public.

Children are attracted to sites, and will enter them outside working hours and many tragic events have occurred as a result. Effective prevention measures include ensuring that the fence has no small openings, and there is no gap under the gates through which a child could enter; and providing security lighting and guarding.

Special attention should be given to lighting at site entrances. This will not only help prevent crime but will also enable a tired driver to locate the site and drive into it safely.

16.2   Lighting for quarries and surface installations at mines

16.2.1   The need for lighting

In the winning of minerals, suitable exterior lighting can increase profitability by enabling the working hours to be extended, thus making fullest use of capital plant such as drag-line scrapers, diggers, hoists, conveyors, grading plant and transport. In the UK some 4000 quarries and opencast workings are operative, producing materials such as coal, china-clay, clay, gravel, gypsum, sand, slate, stone, zinc etc. Many of these workings have inadequate lighting or none at all.

16.2.2   General requirements of the lighting

Most such sites require lighting at the entrances and approach roads, the weighbridge area and for the area around the processing plant. At the workings, the lighting should enable vehicles to approach the active area in safety, bearing in mind that steep inclines and tortuous routes will require *area lighting* rather than roadlighting. Within the working, *cross-flooding* (section 16.2.3) will help operators of draglines, diggers and dozers to drive safely without dependence on vehicle headlights. If there is not sufficient reasonably glare-free general lighting, drivers tend to look only into the area illuminated by the vehicle lights, and may not be aware of what is happening at the sides and behind their machine. The mounting-height of the luminaires must be chosen to take account of the eye-level of the operators relative to the horizontal or vertical worked surface. The provision of good lighting is of particular value when cranes are used within the working or from the lip, as in some stone workings.

16.2.3 Cross-flooding

This is a method of getting light into wide, deep excavations. To reduce risks to plant, and the dangers from positioning electrical cables within the excavation, instead of having mobile lighting units within the working, the light may be projected into the excavation from a number of positions on the lip so that every point in the working area is lit from at least two directions. If sufficient sources are so disposed, the general field brightness will raise the adaptation level of the operatives so as to minimize the glare sensation.

An objection to cross-flooding may be made on the grounds that mobile lighting equipment placed at the lip might be blown over the edge by wind, or that the lip will crumble. However, mobile lighting supports can be provided which will withstand the highest wind speeds likely to be met, and, as a precaution, masts can be lowered during times of high wind. Cross-flooding is practicable for throws of up to around 500 m; greater distances necessitate the use of narrow-beam floodlights which can cause excessive glare and are difficult to place accurately. The use of two-way portable radios is a practical aid to aligning floodlights.

16.2.4 Positioning mobile lighting units

The risk of losing the equipment over a crumbling edge will be much reduced by using masts of such height that the lighting unit can be positioned well back from the edge (Figure 16.1). Lighting towers may be jennylights (section 18.2.3), trailer-lights (section 18.2.4), or skid-mounted towers (section 18.2.6).

16.2.5 Galleried quarries

These are employed in quarrying for stone, ore, slate etc. from an inclined workface. The working plane is often more vertical than horizontal. If the lighting towers are closely offset (i.e. spaced off less than five times their mounting height), dispersive distribution luminaires are the most practical.

Portable or mobile towers may be used for the higher galleries, with some risk that they may tip over the edge if not staked securely. Best practice is where wide 'lands' or horizontal galleries are cut, wide enough to permit lorries, mining equipment, hoists etc to pass the lighting equipment (Figure 16.2).

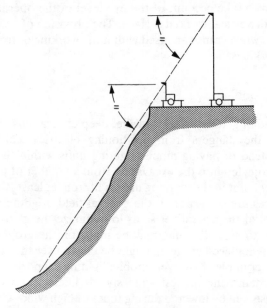

Figure 16.1 *How increasing the height of the lighting support enables the required depression of beam angle to be achieved without bringing the lighting equipment unnecessarily close to the lip of the quarry.*

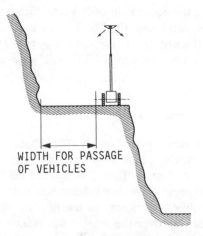

WIDTH FOR PASSAGE
OF VEHICLES

Figure 16.2 *Wide-land method of face-cutting provides safe placements for mobile lighting units, and allows vehicles to pass.*

16.2.5 Opencast workings

Lighting of these workings may help to bring forward completion dates, increasing the profitability on fixed-price contracts, or reducing the risk of penalty for late completion if the work should be delayed by flooding or bad weather. With suitable lighting the operation can continue 24 hours per day. Although local residents may object to the noise and other inconveniences of continuous operation of opencast sites near their homes, 24-hour working may be acceptable to them as a means for the nuisance to be ended sooner.

Trailer-lights are used for this work, as are temporary high masts. Advantages of employing self-powered lighting units on these large working areas are that there are no trailing cables and they can be repositioned frequently to take account of the changing topography of the site. A trailer-light can be readily moved; it takes about 15 minutes to lower the mast and hitch the unit to a tow vehicle, and about the same time to set up the unit at a new position.

16.2.4 Surface installations at mines

Three-shift operation is normal at most mines, and the production below has to be matched by continuous surface operations such as mechanical handling of dross and spoil to tips, grading of the extracted mineral, and transport of the product off the site. The efficiency of these operations is dependent on the provision of suitable lighting for movement and safety on the roadways, in the lorry parks and at railway sidings.

At mines where precious metals and gem stones are won, the security aspects of the site may be important. In the less developed countries, and especially at isolated mines, the problem may be acute and conventional policing not available. Theft of tools, stores, materials, supplies and mine products may be controlled by clearing a perimeter area, fencing it, and providing guarding and security lighting (chapter 17). The use of 'citadels' within which to store security containers may be essential.

16.2.5 Operational requirements

On rugged sites, electric cables – even if armoured – are liable to be damaged by the passage of vehicles or rock-falls, etc. Unless prohibited by local regulations, armoured concentric-core cable may be the safest means of distribution to exposed units.

Remote switching of mains-powered lighting towers should present no serious problems; mains-borne signalling is sometimes employed. Trailer-lights and skid-mounted lighting units can be remotely controlled. Equipment is now available to start engines by radio signal, with provision for the lamps to be switched on after the engine has reached full speed.

The fuel-tank capacity of self-powered lighting units should be sufficient to run the units for the whole night without refuelling.

16.3   Ports, ship building and repairing

16.3.1   Objectives and constraints

This is an industry sector which has undergone considerable change in the past decade or so. Some ports now handle mainly containerized cargo, and some have got rid of many of their old warehouses. Lighting is required for their immense new outdoor working and storage spaces. Ship building is now largely a factory operation, with major assembled units being brought to the dock for final assembly of the vessel.

Ports and docks tend to have certain factors in common as regards lighting:

- The spaces to be lit are considerably larger than those met in practically any other industrial application.
- 24-hour working is usual.
- All have major security problems relating to such matters as illegal entry into the country, illegal imports and exports, and theft. On all these locations, lighting forms an important part of the security plan.
- The lighting equipment must be constructed and finished to withstand salt spray.

Security lighting for a port or dock cannot be considered separately from the rest of the lighting, but should be integrated with all the other outdoor lighting functions. Where several authorities have control of lighting for different areas, the controls and layout should ensure that switching off one section does not deprive another area of light needed for safety or security. Lighting for entrance roadways, car parks, lorry parks and storage areas should provide smooth transition from one illuminance to another, so that there is no unnecessary strain on drivers, particularly in bad weather.

Glare may be projected great distances, so modifications and additions to port lighting should take account of distant viewpoints. For example, a newly-lighted area might seem satisfactory to those working in it or passing through, while it is projecting uncomfortable glare to distant straddle-truck drivers and crane-drivers whose eye-levels are higher. Remember that railway trains do not have powerful headlamps like road vehicles. Train drivers must not be subjected to excessive changes in field luminance or glare as they pass by a lighted industrial area.

In addition to the usual legal constraints (appendix A), lighting at ports must not project light to seaward 'confusable with a navigation light', nor shall light to seaward cause glare to pilots or be a hazard to mariners. The use of cut-off lanterns which project no light above the horizontal, i.e. Type 'F' (section 6.4) may be desirable. In some situations it can be arranged for light to flow inland from the shore side.

16.3.2 Cargo terminals

The essential need is to get light onto the ground, irrespective of how it is obstructed from hour to hour by containers, cargo, vehicles and port equipment. Any attempt to light quays by the minimum number of masts that will theoretically give cover is likely to give poor results. This is an example of one of the most common mistakes made by lighting designers, that is to try and light the empty space, instead of lighting the space as it will be occupied in use. The normal rule of thumb to ensure that light comes to every point on the ground in substantial proportions from at least two directions may not be satisfactory for a busy cargo terminal with a great many ground obstructions. A better aim would be for every point on the ground to receive light from say, four directions.

The use of light-coloured surfaces to roadways and quays aid visibility, but these are difficult to maintain because of oil spillage etc. Bold yellow or white ground markings aid visibility, and these should be positioned in relation to the lighting towers. Illuminated bollards and kerbs are used to prevent vehicles colliding with lighting towers. The use of light-coloured paint on boundary walls, adjacent buildings etc is very helpful (but not aluminium paint which can cause hazardous reflections).

Valuable economy in energy may be achieved by switching out the main lighting at unoccupied berths, but there should always be a minimum illuminance of say, 5 lx over all areas accessible to pedestrians and traffic during the night hours.

16.3.3 Parking areas

It is reported that some 10–15% more vehicles can be parked safely in a well-lit lorry park as compared with an unlighted or poorly lighted one. When vehicles are in position, they cause extensive shadowing. This can be minimized by selecting mounting heights of the order of 8–14 m rather than using large clusters of luminaires on high masts. The latter may seem to be the elegant solution, but the effectiveness of lorry-park lighting should be judged not from its appearance on a fine dry night, but on a foggy night when the area is full of vehicles. If lighting supports are positioned within the area, they should be positioned along access lanes upon which vehicles should not be allowed to park.

16.3.4 Security lighting at docks

Sufficient illuminance for security purposes should be provided over the whole dock area, i.e. say 0.2–1 lx (which is sufficient for spotting people or vehicles). The general principles of security lighting (chapter 17) may need to be adapted to conditions at docks, for there are large areas which have to be defended even though they are not fenced.

If within the port boundary there is a foreshore which is uncovered at low tide, it may be possible for persons to perform various illegal acts in darkness, e.g., by walking along the foreshore a person may gain entry into the dock area, or may depart from it with illegal or stolen goods. Illegal goods dropped from ships or a quay might be recovered by a collaborator at low tide. To prevent such similar crimes, it may be necessary to provide lighting of the foreshore and shallows. Special techniques may be used, including mounting submersible luminaires on the faces of quays below high water mark. (Note that submersible luminaires may not be watertight, but usually work on the diving-bell principle.) Bouy-lights have been used, as have flood-lights mounted on barges or pontoons moored offshore and powered by underwater cable.

Although docks and shipyards do present real difficulties in devising effective security lighting, almost invariably it is possible to find an economical and effective counter to each threat by correctly applying the basic techniques of security lighting. If there are special and unusual risks, then some ingenuity must be exerted to deal with them. For example, it has been known for trailer-lights (section 18.2.3) or jennylights (section 18.2.4) to be mounted on vessels or barges to cover a temporary risk.

16.3.5 Crane operation

The crane driver and banksman need to see the load, the pick-up point and setting-down point clearly, without being confused by glare from luminaires, nor by glare reflected from wet surfaces or water. An illuminance of 50–70 lx normally suffices for dock work, but extra floodlights may need to be brought in for loading and unloading operations.

Although it cannot always be avoided, floodlights will preferably not be mounted on the crane jib, nor on the tower of a crane that tracks or rotates. Moving shadows can cause confusion, and the hook or the load may be masked by glare when seen from below. Preferably, local area lights will be mounted below the cab level.

Closed-circuit television (cctv), with a monitor screen in the cab, helps the crane driver considerably, and may obviate the need for a second banksman if the working point is out of the driver's view. With cctv cameras of normal sensitivity, an illuminance of at least 25 lx on planes facing the camera is required. The picture can be confusing if there is strong shadowing, and there will be 'ghosting' if the camera can look directly at a light-source. The monitor screen should be mounted in the cab where the driver can see it easily, without masking reflections on the screen. A polarizing enclosure on the luminaire in the cab may reduce reflections, but perhaps of greater utility is a dimming control for the lights in the cab.

16.3.6 Lighting at shipyards

The modern shipyard more closely resembles a large conventional factory than a dock. Its covered areas used for prefabrication of ship parts should be lighted according to current practice for industrial interior lighting. However, for the forming of hull sections, and the machining of large components such as propellers, systems of local lighting to give enhanced vertical illuminance on the tasks may be needed. Some subassembly of large assemblies is performed out of doors to gain the advantage of ease of crane usage and freedom from height restrictions.

Fitting-out bays and stores are notorious for 'shrinkage' of stock by pilfering, such petty crime reaching proportions that may threaten the viability of the concern. Stringent security measures, including the division of large areas into zones by internal fences and providing security lighting, are therefore essential (chapter 17).

Because the ship constructor is invariably working in an 'earthy environment', it is best if all lighting equipment within hand-reach height is operated on 110 V (future regulations will require 100 V), or even 50 V. Supplies to these circuits should be derived from a double-wound step-down transformer having an earth screen between its windings, and with its secondary winding solidly centre-tapped to earth. Special care must be taken to protect mains-voltage supplies to transformers and other plant by use of armoured cables or cables in rigid or flexible conduits. It is sound practice to use 'all-insulated' or 'double-insulated' portable lighting equipment in these situations, with additional protection by earth-leakage circuit-breakers.

For work on hulls in dry-docks and slipways in tidal waters, the luminaires should be capable of being submerged in salt water without risk of failure or hazard. Luminaires constructed on the 'diving bell principle' may be employed, but these are not suitable for continuous immersion as temperature cycling will eventually cause the luminaire to fill up with water. A practical point: submersible luminaires are heavily constructed, but may in fact be buoyant, and therefore must be very securely fixed – especially if subjected to regular tidal immersions.

16.3.7   Contingency planning

Thought should be given to the particular risks in the dock environment, such as a vessel on fire at a berth, leakage of noxious chemicals or petroleum products, or the dock being the scene of sabotage or terrorist activity. Most urgent interventions in emergencies at docks will be in relation to fire, rescue and evacuation of personnel. As with all contingency planning, efforts should be made to review all possible events which could lead to injury or loss of life, especially in darkness under mains-failure conditions. Not only should systems of interior emergency lighting and illuminated signs (chapter 14), and exterior emergency lighting and pilot lighting (chapter 19) be installed, but additional resources such as mobile lighting (chapter 18) and stand-by power supplies (section 22.5) should be in readiness for emergencies.

16.4   Petroleum and chemical industries

16.4.1   Introduction

All considerations for lighting petrochemical plant must be dominated by the dangers associated with the products handled, that is fire, explosion and corrosion hazards. Lighting equipment must be selected to suit the zone classification (section 7.10), and a high degree of reliability must be built into the electrical system (chapter 22). Installations must satisfy the quality and safety standards of the client, the insurers, and the Petroleum Officer or other special officer of the HSE Inspectorate.

A recurring problem is the difficulty of installing additional equipment or modifying the installation once the plant is in operation. Plants may operate continuously for many months between shutdowns for essential maintenance, and it is only when the plant has been purged of flammable materials that it will be possible to use welding equipment, or to introduce mobile plant (cranes, access equipment etc.) unless this is constructed for operation in the particular hazard Zones.

Because flame-proof and protected equipment is costly, skill must be exercised in designing layouts that will achieve the lighting objectives with the minimum of redundancy and unnecessary expense. Paradoxically, it may be necessary to duplicate some lighting services, or provide some overlap of facilities, to give greater reliability and ensure that under all circumstances there will be sufficient lighting for emergency escape use or for essential standby operations (especially for processes that cannot be shut down at short notice).

The design engineer specifying lighting equipment must work within a clear brief from the client as regards such matters as the nature and extent of hazardous Zones, for their definition and extent is the client's responsibility (section 7.10.1). The proposals should include safe methods of relamping, and how access to lighting equipment will be gained. Routine testing of electrical plant can only be carried out under a 'permit-to-work' at a time convenient to the operating schedule.

16.4.2 Jetties, pipeline terminals

The guidelines given for port lighting (section 16.3) are generally applicable to depots and locations where petroleum and chemical substances are handled. At depots located on the coast, particular attention should be paid to the construction and corrosion-protection of lighting supports since they will have to resist salt corrosion, as well as possibly being subjected to high winds. The latter consideration may prompt investigation of the use of lower mounting heights for luminaires. However, note that hazardous Zones are defined in the vertical plane as well as the horizontal, so that lower mounting might necessitate employing a higher degree of environmental protection.

At coastal pipeline terminals, the above considerations apply, and the installation could also be a target for sabotage, so that a high standard of fencing, patrolling and security lighting may be required. The matter is complicated when it is necessary to install security lighting within hazardous Zones. Pipeline terminals may have extensive port installations (section 16.3) with facilities for major road vehicle movements. Some have helicopter landing facilities (chapter 20). All the lighting requirements have to satisfy the constraints of zoning and the restrictions on showing of light to seaward (appendix A).

16.4.3 Tank farms

Tank farms, whether remote from other plant or associated with jetties, pipeline terminals, refineries etc., present their own special problems for lighting. Security and operational staff need to be able to patrol throughout the installation safely at night, often necessitating walking or driving a vehicle along the top of bund walls around storage tanks – so that the patrol can see down into each bunded area.

For petroleum products giving off vapours denser than air, and in tank farms storing crude oil, it is usually possible for the client or the insurer to specify the minimum permitted height of luminaires above any gas-air concentration that could cause fire or explosion. This minimum height may be as little as 8 m or as great as 30 m according to the products stored and the local topography. All electrical equipment below that minimum height will have to be in *proof* enclosures, so it becomes more economical (if slightly less convenient for maintenance) to locate control gear for HID lamps aloft rather than at the bases of columns.

Locating lighting towers in a tank farm can present difficult problems of geometry, for it is an objective to throw light on all the peripheries of the cylindrical tanks, and to ensure that the sloping walls of the bunds are adequately lit, both for operating reasons and for security. The problem is made more difficult by the fact that lighting supports cannot usually be mounted on top of the bunds, these usually being formed of earth and not firm enough to support a tower. An illuminance of around 5 lx usually suffices for tank farms, but a lower illuminance may be acceptable if the district brightness is low.

It is common practice for planning authorities to insist that landscaped hillocks or bunkers be raised inside the perimeter fence of petroleum installations, these being intended to hide the plant from nearby roadways and residences. Sometimes the operator is required, for aesthetic reasons, to plant coppices of conifers or evergreen shrubs. These practices greatly complicate the creation of a sound security lighting installation, for they prevent the perimeter fence being supervised from within the site. This matter is discussed in section 17.9.4, and the use of additional internal fences described.

16.4.3   Refineries and chemical plant

The guidelines in this section are generally applicable to refinery and chemical plant lighting practice, but additionally it should be noted that exterior plant may have to be lit mainly by small local luminaire which will put light on vertical surfaces and adequately light walkways and ladderways, with proof luminaires and electrical installations used throughout.

Under conditions of breakdown of plant or other emergency, it may be necessary to provide temporary lighting to enable repairs to be carried out or rescue effected. Special forms of trailer-lights (section 18.2.4) adapted for use in flame hazard zones are invaluable. Light can also be projected into hazard Zones from safe areas. The ability to bring in additional lighting, independent of all local electrical supplies, is an essential component of the contingency plan for dealing with emergencies and possible disasters at such locations.

16.4.4   Offshore structures

The lighting and electrical installations for oil-rigs and drilling platforms have to achieve higher standards of reliability and safety than is required in practically any other application, the equipment having to function within the constraints of restricted space, tight schedules for supply and construction, and with minimum opportunities for carrying out maintenance work or modifications to the installation. The plant operates in an environment that combines the risks of fire or explosion from the petroleum or gas being extracted, with the corrosive and destructive effects of the sea and the weather.

In a very small space, a drilling rig must be provided with lighting for safe movement, repairs and emergency actions in all weathers. It may be required to project light to seaward for various purposes including diver operations, loading and unloading of vessels by crane, docking of tenders, rescue operations etc. The exterior installation also usually includes a helicopter landing pad (section 20.7), but of dimensions much smaller than those used on land.

Any attempt to refer to all the relevant sections of this book which are applicable to offshore installations would result in a tedious listing of a large proportion of its entire contents, and will not be attempted. By their complexity, multifunctional use of space, and environmental constraints, offshore rigs demand the best skills of illuminating engineers and electrical engineers, as well as the achievement of a degree of perfection and meticulous attention to specification that is unique in lighting practice.

17 Exterior security lighting

17.1 Applications and objectives

17.1.1 Introduction

Security lighting is exterior lighting which is operated every night from dusk to dawn, the main purpose of which is to protect premises, property and people from criminal attack.

Almost any kind of exterior lighting will enhance the night security of premises, but lighting that is designed primarily to defend against intrusion or night attack will perform this function more effectively, and the desired results may be achieved at lower capital cost, with lower operating and maintenance cost, and with greater reliability.

The security lighting methods described in this chapter have been extensively and increasingly used in many countries since the late 1960s, and have proved very effective. The techniques described here are endorsed by the crime prevention officers of many UK and overseas police forces.

In this chapter, we are concerned mainly with exterior lighting, but the subject includes the lighting in checkpoint huts and security control centres (section 17.4). Some aspects of the subject are dealt with in other parts of the book. Lighting provided within buildings to enable security guards or fire patrols to move about safely and detect signs of intrusion or fire, (sometimes termed *police lighting*) is discussed in Chapter 12. In many factories, interior lighting which has a security function is identical with that which is provided for emergency and escape purposes (sometimes termed *safety lighting*), and this is discussed in Chapter 14. Some exterior lighting which is provided as *exterior emergency and pilot lighting* also has an important security function, and is discussed in chapter 19.

17.1.2 Scope

This chapter discusses security lighting for industrial premises, but specifically excludes forms of security lighting used to defend military premises and sensitive targets against forcible attack. (A forcible attack is one made openly by day or by night, often accompanied by threatened or actual violence against persons, often with disregard for any alarm which may be triggered, and carried out with the intention to steal, do damage or personal assault etc., regardless of the consequences.) Also excluded from this chapter are applications of mobile lighting by the police and military in dealing with dynamic security situations, for example, where there is confrontation with rioters, or where hostages have been taken. Such applications lie outside the needs of users and providers of industrial lighting, and are not dealt with in this book.

Security lighting as discussed in this chapter is provided at industrial premises primarily to aid defence against surreptitious criminal attack by night. In such an attack there is usually no intention on the part of the intruders to confront the defenders. However, a surreptitious entry may be the prelude to an attack with violence on the occupants.

17.1.3   Lighting as a defence against night crime

A review of the types of crime which may be perpetrated against business premises and the protective measures that may be taken is given in another book by the author[59]. Defence against many types of criminal activity at industrial premises is aided by security lighting in conjunction with other appropriate security measures.

Typical crimes involve surreptitious night-time entry, usually to steal. Intruders are sometimes the firm's own employees, ex-employees or contractors. Entry may be accompanied by destructive acts. Many fires at business premises are associated with criminal entry, started either maliciously, or to conceal fraud or stock shortages, or accidentally (e.g. by ignoring 'No Smoking' notices).

Security lighting also helps in protection against fire in other ways. If there is adequate light to permit the defenders to patrol about the premises, they are better able to prevent events which could lead to fire, and may discover fire earlier, and thus prevent its spread. Fires, whether of accidental or malicious origin, may smoulder for hours unnoticed before breaking out in a violent conflagration. If there is security lighting on the site, smoke from such a 'dark fire' may be detected earlier.

Illegal entry by a visitor, member of staff or contractor's employee may be effected by the individual remaining concealed in the premises at the end of the day. Security lighting will help the defenders discover people breaking in or breaking out.

Crimes by employees range from petty theft (concealing small items on their person or in their cars) to major premeditated crimes, possibly performed in collusion with outsiders. Companies commonly suffer losses through *vandalism* (graffiti, stone-throwing at windows, damage to parked vehicles), *fly-tipping* (unauthorized dumping of waste materials on private land), and *malicious damage*. The last of these includes not only the planting of bombs and incendiary devices, but also aggravating and costly occurrences such as the opening of drain-cocks on fuel tanks, the damaging of installations and costly machines such as cranes, the opening of weirs and sluices to cause flooding, and the deliberate pollution of reservoirs and waterways.

Crimes involving the theft of cars, and of heavy goods vehicles and their loads from industrial premises are common. Suitable lighting enables parking areas to be properly supervised, and makes it more difficult for a thief to force a gate or breach a fence undetected and drive a vehicle away.

Security lighting aids supervision at ports (section 16.3), helps to control movements on construction sites (section 16.1), and is vital to the security of petroleum and chemical plants (section 16.4) and at mines where precious metals and gems are extracted (section 16.2).

All kinds of businesses have reason to be concerned about their night security, and should consider the use of security lighting as part of their defence plan. Effective control of entry into their premises is vital for every business, for the viability of companies depends on the security of their records and computer equipment. Unauthorized entry may involve theft or destruction of data, and possibly theft or sabotage of equipment.

Since the use of security lighting became common in the 1970s, criminals have come to recognize that lighted premises present greater risks of detection and apprehension. This is reflected in insurance premiums, which may be lower for premises so protected. Indeed, the provision of suitable standards of lighting, fencing and guarding may be made a condition for insuring high-risk premises. There have been some attempts by criminals to tamper with lighting circuits, and it is now good practice to integrate monitoring of the lighting into electronic security supervisory and detection systems.

While the principal objectives of security lighting are to prevent loss due to criminal acts, it may serve other purposes, such as assisting in the safe movement of persons and vehicles on the site, and aiding the performance of work. The floodlighting of buildings may aid prestige or publicity. Not least, security lighting will improve the working conditions for night security staff, reducing dangers and the feeling of isolation that make it difficult to recruit and retain the services of good security staff.

17.1.4 Development of security lighting

Security lighting is an application of science-based concepts involving understanding the capabilities and limitations of the dark-adapted eye. It exploits the phenomenon of glare experienced at low illuminances, and an appreciation of how objects are perceived in low-illuminance conditions. The technique takes account of certain physiological attributes of the eye, e.g. night myopia (section 1.7.5), and dilation of the pupil in subjects experiencing excitement or fear (section 1.1.1) which makes them more susceptible to glare.

Practical techniques take account of reflection from fences, and the fact that revealment of an object against a dark background is enhanced by vertical illuminance. The techniques were developed during 1969 to 1972 largely by the author with the aid of J.E.Baker, D.W.Durrant and Dr P.Boyce, under the auspices of the former Electricity Council. Developments since then have been largely to improve the performance of installations, to achieve the required results with lower capital and operating costs, and lower consumption of energy. In the 1980s development continued largely by field experience and practical experimentation, enabling the economical lighting of large areas, and the revealment of intruders at great distances with the mininum amount of light. The lighting techniques described in this chapter can generally be applied empirically with only a limited need for calculation.

17.2 Principles of security lighting

17.2.1 Capability

Security lighting systems for industrial premises provide enhanced night security by:

- deterring the intending criminal;
- revealing the criminal before, during and after the attack on the premises;
- in many cases, providing some degree of concealment of the defenders from the view of the attacker.

Additionally the lighting may facilitate the safe patrolling of the area by the defenders.

17.2.2 Lighting in the security system

Security lighting has only limited value unless it forms part of a system comprising:

- *good physical defences for the premises* (e.g. strong fences, walls, gates etc) to slow down the attack and to make it more difficult and more time-consuming for the attacker;
- *defenders* (e.g. the normal occupants of the site, security guards, police and members of public who can see into the site from outside) to observe and respond when an attack occurs or is threatened;
- *lighting*, which preferably will be designed along the guidelines given in this chapter, and which may also serve the purposes of amenity, safety, publicity or décor.

It is the combination of 'lights, fences and men' that gets the desired results.

17.2.3   Cost-effectiveness

The cost-effectiveness of security lighting arises because:

- it has an excellent deterrent effect upon criminals;
- it enables the supervision of large areas and long perimeters with a small defending force;
- it reduces the need for costly physical defences (e.g. high brick walls);
- has long life;
- together with fencing, it has low maintenance and operating costs compared with far more sophisticated systems of defence.
- it can provide the illumination for economical cctv systems and video equipment.

17.2.4   Physical defences

No premises are immune from attack – even if they contain nothing that could conceivably be of interest to any intruder – and no premises are physically invulnerable to attack. Given enough know-how, time, resources and motivation, even the strongest defences can be overcome. All that physical defences can do is to make entry more difficult and risky for the intruder. If the perimeter defences make the crime take longer, the criminal is exposed to greater risk of discovery. Many premises are very secure by day, but highly vulnerable by night if not provided with security lighting and suitable guarding. The effectiveness of solid walls versus chain-link fencing or palisade fencing is discussed in section 17.3.

17.2.5   Philosophy

It is better to have a low-technology security system that positively discourages crime and actually aids the handling of an intrusion situation, than to have a sophisticated system of detection that rings a bell to tell you that an attack is taking place.

Lighting is clearly a deterrent. (Convicted persons, interviewed by the author, said that they would always prefer to break into dark premises rather than into those with security lighting.)

Security lighting is a positive asset in dealing with an intrusion through the perimeter. If the site is not lit, and intrusion is discovered or is detected (e.g. by high-sensitivity low-light cctv monitoring, by radar, or by active or passive infrared detectors etc), it will still be necessary to switch on some lighting in order to deal with the situation. It is very difficult to chase criminals in the dark, even with a torch!

It is always better to have the security lighting on all night, from dusk to dawn, preferably controlled by automatic switching so it cannot be forgotten or deliberately left off by a corrupt security guard. *Triggered lighting*, i.e. lighting that is switched on by a detection device, has many disadvantages (section 17.2.6).

17.2.6 Disadvantages of triggered lighting

Triggered lighting (also called *trip lighting*) is of great value for switching on lights at the back and front of a domestic dwelling, both for safety and security. Thousands of such installations exist, generally using passive infra-red (pir) detectors incorporated into small bulkhead or floodlighting luminaires. It must be clearly stated that such lighting has no place in the defence of industrial premises.

Triggered lighting can easily be abused by intelligent criminals, and the system outwitted. They may exploit the weaknesses of such systems by deliberately tripping them to bring the lights on, thereby discovering:

- the *reaction time* of the defenders (how long does it take the defenders to get there?);
- the *reaction direction* (from which direction do they usually come?);
- the *reaction strength* (one elderly guard on foot? or a posse of strong young guards in a Landrover?);
- whether the defenders notify the police as part of the routine of any alarm (or if a central security station link is employed).

A criminal may wear out the response reflex of the defenders by such *nuisance tripping* until they tire of responding, or decide that the lighting and/or alarm system is faulty, and may switch the system off or ignore its operation, giving the criminal the opportunity to enter undetected. Criminals in collaboration can perform nuisance tripping at several points on the perimeter, and can tire and confuse the defenders and divide the reaction force between a number of incidents. For these reasons, security lights should be switched on all night and every night, and the operation of an alarm should not be signalled to the intruder.

It is sometimes said that trip lighting is necessary where the defended area is near a wildlife conservation area or a 'lovers' lane' in which casting of light would be an intrusion. This is fallacious, for generally it is possible to devise lighting that causes very little nuisance outside the perimeter. There is another fallacious argument suggesting that dark premises would be less likely to attract the interest of criminals, on the basis that defensive measures imply there is something there worth stealing. Any Crime Prevention Officer will confirm that criminals do not pick their targets as casually as that.

One final reason for opposition to trip lighting is that the need for instant full light output restricts the user to tungsten filament lamps, tungsten-halogen lamps and fluorescent tubular lamps, preventing the use of HID lamps which are so much more economical to operate.

17.2.7   Revealment effect

Security lighting is designed to project light in the possible directions of an approaching intruder, revealing the intruder to the defenders by his illuminance contrast with the background, or by his silhouette against a brighter background. Revealment is discussed in section 17.3 in relation to perimeter lighting, in section 17.5 in relation to area lighting, and in section 17.6 in relation to building floodlighting.

17.2.8   Concealment effect

The concealment effect in security lighting installations is achieved by exposing moderately high luminances towards possible directions of an intruder's approach, so that the intruder is exposed to this brightness while remaining dark-adapted, and thus suffers some embarrassment of vision. This effect is utilized to give a measure of concealment to the security guard while on patrol inside the perimeter fence (section 17.3), and is also employed to conceal the guards within a checkpoint hut (section 17.4).

17.2.9   Deterrent effect

Because of the revealing power of the lighting (section 17.2.7), would-be intruders know that their presence may be detected even before they attempt entry into the premises. They know that, if they are seen, they may be positively identified if known to the guards, or may be accurately described to the police.

Because of the concealing power of the lighting (section 17.2.8), intruders know that, even before starting the crime, the guards may have had them under observation without them being aware. The lighting denies them the advantage of knowing the number or location of the defenders, they do not know if they have been seen. Intruders are thus very vulnerable, and are at a visual and tactical disadvantage. In many cases they are deterred from attempting the crime, and may instead direct their attention to other premises not equipped with adequate security measures.

Other than statistically, it is difficult to prove the efficacy of security lighting; effectiveness results in no attacks. Alarm systems produce statistics of false alarms and aborted attacks, but security lighting tends simply to produce peace and quiet.

17.2.10 Security lighting techniques

Five basic techniques are used which are based on traditional lighting methods. They are discussed in the following sections.

Perimeter lighting

This is lighting of the fence line, enclosing wall or boundary, commonly with outward facing luminaires arranged to cause glare to the intruder (section 17.3).

Checkpoint lighting

This enables people and vehicles to be checked at a point of entry or egress. This may include special lighting within the checkpoint hut and special window arrangements to enable the security guards within the hut to see out without themselves being visible from outside (section 17.4).

Defensive area lighting

This is the lighting of open spaces around and between structures from luminaires mounted on local structures, or on poles or towers (section 17.5).

Defensive building floodlighting

This is floodlighting of buildings, plant, walls etc., to reveal people locally within the lighted areas, or to reveal them by silhouette against an illuminated surface viewed from a greater distance (section 17.6).

Topping-up

This is the provision of lighting in areas which otherwise would be shadowed e.g. by deep doorways, under canopies, in narrow spaces between buildings etc.. Topping-up also refers to mobile lighting equipment brought in temporarily to provide lighting for some special area of risk (section 17.7).

17.2.11 Assessment of risk

A security lighting system and fencing system may be devised that will give effective protection against likely threats. In quest of economy, and to ensure that sufficient resources are devoted to proper protection, the designer may pose the following three questions:

- Are the premises easy to enter by night? The designers should consider their situation, the amount of local activity at night, the district brightness, the strength of the fences and other means of keeping out intruders, the nature of the night occupancy, and the standard of night guarding.
- Do the premises contain property attractive to criminals? Cash is the most attractive prize, while gold, silver, mercury and gems etc., are easy to carry away and difficult to identify later. Consumer goods (tobacco, alcoholic drinks, fashion clothes, televisions and videos etc) find a ready market to the unscrupulous. Some bulk foodstuffs are easy to dispose of, and petrol, tyres and lorries loaded with anything at all are also attractive to thieves. Anything moveable is stealable.
- Could a break-in be seriously damaging to the business? Uninsured losses can be crippling (loss of goodwill because of delayed deliveries). The designer should consider loss of profits or viability after loss of equipment or tools, the risk of fire from intruders and the risk of losing records, firm's secrets or interference with computers.

If the answer to all three questions is 'yes', assess the risk as class A – (extreme risk). Two 'yes' answers merits Class B (high risk) and one 'yes' answer merits class C (moderate risk). These three arbitrary gradings can be described as follows:

Class C – moderate risk. Premises where the physical security is reasonably good, and where the goods contained are not particularly attractive to thieves. The lighting installation will comprise standard robust luminaires, installed according to normal electrical practice, but carefully sited. Lights to be switched from a central point not easily accessible to visitors.

Class B – high risk. Premises that are fairly easy to enter by stealth at night, and/or contain goods attractive to thieves. Luminaires should be selected to withstand considerable violence from stones and airgun pellets, and the mountings will be robust. Wiring to be in heavy-gauge steel screwed conduit or single-wire armoured cable, concealed from view. Switching by time switch or photoelectric controllers. Key lights to have stand-by alternative supplies, such lamps being of types that relight without appreciable delay, e.g. GLS, T-H, MCF.

Class A – extreme risk. Premises which are easy to enter by stealth at night, or are in isolated, quiet places, what ever they may contain. Also, premises which are quite well defended, but which contain goods that are particularly attractive to thieves. Installation recommendations as for class B to be followed, with additional precautions to prevent sabotage. Lighting controls to be linked into a security monitoring panel to signal any interference. Switching by photoelectric controllers local to the luminaires. Vital lights to be operated in the 'maintained mode', with 3-hour battery backup duration. Preferably the perimeter will be defended by palisade fencing with intrusion detection facility (section 17.3.4).

A main factor in determining the likelihood of intrusion is the ease with which entry can be effected. Vagrants will get into premises that are easy to enter, looking for shelter or something to steal; the fire risks from such penetrations are great. A bullion vault is unlikely to be breached because its defences are strong, but the losses due to successful entry could be great. Small business premises are frequently entered, though the reward to the criminal may be negligible; such penetrations are frequently disastrous to the business. Crimes of opportunity occur because the opportunities are provided. Good fencing, security lighting and efficient patrolling greatly reduce the risks.

A useful reference when planning security lighting schemes is the publication *Essentials of Security Lighting*[61].

17.3   Perimeter fencing and lighting

17.3.1   Fences and walls

A perimeter fence or wall will slow down intrusion and make it more difficult, but by itself may not prevent intrusion. The presence of guards is essential, and they can only function efficiently at night if suitable lighting is provided. With lighting, a small number of guards can defend an extended perimeter.

High solid walls, particularly if fitted with anti-climbing obstacles, can be effective barriers, but their cost is great. In some cases, use may have to be made of existing solid walls, with new security features added (section 17.3.8). Chain-link fencing or palisade fencing is generally cheaper and more effective. Planked (boarded) fencing is usually best replaced with chain-link or palisade fencing. Chain-link and palisade fencing may be topped with rotating spikes, barbed-wire or barbed aggressive tape.

Both chain-link and palisade fencing have the advantage of being 'transparent'. The defenders can see through them to supervise the approaches to the site, and – if there is no objection from adjacent occupiers or the local authority – light can be projected through, as well as over, the fence.

17.3.2   Typical arrangements

Perimeter defence arrangements will typically have the following features (Figure 17.1):

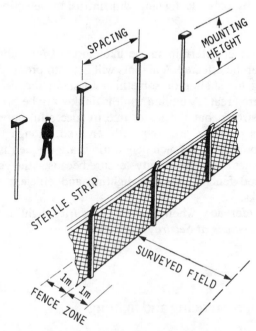

Figure 17.1 *Typical perimeter lighting installation with 3-m or 5-m perimeter lights. The patrol path is well behind the array of lights.*

- The *defended area* is the whole of the site within the perimeter fence.
- The *surveyed field* (or 'surveillance zone') ideally would extend for a minimum of 20 m, and preferably 50 m outside the fence, would be reasonably flat, and clear of any obstruction (e.g. trees, bushes, parked vehicles) which could conceal anyone from view from within the defended area. Preferably light from the perimeter lighting system will pass through or over the fence and illuminate the surveyed field. If the surveyed field is obstructed (i.e. has cover for intruders), or if there would be objection to light being projected into it, a double fence line may be employed.
- The *sterile strip* is a clear zone immediately inside the fence, say 6 m wide, and extending back to the array of perimeter lights. A sterile strip between double fence lines is sometimes called a *no-man's land*, and should be broken into short sections by transverse fences called *septa*.

- The array of outward facing *perimeter lights* will be located typically 3–5 m back from the fence. Perimeter lights may typically be any kind of small floodlight luminaires mounted on columns 3–5 m high, or may be low-mounted glare-lights mounted at 0.7–1.2 m above ground (Figure 17.2).
- The *fence zone* is the strip of land 2 m wide centred on the fence. It is important that this area should be kept quite clear of all obstructions such as vegetation, rubbish or windblown sand etc.

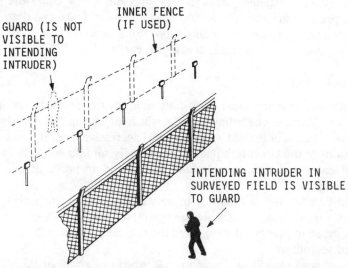

GUARD (IS NOT VISIBLE TO INTENDING INTRUDER)

INNER FENCE (IF USED)

INTENDING INTRUDER IN SURVEYED FIELD IS VISIBLE TO GUARD

Figure 17.2 *General arrangement of single or double fence-line with glare lighting.*

- If there is a *double fence line*, the inner fence will be spaced 1 m back from the line of the perimeter lights (Figure 17.2).
- The *patrol path* lies well behind the array of lights, or, if an inner fence is provided, behind the inner fence. Important: in the latter case, to prevent the *sleeping sentry syndrome* occurring, the patrol path should be within 1 m of the inner fence, or 3 m or more from it (section 1.7.5).

Additional fences may be needed if the defenders cannot see the entire perimeter fence from within the defended area because of obstructions, or if 'citadels' are created to protect vulnerable points (section 17.9.4).

17.3.3 Fence construction

Types of fencing used for industrial site perimeters are discussed in the following sections.

Wooden fencing

For temporary installations (e.g. construction sites), chestnut paling fencing, reinforced with strong arris rails, or wooden palisade fencing, will suffice if the security risks are not high. Chestnut paling has the advantage that it can be rolled up for easy transport to the next site. If properly erected and topped with barbed wire or barbed aggressive tape, wooden fencing will provide a reasonable degree of discouragement to unauthorized entry. However, chestnut paling will be readily distorted or smashed by collision with vehicles. If not very well installed, children and dogs can get in under the pales. A rubbish fire injudiciously placed too near a wooden fence can destroy it.

Wooden paling fences (including those using concrete posts) are specified in Part 4, and wooden palisade fencing in Part 6 of *BS 1722*[62].

Chain-link fencing

The fence construction most commonly used for the protection of industrial sites is 'anti-intruder chain-link fencing' which may be up to 2.8 m high topped with three strands of barbed wire or barbed aggressive tape. It is essential that the bottom of the chain-link panels are securely set into concrete footings to prevent entry under the panels, and not merely anchored at two or three points per panel.

Chain-link fencing is available in very heavy gauge wire which will defy ordinary wire-cutters. Preferably the chain-link will be PVC-coated in black or dark green to prevent glare-back to the defenders which will handicap their outward visibility.

Anti-intruder chain-link fencing is described in Part 10 of *BS 1722*[62].

Metal palisade fencing

Steel palisade fencing provides considerably greater strength and durability than either wooden fencing or chain-link fencing, and its higher cost is justified for high-risk installations. Because of the weight of the components, it is essential that the metal or concrete posts are set in concrete foundations of sufficient mass and extent in relation to the ground conditions to ensure that the fence will not fall over on impact by a vehicle. Metal palisade fencing should be painted black or dark green on the inner face to prevent glare-back to the defenders from the perimeter lighting, and may with advantage be painted a light colour on the outside.

Steel palisade fencing is described in Part 12 of *BS 1722*[62].

17.3.4 Detection systems on fences

The technology of fence alarms lies outside the scope of this book, but is described in another book by the author[59]. Mention will be made here of one product which has some exceptionally good features; this is 'Sabretape', which is marketed by Pilkington Security Systems (appendix G). Sabretape is a barbed aggressive tape having the same general function as barbed wire, but without any unbarbed places which may be grasped. Along its surface there is a 50-micron glass optical fibre carrying coded light pulses. The tape has weak places where it will break if stressed. If the tape is cut or is broken at one of its weak places (for example, by someone trying to scale a fence) an alarm condition is established.

An excellent system is afforded by combining Sabretape with a robust steel palisade fence, for example that supplied by Prima Security and Fencing Products Ltd (appendix G), in which the steel pales are pierced so that the Sabretape can be threaded through to form a formidable obstacle to intruders. If required, rotating steel spikes can be added to the top of the palisade fencing.

17.3.5 Double fence lines

The distance between the two fences should be at least 6 m, or 8 m if the ground is available. Distances less than 6 m are thought to be capable of being bridged by an intruder using a ladder or pole.

17.3.6 Perimeter lights

It is not usually necessary to do more than very simple calculations to design perimeter lighting. All that needs to be done is to select the type of equipment and follow the maker's instructions regarding mounting height, spacing and stand-off from the fence. Some layouts using tungsten-halogen lamps in luminaires of type 'Z' distribution (section 6.4.1) and the results of their field testing are given in Table 16.

In the tests reported in Table 16, the performance of the luminaires tested was significantly affected by the accuracy of aiming (section 15.6). If the land outside the perimeter fence is not level, the performance of perimeter lights will be affected and it may be difficult to get good effects with low-mounted glare lights (section 17.3.7). If the ground falls steeply outside the fence, it may be necessary to use higher mounting (Figure 17.3A), and if it rises, the aiming will have to be raised in order to get a good throw (Figure 17.3B). The objective is to arrange for the main flow of light from the perimeter lights to be projected parallel to the ground in order to obtain the maximum revealment effect. The illuminance on the ground in this case is almost of no significance to the effectiveness of the system.

In typical installations the ground will be soft, and it will not be practicable to use ladders etc to gain access to the luminaires for maintenance. Therefore

Table 16   *Perimeter lighting arrays using T–H lamps*

Lamp (W)	Mounting height (m)	Spacing (m)	Aiming distance (m)	Revealment effect	Concealment effect
500	3	10	15	B	A
300	3	10	10	C	B
500	5	10	25	B	B
300	5	10	25	B	C
500	5	15	25	C	B
300	5	15	25	D	C

Revealment and concealment effects are expressed as five grades of satisfaction: A, excellent; B, superior; C, good; D, moderate; and E, poor. In the above examples, the stand-off from the fence is 10 m, but if this is reduced to 6 m the revealment effects will be reduced by one grade. Gradings will be similarly reduced if the mesh of the fence is of galvanized finish, and not dark and fairly non-reflective. The concealment effect will be reduced by one grade if the ambient illuminance on the site is between 1 and 5 lx; if over 5 lx, the concealment effect should be tested.

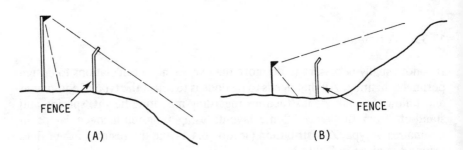

Figure 17.3 *Matching the mounting height and aiming of perimeter lights to the rise or fall of land outside the fence.*

base-hinged luminaires are preferred, these being of a kind that can be lowered by one person (Figure 17.4).

17.3.7   Low-mounted glare lights

In this type of perimeter lighting installation, less importance is given to the ability to detect people in the surveyed field than to projecting glare towards them to give a measure of concealment to the defenders in the defended zone.

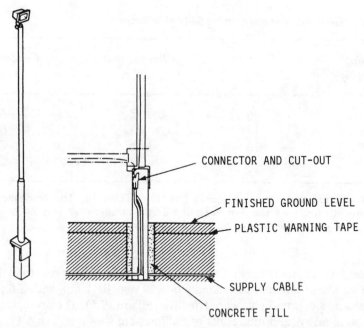

Figure 17.4 *Base-hinged 3-m or 5-m column for perimeter light, capable of being lowered by one person.*

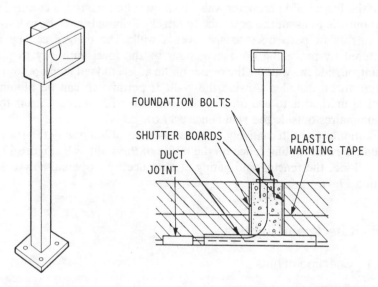

Figure 17.5 *Low-mounted (1.2 m) perimeter glare light housing one 50-W MBF lamp.*

Table 17   *Glare-lighting arrays using 50-W MBF lamps*

Mounting height (m)	Spacing (m)	Revealment effect	Concealment effect
1.2	6	C	A
1.2	12	D	B
1.2	18	D	C
3.0	6	B	C

For explanation and conditions see footnote to Table 16. The revealment and concealment effects for the 3-m mounting height arrangement may be one grade higher if the ground outside the fence slopes upward.

Practical tests at an installation where the district brightness was low, using a simple bulkhead luminaire containing one 50 W MBF lamp, with a flat glass front and no sideways light, were effective (Figure 17.5). The arrangement of the lighting arrays and the results of the tests are shown in Table 17.

17.3.8   Solid perimeter walls

If the site has a solid perimeter wall which must be retained, a criminal could loiter outside to await the best time to attack. Criminals have been known to use mirrors or periscopes to see over a wall. The danger can be partly countered by prevailing on a neighbour or the local authority to provide lighting outside the wall, or the owner of the adjacent land might permit some lighting to be installed outside the wall. If permission can be obtained, a practical method is to use roadlighting columns with outreach arms to hold the luminaires outside the wall (Figure 17.6).

Additionally or alternatively, a better method of improving the security is to add a new chain-link fence on the inside of the solid wall (Figure 17.7). If this is done, the fencing and gating at entrances must be suitably designed (section 17.4).

17.4   Checkpoint lighting

17.4.1   Checkpoint huts

The provision of a strong perimeter fence may increase the chance that a criminal will attempt to use a normal gate to enter the premises, either

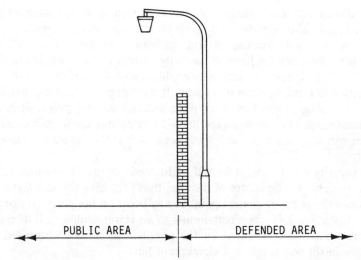

Figure 17.6 *Lighting the outer surface of a solid perimeter wall by use of a roadlighting lantern on standard with outreach arm.*

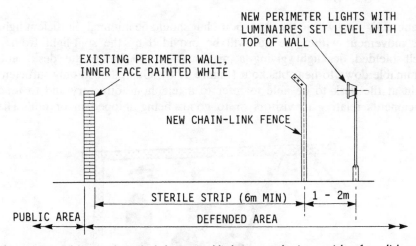

Figure 17.7 *Adding a chain-link fence and lighting on the inner side of a solid wall.*

surreptitiously, or by trick. It is therefore necessary to design the checkpoint hut and its environs to ensure that the security guards can observe and control all movements of people and vehicles into and out of the premises.

It is preferable if the functions of a checkpoint hut are limited to the purposes of providing shelter and accommodation for the security guard/s responsible for control of movements through that gate, and that it is provided with minimal communications, e.g. nothing more than a telephone and emergency alarm button, and possibly cctv monitor screens.

The practice of locating alarm indicator panels, mimic diagrams, key-boards, switches for lighting, security system controls etc in checkpoint huts is ill advised, because such may enable a visitor to quickly acquire much information that might assist in a crime. If it is not possible to arrange for this equipment to be located in a separate security control point, at least the equipment should be arranged so that visitors cannot see it from outside the hut, nor can they see it or easily get access to it if they enter the hut on any pretext.

The security control point should be located not on the perimeter, but as close as possible to the centre of the site, thus providing the shortest average route from the security control point to any point on the perimeter or to any 'citadels' (section 9.4) when responding to an alarm situation. If a rest-room or catering facility is provided for the guards, this should be located at the security control point, not in a checkpoint hut.

17.4.2   Lighting within the checkpoint hut

General lighting within the checkpoint hut should be minimal. Sufficient light for movement within the hut will be provided by the spill-light from a well-shielded desklight giving a maximum of 300 lx on the desk, and dimmable down to near blackout (Figure 17.8). Guards need only sufficient light at the desk to be able to refer to a telephone directory and to read documents relating to visitors or to goods being shipped in or out. The

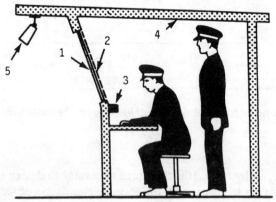

Figure 17.8 *Lighting in a checkpoint hut. 1, glazing is tilted to reduce reflections; 2, wire-mesh or perforated metal screen; 3, shielded luminaire with dimming facility and emergency battery power; 4, ceiling and upper walls are painted subdued colour; 5, canopy lights.*

objective is to keep the guards' adaptation level as low as possible, so that they are able to keep observation on the lighted area outside, and so that they have to make the minimum adaptation adjustment on stepping outside. The lighting in the checkpoint hut should have high reliability, and will preferably be operated as maintained emergency lighting (section 14.3). Preferably the desk will be located under the window which will face in the direction needed to supervise the entrance.

17.4.3 Checkpoint hut window

The window should be tilted outward at the top at an angle of about 7° to reduce reflections. The ceiling and upper walls of the hut should be painted a matt subdued colour of about 25% reflectance to limit window reflections (Figure 17.8).

Various methods of screening a checkpoint hut window are used to prevent a person outside seeing in (and thereby learning how many guards are present, and perhaps ascertaining if they have been seen by them). Half-silvered glazing and glazing partially obscured with mirror stripes are effective. To reduce risk of injury to the guard if the window should be smashed by a missile or explosion, the window may be glazed with a robust plastic sheet material rather than glass.

A simple way of reducing inward visibility is to glaze the window in very small panes, with glazing bars 25–40 mm wide and painted white on the outside and matt black on the inside. Another proven method is to place a sheet of perforated steel plate inside the glazing, the plate being painted white on the outside and matt black on the inside. This giving protection against smashed glazing and a strong 'one way vision' effect.

17.4.4 Location of the checkpoint hut

The checkpoint hut should be carefully sited to ensure that the occupants have a clear view of all the approaches to the checkpoint. It could be best to locate windows on two adjacent walls and position the hut diagonally, or provide a bay window, so that the guards can see clearly into and out of the site. However, if the checkpoint hut has windows on two or more walls, e.g. if it is at an island location between entrance and exit roadways, care has to be taken to ensure that the security guard within the hut is not silhouetted against a window. It is desirable that prowlers outside should not be able to see easily if the checkpoint hut is occupied, nor how many guards are present.

The hut may be provided with an external canopy to give weather protection to the guard, and under-canopy lights provided (with local switching).

The layout for a checkpoint hut and entrance should preferably provide space for a vehicle to draw in off the public road before reaching the first check-bar or gate. The layout should take account of a double fence-line or an additional fence placed inside an existing solid wall (section 17.4.8). The entrance should provide a *lock* (the strip of road between two check-bars or gates) long enough to hold the longest vehicle and trailer that is likely to enter (Figure 17.9).

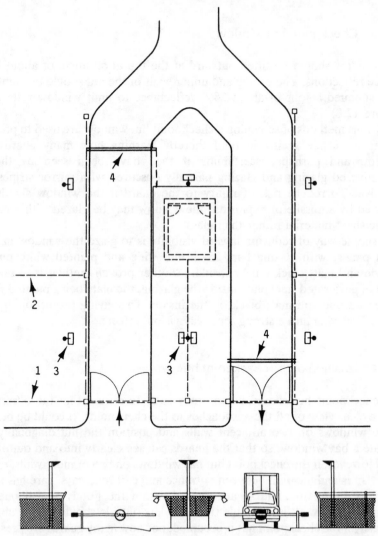

Figure 17.9 *Layout for a major site entrance. 1, outer perimeter fence or solid wall; 2, inner perimeter fence; 3, lighting columns; 4, check-bars or gates (to be in reverse positions in countries where traffic travels on the right).*

Check-bars may be counterbalanced and hand-operated, or hydraulically operated from outside or from within the hut. Gates take longer to open and shut than check-bars, and occupy considerable linear space along the roadway, thus making the installation larger and more costly, but they offer the means of totally enclosing the lock so that it is not possible for someone to slip through while the guard is inspecting a vehicle.

When checking vehicles, a good stratagem is for one guard to be outside, while the other remains in the locked checkpoint hut with access to the telephone or emergency-button in case of trouble outside.

Note: for some large sites, additional checkpoint huts may be required for *pounds* (vehicle quarantine areas), at *zonal barriers* or at *citadels* (section 17.9).

17.4.5 Lighting of checkpoint entrance

The lighting at a site entrance should enable an arriving driver to recognize the premises and see clearly how to bring in a long or articulated vehicle. The lighting within the lock must enable the guard to search vehicles, looking for illegal goods and for 'false ballast' (for weighbridge frauds), the latter often consisting of polythene water containers or other heavy items concealed between the longerons under the vehicle.

The lighting should enable the guard to see into open trucks, inside vans and lorries, into driving cabs, and under vehicles. To aid under-vehicle search, a mirror may be used, usually of convex design and sometimes placed on a wheeled frame with a long handle so it can be pushed under the vehicle. Some mirror devices incorporate a light, but these are not satisfactory; if battery-operated the power and reliability are low, and if mains-operated they are not without danger even if operated on reduced voltage. The flexible cable is inconvenient to handle in the dark and wet.

Lighting for inspection under vehicles can be achieved by recessing suitable robust luminaires into the road surface (e.g. airfield runway lights); but these are costly, and a great deal of maintenance is required. A much simpler and more effective method which is widely employed is to paint the road surface within the lock with a white durable paint (e.g. epoxy resin paint) so that light from the adjacent luminaires is reflected up under the vehicle. The lighting should be ranged on both sides of the vehicle.

If a two-lane entrance is used (Figure 17.9), it is essential to provide a central fence to separate the carriageways, and either have two checkpoint huts (as at some very busy establishments) or a central checkpoint hut between the two carriageways. It is desirable to have a row of luminaires on posts along the centre reservation so that dense shadows are not formed when large vehicles are stopped in both lanes.

For most installations, luminaires mounted at 5 m will give good results. If the checkpoint lighting is to be kept on continuously, these may be HID lamps. Tungsten–halogen lamps may be used if the lighting for vehicle inspection is only switched on when required.

If high-sided vehicles pass in and out of the premises, 7 m or 9 m mounting heights for the adjacent luminaires should be adopted, and it will be necessary to provide a wheeled ladder-tower for inspecting the tops of vehicles.

17.5   Defensive area lighting

17.5.1   Strategies

Moonlight versus searchlight

Security lighting techniques use proven concepts involving understanding the process of vision at low illuminances, the behaviour of the dark-adapted eye, and the phenomenon of glare. Just as it is clearly established that a low level of general area lighting always produces better visibility for the guard than patrolling in darkness with a torch, so it is established that moonlight levels of area lighting are always better than 'Colditz' (beam searchlight) installations which are heavily weighted in favour of the intruder. The intruder can usually outwit the defenders by following the simple rule: 'When the beam of light moves – freeze; when the beam is stationary – move!' This advice, given by a former prisoner of war who escaped from a prison camp was confirmed as sound by the author who conducted some simple experiments. Operating a powerful 'scheinwerfer', he found that his eyes adapted to the beam-spot brightness, and that it was quite impossible to see an 'intruder' who kept still while the beam was moving. Later, using the same 1 kW of energy in several small floodlights to create moonlight conditions, he found it easy to spot 'intruders' at great distances and over a considerable area.

Working sites

If the interior of a site is in virtual darkness, the lighting arrangements at the perimeter will be more effective (section 17.3). If the site is a working area at night, area lighting will be needed to enable work to be performed (chapter 15). On some large sites, perimeter lighting will be used, and only subdued lighting provided elsewhere. However, areas well away from the perimeter may have normal area lighting without detracting seriously from the effectiveness of the perimeter lighting.

If consideration shows that there will be more benefit from lighting the central areas than from providing concealment of the guards, then the choice may be to light the area. In control of certain sensitive locations, e.g. nuclear power stations, the tactical advantage of knowing for certain the present location of an intruder may be more important than concealing the guards on patrol. Very rarely there may even be justification for having maintained perimeter lighting with switched or trip-switched lighting of the interior, with all its disadvantages (section 17.2.6). Each location must be studied and the best plan worked out within the terms of the brief.

The methods and installation techniques for area lighting for security purposes and those for ordinary industrial purposes may be identical, though the illuminances used in the latter will be much higher (appendix C). Intermittently-operated lighting will use tungsten-halogen lamps which have lower efficacy and higher operating costs than HID lamps.

Sites surrounded by open country

Sites which abut natural open country or farmland may be particularly at risk from intrusion. It may be permissible for the occupier of the site to project light over areas of land outside the perimeter. The modern techniques of defensive area lighting may be particularly applicable to such situations where the surveyed field has a very low district brightness. If the land is unobstructed by trees, bushes etc, and is reasonably flat, it is quite feasible to set up a lighting system using minimal amounts of light projected from low-mounted floodlights that will enable the ready detection of intruders in the surveyed field at distances of 250 m, or even up to 500 m (section 17.5.2).

17.5.2 Revealment effect

Security lighting is designed to project light in the possible directions of an intruder's approach so that the intruder is revealed to the defenders by luminance contrast with a dark background, or by silhouette against a brighter background (section 17.6).

The illuminances employed in security lighting are quite low (Table 21, appendix C), and are graded in relation to the district brightness and the degree of risk. In Table 21, higher illuminances in each range which were proposed when the recommendations were first formalized in 1969, have been abandoned, and the lower illuminances in each range, which were formulated in 1975 following practical experimentation, have been included. The suitability of these figures has since been confirmed in practical trials carried out by the author.

Experience shows that very good revealment can be obtained by projecting light in a substantially horizontal direction. Effective results can be gained even if the illuminance on the horizontal plane is as low as 0.2 lx. Subjective assessments of revealment made by observers more closely relate to the vertical illuminance (E_v) than with the horizontal illuminance (E_h), a finding that holds good at very low illuminances and for trying to reveal individuals at extreme range[60].

It is possible to reveal a dark-clothed person at distances of up to 500 m in open country without ground cover. In one test performed by the author, with conditions of E_h well below 0.1 lx, but with E_v of about 1.0 lx, there was very good revealment over grass at 550 m.

Such ideas must affect our approach to the design of both area lighting and perimeter lighting (section 17.3), for they can lead to greatly reduced costs for security lighting and reduced energy usage without diminishing the effectiveness of the installation. By employing much lower mounting heights than would be used for conventional area lighting, considerable savings may be made without lessening the revealment effect. Mounting heights of 3–5 m, using asymmetric-distribution floodlights have proved very effective. The foregoing statements hold good under clear-air conditions; revealment will, of course, be much reduced when there is absorption and scattering of light due to mist, rain etc.

17.5.3   Area lighting installations

In designing area lighting for the central areas of an industrial site, it is tempting to mount luminaires on buildings and structures to save the cost of columns. This has the following practical disadvantages:

- the buildings themselves may not be adequately illuminated;
- vulnerable features (fire-escape stairs, roof catwalks, recesses) may be left in darkness;
- the available mounting heights and spacings may be far from ideal;
- getting access to the luminaires for servicing may be more difficult and costly than dealing with luminaires on easy-climb towers or hinged masts.

17.5.4   Illuminances for defensive area lighting

If the area lighting is provided mainly for the purposes of work, then the illuminances should be chosen to suit the tasks (Table 20, appendix C). However, if the area lighting is provided for the purposes of security, the illuminances should be chosen to suit the district brightness (Table 21, appendix C). If the area is used for work part of the time only, switching should be installed to enable the working illuminance to be reduced to the security illuminance when work is not proceeding.

If cctv is to be installed, it is important to ensure that the lighting layout and camera positions are considered together. Cameras should be located so they do not look into the beams of floodlights. Camera function is dependent on the luminance of surfaces rather than their illuminance; thus it is the value of the vertical illuminance that determines the picture quality. An illuminance of about 100 lx on building surfaces will enable a vidicon cctv camera to register an excellent picture, but if the building surfaces are of light colours, a lower illuminance may suffice. A very much lower illuminance may suffice if a more sensitive type of camera tube is employed, e.g. Newicon or Ultracon. Early liaison with the specialist supplier of cctv equipment is advised.

Note the need for emergency lighting in outdoor areas (Table 19, appendix C). In critical security installations, the entire area lighting installation, or at least some key lights within it, should be provided with some form of mains-failure back-up (section 22.5).

17.6 Defensive building floodlighting

17.6.1 Purposes

In security lighting installations, floodlighting of the vertical surfaces of buildings and structures is used for two purposes:

- to reveal anyone attacking a building, e.g. trying to force an entry, mounting a fire escape, being on a roof catwalk etc; and
- to reveal an intruder positioned in an unlit area between the floodlit building and the defender's position, i.e. to reveal the intruder in silhouette against the bright background of the floodlit surface.

17.6.2 Techniques

Decorative floodlighting often uses *close-offset* floodlighting, i.e. floodlights placed close to the vertical surface to be lit, the luminaires sometimes being recessed into pits or otherwise concealed from general view. The method preferred in defensive floodlighting is for the luminaires to be mounted on poles or towers standing a little distance back from the structure so that the light flows towards the critical lower parts of the target surfaces in a substantially horizontal flow. In such an arrangement, there will be minimum shadowing due to features on the face of the building or structure. It also gives better revealment of an intruder who enters the lighted area between the luminaires and the structure, who may cast a shadow on the building face, and an intruder between the building and a distant point of observation will be seen in silhouette.

To aid revealment, it would be preferable for the building faces to be painted white, or at least a very light colour. Various systems of painting bars and stripes to increase visibility do not seem to be effective; the best 'anti-camouflage' treatment is matt white paint.

17.6.3   Floodlighting in zoned sites

The rationale of zoning a site (section 17.9.4) is not only to prevent unauthorized movements from one part of the site to another, but also to make the person easier to detect and apprehend. In situations where there is zoning by an internal 'transparent' fence between the observation point and a floodlight structure, whether the fence is dark coloured or not, it may be impossible for the guard to determine with certainty whether an intruder is situated on the near side or far side of the fence. However, if the floodlights are positioned on the guard's side of the fence and close to the fence (Figure 17.10), it will be found that intruders between the fence and the building will be directly illuminated and seen against the white of the building face (and their shadow may be seen), and intruders on the guard's side of the fence will be seen in silhouette. Thus in each case the location of the intruders can be determined.

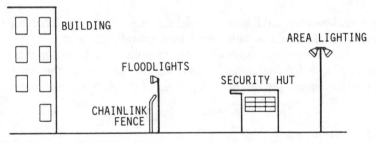

Figure 17.10 *Use of floodlighting of buildings to aid supervision of zoned site.*

If there is lighting on both sides of such an internal fence, it may be arranged to light the areas on either side of the fence by different illuminants, say HP sodium lamps on one side and MBI lamps on the other. The location of a distant intruder may then readily be determined by the colour of lighting under which they are seen. This technique is of value for supervision of adjacent areas zoned for public access and private access respectively as is often the case at airfields, and on those difficult locations where a public footpath runs through the defended site.

17.6.4   Spill-light

A considerable proportion of the light projected onto a light-coloured building or structure is reflected in a diffused manner, and this light flow should be taken into account when planning a security lighting scheme. This effect can raise the ambient illuminance at considerable distances from the building, and may appear as an asset or a disadvantage. At low field illuminances, reflected light from other parts of the installation, spill-light from adjacent premises, light from illuminated signs, and even light shining out of windows can all change the conditions of illuminance and thus affect the visibility of intruders and the concealment of the defenders.

17.7 Topping-up

17.7.1 Within a security lighting installation, there may be areas shadowed by structures, or places into which light cannot penetrate, e.g. under canopies and into deep doorways. Additional lighting is directed into such places to leave no shadowed place where a person might hide, or a vehicle or stolen goods might be placed, or where a crime (such as transferring goods to a vehicle, or forcing a shadowed door or window) could be committed. The process is called *topping up*. Topping-up should not be an afterthought, but the need for it should be identified during the survey (section 17.9.2), and the equipment for it specified.

17.7.2 Techniques

Local topping-up

Examples:

- Placing vandal-resistant bulkhead luminaires within deep shadowed doorways.
- Arranging that under-canopy lights are kept on at night as maintained lighting.
- Where a building is close to the perimeter fence, installing one or more 'glass-brick' windows on the blank side, with lights within the building to illuminate the space between the building and the perimeter.

Remote topping-up

Examples:

- Fixed narrow-angle floodlights aimed at vulnerable parts of plant and under equipment and cranes etc. (This is a useful technique for flame

hazard zones if the floodlights can be positioned in a normal-atmosphere zone.)
- Topping-up areas which are difficult to light from towers or masts, perhaps mounting the luminaires on buildings or structures and thereby saving the cost of a lighting support.
- Fixed narrow-angle spotlights aimed at vulnerable gates.

17.7.3   Emergency topping-up

When the integrity of the perimeter fence has been breached (for example, because a vehicle has crashed through it), or a serious security risk exists on the site (for example, because valuables have had to be brought out of safe buildings because of fire or other emergency), it may be necessary to effect some emergency topping-up, i.e. to install some temporary means of lighting local to the point of need. This may conveniently be achieved by the use of self-powered mobile lighting equipment (chapter 18).

17.8   Guarding of lighted sites

17.8.1   The defenders

The term 'defenders' is used to describe either the ordinary occupants of the premises, or security guards, watchmen or fire patrolmen employed specifically for protecting the premises. Guards may patrol the perimeter either inside the fence, or possibly outside (if there is access). The lighting has to be adapted to suit whichever method is used. Supervision from outside includes observation by passing police patrols, and random observation by neighbours and members of the public.

Mobile security patrols may visit a site without entering it, but it is more usual for them to enter and make detailed checks of vulnerable points. In some cases the mobile patrol enters and searches as routine, with additional random visits for inspection from outside. The timing and number of visits by mobile patrols per night should be random and variable. If there are on-site and mobile patrols, it is desirable that they are able to communicate by portable radios.

17.8.2   Instructions to security guards

The following instructions may be issued to security guards during training.

Instructions to security personnel – working with security lighting

1. Although it may be necessary for you to carry a torch or handlamp, try to make minimum use of it. Shining your torch when you are on patrol tells any hidden intruder exactly where you are and what you are doing. Intruders know that if the beam of your torch does not fall on them, you probably do not know they are there, and they will have the tactical advantage. Further, when you look at the bright patch made by your torch, your eyes will become adapted to that brightness, so it becomes difficult – if not impossible – for you to spot a dark-clothed intruder in the shadows.

2. When in the security hut in the dark hours, use only the essential minimum lighting. This will enable your eyes to adjust to a lower brightness, so that you will be able see out of the windows more easily; and, when you step outside, you will adapt more quickly to the lighting level there.

3. When you go outside at night, give your eyes a few moments to become adapted to the darker conditions, and then you will see far more than if you had switched on your torch. Rely on the security lighting, and do not impatiently start using your torch for ordinary movements about the site.

4. Security lighting works best as a component in a three-part system consisting of:

- the physical defences (fences, locks, bars etc);
- good supervision by the guards; and
- the lighting.

The lighting alone will not defend the premises well, neither will the physical defences if you are not there, and neither can you defend the premises well without the physical defences. At night, with the best physical defences and the most efficient guards, the premises would still be vulnerable if suitable security lighting were not provided.

5. Security lighting helps you defend the premises, its contents and the personnel in three ways:

- by revealing intruders when they approach your defended area;
- in many situations by providing you with concealment behind the glare of the lights; and

- by a combination of these factors, to deter all but the most determined criminal.

6. Darkness is not your ally – it always helps the intruder. In the dark, on a quiet night, you might hear a pin drop at two paces away. But, with a good security lighting installation, you can see a person dropping the pin at 200 paces.
7. At dusk each evening, check the lighting and check your torches. Report any failed lamps. Report also any signs of damage to the security lighting system; criminals sometimes try to sabotage the lighting to aid their attack later.
8. Use the lighting tactically, and only switch on your torch when you really need it. Think about how someone may try to break in. What route would they take? What would they be after? When are they most likely to come? Approach each target area so as to surprise any intruder who may be there. Come from behind the glare of the lights; conceal yourself behind the lights; give the intruder all the disadvantages of your cunning tactics. The lights are there to help you.
9. Patrol like a fox – never going by quite the same route twice. Vary the times of your patrolling; vary the number of patrols over each part of your patch each night; come to each vulnerable area from a different direction each time. Stand still and listen for a few moments; listen and watch. Imagine how the lights would be affecting you if you were an intruder.
10. Security lighting must be on all night, every night of the year. If it gets switched off during the dark hours, it is not security lighting. Do not try to save a little electricity by switching-off security lighting – it is a well-designed system that uses energy wisely and for a good purpose.

© *S.Lyons 1991 The foregoing 'Instructions to Security Personnel – Working with Security Lighting' may be photocopied or reproduced without fee for use in training security staff.*

17.9   Security lighting surveys and scheme planning

17.9.1   Specification

All the guidelines on the specification and procurement of a lighting installation given in chapter 24 apply to security lighting schemes. If a security consultant is involved, the specification for the lighting should be developed in liaison with him, and early consultation with the insurers could be invaluable. The specifier should know what are the threats (e.g. theft of goods, materials or cash; political extremist attacks; civil unrest; employee theft or collusion; break-in through the perimeter fence; vandalism; fly-tipping etc.) Then the degree of risk should be assessed[59]. Such considerations will enable an outline lighting specification and final lighting specification (section 24.2) to be compiled. The design criteria must include consideration of the possible strategies for defence; for example, whether perimeter glare-lighting to give complete concealment of the patrolling guards (section 17.3) is possible or required.

In the case of green-field sites, early collaboration with the architect, and cross-briefings between the lighting engineer, security consultant and the architect are of great value.

17.9.2 Surveys

Surveys of premises for security lighting are usually best carried out in collaboration between the engineer and the senior person in the client organization responsible for security. Consultation with the Crime Prevention Officer of the police may yield valuable advice about the local conditions and risks relating to adjacent premises.

During the survey, all the usual physical constraints and legal constraints (section 24.1) must be recognized. The survey should be carried out along the lines set out in section 15.5, with attention to the following additional matters:

- Obtain a scale plan (preferred scale 1:500). Ensure it records all structures and all other lighting. Note the rise and fall of the land, especially at the perimeter. Note the nature of the 'surveyed field' outside the perimeter, and what risks may be associated with each part, e.g. overhanging trees, vagrants camping nearby, access to the fence by public and private roads or footpaths and proximity of lighted roads and lighted premises. Consider the lighting in relation to possible hazard or nuisance to railways, to road users, neighbouring properties and their activities. Study the approaches to the entrances, and consider the design of the lighting, fencing and gating as one entity.
- Consider the nature and suitability of all perimeter fences. Note their height, strength, condition. If mesh fences, are they fitted with barbed wire on top, is a strong top straining wire fitted? Is the bottom edge cemented down? Does the fence comply generally with an appropriate specification? Is it correctly installed and properly maintained? Note whether mesh or palisade staves have been galvanized, or have been treated to reduce inward reflection. Note colour of support posts.
- Put yourself in the position of someone who wants to break into the

premises. Consider how you would get goods out if you were a dishonest employee.

- Go right round the perimeter, inspecting every part of it from inside; then again from outside if possible. These inspections should be carried out first by day, and then repeated at night.
- Consider the sight-lines of the security guards from various vantage points, and along any perimeter patrol path. Start to visualize the strategies of defence in terms of revealment and concealment.
- Consider any temporary risks that might occur: material piled up near the fence that may provide cover; cranes parked near fences; vehicle access to remote parts of the perimeter; pools and streams which may become frozen in winter and give easy access; foliage in summer that will provide concealment for an intruder and obstruct the view of the guards.

Treat all this information as highly confidential, since it would be invaluable to someone wishing to commit a crime against the premises.

17.9.3 Planning

It is not possible to stipulate how schemes should be compiled, for there are wide differences between sites as to topography, threats and risks.

A problem area may be a site with railway sidings. The spaces under rolling-stock tend to be deeply shadowed, affording a hiding place for intruders, and somewhere to cache stolen goods etc. The use of lower mounting heights for area lighting luminaires (e.g. 8–14 m) will tend to reduce this shadowing, but when trains are parked on adjacent tracks it is very difficult to get light under them; the undersides of carriages and waggons are dark, and the ground clearance is only 500–700 mm. Some improvement in visibility is obtained by spraying the sleepers and track ballast with white paint or limewash. At really critical positions, well-glass or bulkhead luminaires may be mounted on trackside posts 400–500 mm high. Such will permit the security guard to inspect under trains as routine, and will have a strong deterrent effect.

Where there is a perimeter roadway, this might be lit in a way that will contribute to security. A light-coloured concrete road surface, if suitably illuminated, can provide a luminous background against which an intruder in an unlit area can be seen in silhouette.

17.9.4 Additional fences

There may be requirements for additional fences to be used within the area enclosed by the perimeter fence, and special lighting for these may be required. Typical applications are where the view of the perimeter is obscured, for pounds, for zoning and for citadels.

Obscured view of perimeter

If any kind of obstruction near the perimeter prevents the guard seeing part of the perimeter fence from within the site, an additional fence on the inward side of the obstruction may be required, with both the inner and outer fence zones suitably illuminated. As an example, a local authority required the occupier of a site to erect a landscaped tree-planted mound within the site perimeter, the objective being to hide some plant from being seen from the nearby village. The mound prevented a length of perimeter fence from being supervised from within the site, making the site vulnerable to penetration. Security was restored by erecting a further length of fence on the inner side of the visual obstruction (Figure 17.11).

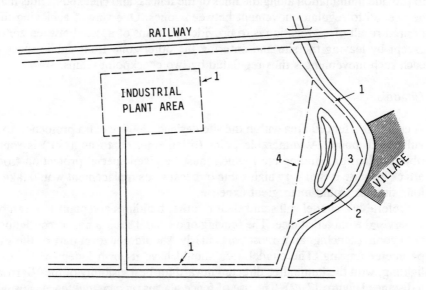

Figure 17.11 *Additional fencing used to protect a site following loss of visibility due to a landscaped mound being placed inside the perimeter on the instructions of the planning authority with the object of hiding the industrial plant from the village. 1, original fence lines; 2, new mound; 3, area now concealed from view and vulnerable to attack; 4, additional fence to prevent entry through the concealed area.*

Pounds

These are quarantine areas in which vehicles may be parked and kept secure and under surveillance by the security staff. For example, a vehicle which arrived outside normal working hours would be placed in the pound if it was carrying a valuable load. Pounds must be securely fenced using 'transparent' fencing (section 17.2.3), and provided with lighting to enable a high standard of supervision.

Zoning

On large sites, thieving and employee crime may be controlled by dividing the site into a number of zones by internal fences, movement of persons and vehicles between zones being regulated by the security staff. Various kinds of frauds and thefts involving the company's own transport and visiting transport can be prevented by requiring that a visiting vehicle shall leave by the gate through which it entered, and if visiting and own company transport are kept apart. Each zone that is visited by road transport should have a weighbridge. Even if the general areas of the zones are not lit at night, it will be necessary to provide illumination along the lines of the fences, and checkpoint huts may be needed to regulate movement between zones. One way of achieving the desired result of segregation is to provide no means of access between zones except by leaving the site and entering the other zone from public space – each such movement is thus regulated by two checkpoint teams.

Citadels

A citadel is a fenced area within the site designed to give extra protection to a vulnerable point. A vulnerable point (often referred to as a VP) is some object (plant or building etc.) which must be given special protection from attack because it is of very high value or because its replacement would take a long time or would incur great expense.

Preferably a citadel will stand clear of other buildings and plant so it can be supervised from a distance. The fencing of a citadel should not be contiguous with zoning fencing, and, most particularly, should not form part of the site perimeter fencing. The citadel area should have its own independent area lighting, with the luminaires directed inward for best visibility of the VP from a distance (Figure 17.12). The use of fence alarms on a citadel fence is usual. It is not unusual for a large site to have more than one citadel.

(The tactical principle of the citadel is also employed inside buildings, particularly in cash-handling areas. If there is an inner area to which threatened staff can retreat during a violent attack, this is termed a *redoubt*.)

17.9.5   Contingency plans

It would be wise to make a contingency plan to deal with any special security risks which may occur. For example, considering how to deal with an urgent

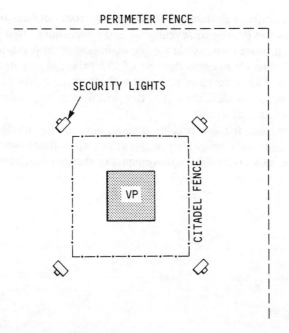

Figure 17.12 *Lighting for a 'citadel'.*

need for mobile lighting (chapter 18). At the level of modest expense, on larger sites it would be wise to provide one 'jennylight' to provide temporary topping-up lighting at any point should there be local equipment failure or a problem to be dealt with (section 17.7.3).

17.9.6   After the installation is completed

When the security lighting has been installed and is operational, the following should be carried out:

● Make a further complete inspection by day and then again by night to ensure that the lighting will provide the security that is planned. Again, go right round the perimeter, inspecting every part of it from inside, then from outside if possible. The security consultant and the Crime Prevention Officer of the police should be invited to join in with this inspection.
● If any weakness, dark patches, places of concealment etc are found, take steps to remedy them at once, perhaps adding additional topping-up lighting or adding further floodlights to towers etc.
● Ensure that the security guards are trained in how to keep surveillance of the premises in conjunction with the security lighting (section 17.8.2). Let

them experience the effects of glare-lighting from outside, and confirm for themselves the degree of revealment and concealment that is provided.

- Ensure that those responsible for site management appreciate that, if there are any future changes to the use of the premises, changes outside the fence, or any later development that will affect the lighting, the fencing or the guarding, modification to the installation to suit the changed circumstances is essential.
- Make provision for the regular maintenance of the installation and, in particular, for bulk relamping at the appropriate times (section 23.4).
- Notify the insurers of the improvements to security that have been carried out.

18 Exterior mobile and portable lighting

18.1 Applications

18.1.1 Routine and emergency use

Typical applications of the types of exterior mobile and portable lighting equipment described in section 18.2 include the following:

- routine use to provide light at work locations where permanently installed lighting would be impracticable, for example on building and construction sites (section 16.1), and at quarries and mines (section 16.2).
- use in any emergency where other lighting is not available, including use during failure of mains power to indoor or outdoor lighting (chapter 14).
- for extended testing of emergency lighting, particularly in premises that are continuously occupied (section 14.4.3).
- in security lighting installations, in which mobile lighting may be placed in position to provide additional defence, e.g. if a breach in the perimeter fence is found, or in dynamic situations e.g. any temporary or unusual security risk or incident for which extra lighting would be a help.

18.1.2 Power supplies

As described in section 18.2, many items of mobile or portable lighting equipment incorporate means of generating power from an alternator driven by a diesel or petrol engine. Such power supplies may be used to supply other lighting equipment (including that used indoors), and also to provide supplies for power hand-tools etc. (section 22.5.4). Where a mains power supply is available, this may be used in place of the inbuilt generator-set in the mobile equipment allowing operation of the lighting without running the engine. This may be done to conserve fuel, to comply with environmental constraints, or to operate in silence (e.g. on sites close to residences or a hospital).

In order that the above conditions may be accommodated, it is recommended that inbuilt generators of mobile and portable lighting units be able to generate at the same voltage as the public supply in the country of use, and should have an integral double-wound step-down transformer to 110 V (future regulations will require 100 V) having its secondary winding solidly centre-tapped to earth (section 16.1.4). This will provide a supply for task-lights etc normally within hand-reach height of operatives, and will also ensure that power can be taken from a mains supply when needed and available.

Important: It is recommended that all mobile and portable lighting units should have all their non-voltage metalwork bonded, and that an earth terminal be provided to which a local earth-spike can be connected[18, 27]. Note that electrical equipment and prime movers may only be taken into designated hazardous zones if constructed in compliance with the protection required for such Zones (chapter 7).

18.2   Types of equipment

Any of the types of equipment described in this section may be specially made for, or adapted to, special conditions of use, e.g. for use in the tropics or in cold climates, at high altitudes, or in the presence of dusts etc.

18.2.1   Portable task-lights

These generally consist of a folding tripod assembly, operating at 110 V (future regulations will require 100 V) derived from a mains supply or from a local generator. Intended primarily for outdoor use, they may also be of value indoors (section 13.1). They are usually fitted with spiked feet, but may be fitted with plates for use on soft ground.

Portable task-lights may carry one or two luminaires containing compact-fluorescent lamps, with the facility to raise the luminaires to around 2 m above floor level and direct the light to where it is needed on the task. Larger units carry say, 2 × 500 W linear tungsten–halogen lamps for lighting larger task areas.

A portable and mechanically adaptable task-light bearing an enclosed fluorescent-lamp (1200 mm or 1500 mm) luminaire is very useful, serving as a 'plasterers lamp' and for illuminating other tasks requiring a flat surface (screeding, terrazzo etc). Bringing the lamp parallel with the surface being worked reveals any undulations.

18.2.2   Portable generator-sets

Small petrol-engine-driven alternator sets, mounted within a tubular frame for convenient handling, are used to power portable task-lights (section 18.2.1).

18.2.3 Jenny-lights

These are small petrol-engine-driven generator sets similar to those described in section 18.2.2, but combined with an extendible mast typically carrying 2 x 500 W tungsten–halogen lamps at 5 m mounting height (Figure 18.1). Such units can be carried by two people, and when folded down are small enough to be stowed in the boot of a car.

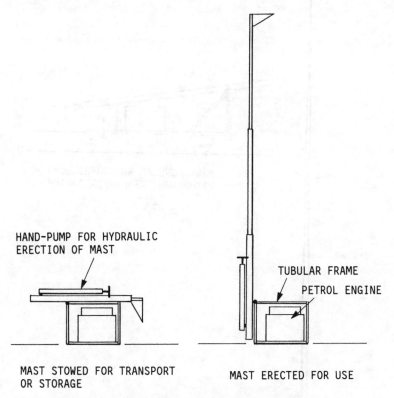

HAND-PUMP FOR HYDRAULIC
ERECTION OF MAST

TUBULAR FRAME

PETROL ENGINE

MAST STOWED FOR TRANSPORT
OR STORAGE

MAST ERECTED FOR USE

Figure 18.1 *A typical 'jennylight' (portable self-contained lighting tower), extending to about 5-m, and carrying 2 × 500-W T–H lamp luminaires. Powered by a petrol engine. Can be carried by two people.*

18.2.4 Trailer-lights

These are mobile lighting towers, powered by a diesel or petrol engine (Figure 18.2). The engine and mast etc are mounted on a trailer for towing. In use, extendible outriggers are employed to provide stability. The mast may be hand-wound up to height, or wound up by a power winch driven by the engine. Typical models extend to 8 m or 19 m. The mast may carry typically 4

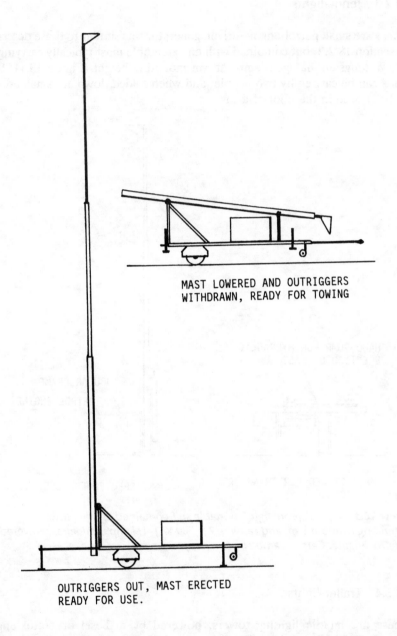

MAST LOWERED AND OUTRIGGERS
WITHDRAWN, READY FOR TOWING

OUTRIGGERS OUT, MAST ERECTED
READY FOR USE.

Figure 18.2 *Typical mobile lighting tower (trailer-light), powered by a diesel or petrol engine driving alternator. Mast is erected by hand-wound or power winch. Typical models erect to 8-m or 19-m. Mast may carry 4 × 1-kW SON or 4 × 750-W MBFR. Can be powered from mains if available.*

× 1 kW SON lamps, but some have greater lighting loads. Note that these units can provide power for other lighting or hand-tools. For silent running, or for continuous operation without having to refuel, trailer lights can be powered from a mains supply if available.

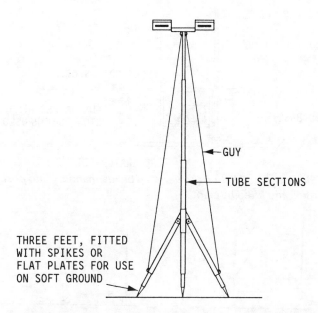

GUY

TUBE SECTIONS

THREE FEET, FITTED
WITH SPIKES OR
FLAT PLATES FOR USE
ON SOFT GROUND

Figure 18.3 *Typical tripod unit. To carry up to 4 × 1-kW T–H lamps. Can be erected by one person; small enough to be carried in boot of a car. Powered by local generator set or mains.*

18.2.5   Portable demountable towers

These units can be an economical form of temporary lighting for situations such as constructions sites, for they can be brought to the point of use, and after use dismantled and returned to store or transferred elsewhere. Portable tripod towers, with heights up to around 6 m, may be braced with steel guys (Figure 18.3). They can be carried in the boot of a car when dismantled, and are capable of being erected by one person. Such portable towers may be powered by a separate generator set or from the mains supply. Greater resistance to wind can be achieved by use of additional guys secured by ground screw anchors.

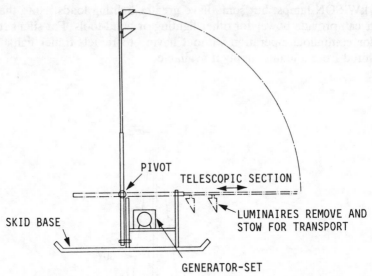

Figure 18.4 *Skid-mounted lighting tower. Can be manhandled into position. Is moved by truck, and lifted by crane.*

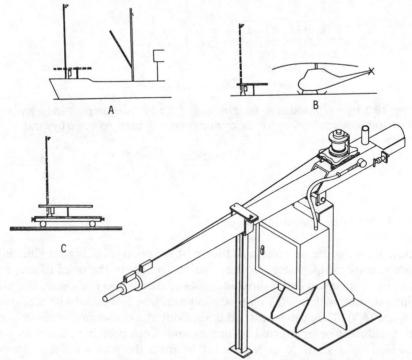

Figure 18.5 *Typical demountable 8-m tower for such applications as (A) on board ship, (B) at helicopter bases, (C) mounted on railway flat-car.*

18.2.6   Skid-mounted towers

These units may be equipped with a generator set to supply the luminaires at the tower head. Those of simple construction have a twofold hand-erected 5 m mast carrying say, 4 × 1 kw tungsten–halogen floods. Others, with a telescopic mast of 8 m or higher, may carry a similar head load to a trailerlight (Figure 18.4). Skid-mounted lighting towers are popular with civil engineering contractors, for they are rugged and can be dragged about the site by tractor. A crane is usually available on sites to lift the unit off the transport. Skid-mounted equipment is very suitable for quarries (section 16.2) for skids are not easily overturned, and have no tyres to be punctured by sharp rocks.

18.2.7   Demountable towers

These units are towers with telescopic masts of 8–10 m of the same type as used on trailerlights, but provided with a base-plate which can be secured to the ground, or, for example to a railway flat-car, or the deck of a ship, raft or lighter (Figure 18.5). This type of equipment is also used on pontoons, rafts or lighters during dredging and engineering operations in the construction of dams etc.

19 Exterior emergency and pilot lighting

19.1 Legal responsibilities of occupier

The *Health and Safety at Work Act*, 1974[42] and other legislation[45] impose duties on occupiers of premises which go far beyond the basic common law duty to provide for the reasonable safety of employees and other persons who enter their premises. The word 'premises' in that Act, and in the common law, does not mean only the buildings; it also includes all exterior spaces. In other words, the occupier has a clear legal duty to provide 'sufficient and suitable' lighting over the whole of his premises, including all outdoor areas accessible to the occupants.

Similarly, the *Fire Precautions Act*, 1971[51] requires that the means of escape to a place of safety 'shall be capable of use at all material times' – this phrase being construed to mean that sufficient and suitable lighting for escape shall be provided, and this surely should apply to outdoor areas. It is often wrongly assumed that, in an emergency, persons will be safe if they can find their way out of the building. However, outdoor areas may contain many kinds of hazards, and the need for 'lighting for escape' may extend a considerable distance from any building.

An outdoor industrial area may be a dangerous place. It may, for example, contain chemical storage vats or gas cylinders which could cause injury to persons should there be a fire or an accident with a vehicle or a crane. In such circumstances, people in the vicinity will need to be able to see so they can escape to a place of safety, even under mains-failure conditions at night. Only when lighting is provided that satisfies this need will the occupier's duty to provide for the reasonable safety of persons be discharged.

19.1.1 Current practices

Outdoor lighting installations at industrial premises in the UK rarely incorporate proper facilities for personnel safety in the event of mains failure. In general, industry does not provide lighting levels out of doors that meet the minimum recommendations of the CIBSE for similar tasks performed indoors. Outdoor emergency lighting to facilitate escape is rarely provided, and, when it is provided, it may not be fully effective.

It is clearly a matter of great importance to provide sufficient illuminance under mains-failure conditions to enable a person to move safely across an industrial yard cluttered with potentially hazardous obstructions on a dark night.

19.1.2 Possible future practices

The recommendations of *BS 5266*[47], which are discussed in chapter 14, have little relevance to the need for providing lighting for escape in outdoor areas. For example, if a minimum illuminance level for emergency lighting of 0.2 lx is needed to enable a person to move along a well defined corridor with a level floor, it is apparent that a rather higher illuminance is necessary to enable the person to descend safely in darkness down a series of ladders from a partly constructed building when the lighting has failed.

It has long been the opinion of the author that there is a need for a new standard for outdoor emergency lighting, and in 1980 he published some proposals for this, having in mind the special dangers on construction sites, in chemical plants, and in other hazardous outdoor areas. The subject is still a neglected one, but updated detailed proposals will be set out in a new book, *Handbook of Emergency Lighting*[48]. It is hoped that the law will be amended to make it clear that occupiers have a duty to provide emergency lighting out of doors. Further, it is hoped that suitable standards for exterior emergency lighting will be imposed in due course by EC legislation.

19.2 Practical steps

19.2.1 Pilot lighting

It is recognized that it may be impracticable and extremely costly to try to apply *BS 5266* standards of emergency lighting to large outdoor industrial areas. However, emergency lighting at many such locations might be achieved by installing a system of *pilot lighting*, i.e. a system of strategically placed luminaires which would act as markers under conditions of failure of the normal lighting.

When the normal lighting is not operating, pilot lighting would enable a person to safely traverse a series of clearly marked straight paths through the site which are intended for fire/security patrols, or enable the person to move to an exit or designated place of safety. If such pilot lighting systems were supplied from an uninterruptable electrical supply, a considerable increase in safety over present practices would be obtained. In hazardous areas, the pilot lighting luminaires would be of a suitable protected design.

19.2.2 Escape route marking

For marking escape routes and denoting exits in outdoor working environments (and especially in hazardous ones) the use of intrinsically safe phosphorescent markings and exit signs has been reported (section 5.13), and this practice has much to recommend it. Particularly in places where the public may be present, the use of luminous handrails or ballustrades for guidance along the escape path in darkness and smoke should be considered (section 14.4.7).

19.2.3 Personal lights

It is not practicable in many situations for people to carry with them a personal light, i.e. a torch or handlamp, such equipment being easily lost or broken, and liable to be stolen. It is also difficult to ensure that the batteries are kept at a good level of charge. However, it is perfectly feasible for persons to carry with them one or two chemical light sticks (section 5.13.1).

20 Airfield and helipad lighting

The information in this chapter on lighting for temporary landing strips and helipads will be of value for remote civil engineering sites and for operations in developing countries. The chapter does not deal with conventional fixed lighting installations for airports which must be installed to international standards.[4] The special requirements for security lighting and the control of dynamic situations (section 20.4), as well as the possible need to cope with crashed aircraft or aircraft on fire at night, may necessitate the availability of mobile lighting which can rapidly be brought to where it is needed (chapter 18).

20.1 Objectives and constraints

20.1.1 Lighting requirements for airfields may present conflicting demands on the designer. While the lighting must provide good visual conditions for technical work and for safe movement of vehicles, it must also suit the needs of pedestrians. There may also be a stringent requirement for security lighting to prevent intrusion and to control persons within the airfield. All this must be achieved without detracting from the required visual conditions for pilots taking off and landing, as well as for ground movements of aircraft.

For example, a line of perimeter luminaires or roadlighting lanterns, viewed from the air in mist, could be mistaken for the flarepath. The pilot may quickly correct this error when he picks up other visual information or fails to find the end-of-runway markers, but such misleading aerial pictures add to the fatigue and risks of the pilot's job. To avoid creating a misleading appearance when viewed from the air, cut-off luminaires may be used, e.g. type 'F' floodlights (section 6.4), or hoods may be fitted to luminaires to prevent light being projected above the horizontal. (The advice of roadlighting experts should be sought on the choice of roadlighting lanterns.) When lighting ground that has a gradient, always orientate the luminaires so that the light flows away from the lower ground towards the upper ground, so that distant low-flying aircraft will not be in the glare zone.

20.1.2 Airfield lighting is not viewed from a single height. Not only must the visual pattern be considered from the pilot's viewpoint, but note also taken of variations in the ground level and the eye-levels of occupants. For example, consider the difference in eye-levels of pedestrians, car drivers, drivers of heavy vehicles and fuel tankers, and pilots of taxiing aircraft. Consider also the view of the flight controllers who must be able to see all that is occurring without suffering glare or unnecessary visual fatigue.

Considering the factors presented so far, at least the following criteria for airfield lighting can be identified:

- safety of airborne aircraft;
- safety of aircraft moving on the ground, and taking off or landing;
- safety of vehicles moving on the airfield, roadways and dispersal areas;
- suitable visual conditions for the performance of work on the aprons and at dispersal areas;
- suitable visual conditions for flight controllers;
- suitable visual conditions to meet the security requirements (section 20.4).

Additionally, consideration must be given to possible exceptional situations, such as crashed aircraft or terrorist attack, as well as power failure causing loss of mains supply to normal lighting (chapter 19 and section 22.5).

20.1.3   There are constraints on lighting peculiar to airports. For example (giving the latest data available at the time of writing), the permitted height of a lighting column or tower depends upon its distance from the centreline of the nearest runway. Thus towers of 18.5 m should not be nearer than 311 m, and towers of 30.5 m, not nearer than 396 m to the runway centreline. A requirement for strict cut-off of luminaire beams may be imposed, with a limit set on the tilt angle according to the projected brightness. For latest data, refer to the current edition of the publication *Licencing of Aerodromes*.[4] A useful update on airport lighting technology is provided by two articles in *Electrical Design*.[28, 29]

20.2   Lighting of hangar precincts and aprons

20.2.1   The lighting requirements within hangars are similar to those in other industrial interiors, i.e. an illuminance of not less than 300 lx general lighting is usually required, this to be augmented by local lighting or task lighting (section 9.4). Where the tasks merit it, the general illuminance may be as high as 750 lx.

20.2.2   When pushing or towing an aircraft out of the hangar at night, the driver and crew would be subject to a steep gradient of illuminance if emerging onto an apron or precinct which was unlit or lit to only a very low illuminance. Conversely, in the case of hangars without windows or with limited fenestration, there could be adaptation problems on coming out into bright sunlight, or on entering the hangar from bright sunlight. In each case it is best if the subject passes through a *zone of intermediate illuminance* (section 9.7), which preferably should be 25% of the higher illuminance. If the subject remains in this intermediate illuminance for about half a minute, reasonable adaptation will be achieved so that not too much difficulty will be experienced (section 1.4). The required conditions for day and night can simply be covered by suitable switching of luminaires (section 22.4).

20.2.3 To reduce the number of ground obstructions, apron lighting luminaires may be mounted on the hangars. These produce only an outward flow of light, and this can lead to excessive contrasts and shadowing of the sides of aircraft and objects facing away from the hangars. It may therefore be necessary to position some lighting supports at the outer edge of the apron to direct light towards the hangar, though it is desirable to avoid having such obstructions if possible. Luminaires mounted on the hangar front must be placed above the doors which may be 20 m or more high. Those mounted on the side walls may be lower (but when the doors are opened beyond the boundaries of the building, the light towards the apron will be obstructed).

20.2.4 An average illuminance of 20 lx will be adequate for movements of vehicles and aircraft, while at least 100 lx will be needed for even simple servicing operations, e.g. refuelling. Engineering tasks require 300–500 lx at the point of work. A good method is to light the area to say, 20 lx–100 lx from hangar-mounted luminaires, and then to boost the lighting at task areas by means of mobile (wheeled) lighting towers which may be mains-powered or powered by local generators. Lighting towers and task lights may be used, both in the hangar and externally. In the latter case they may be self-powered from their own generating sets (chapter 18).

   If they can be sited at a suitable distance from the runway, a small number of lighting towers of 30 m height or higher may be used, each carrying as many as 30 luminaires mounted on a circular gallery which can be lowered by an integral power winch for servicing. The luminaires may comprise wide-angle floodlights for general area lighting and security lighting, as well as accurately-aligned narrow-angle floodlights aimed to illuminate distant areas (section 20.1.3).

20.2.5 When no work is being done on the aprons, the exterior lighting level can be reduced to typical roadlighting level (5–10 lx) for pedestrian movement, or to 0.5–5 lx for security purposes only (chapter 17).

20.3 Roadways and car-parks

20.3.1 The requirements for lighting of roadways and car-parks within an airfield complex do not differ greatly from those for other similar locations except for the constraint on upward light (section 20.1.3). Low-pressure sodium-vapour lamps (section 5.9) should not be used for lighting car-parks because of the difficulty people have in identifying their cars under lighting devoid of colour rendering.

20.3.2   As for other outdoor parking areas, an illuminance of 10–50 lx will suffice for safe movement by the public and for supervisory purposes. Luminaires mounted at around 5–8 m will enable some penetration of light into the vehicles; higher mounting will tend to light the car roofs rather than their interiors.

20.4   Security lighting on airfields

20.4.1   The guidelines given in chapter 17 are generally applicable to airfields, but certain constraints must be observed (section 20.1.3). Where perimeter lighting is to be placed along a fence line near a taxiing path, luminaires should be mounted on frangible poles made of glass-fibre or wood which will readily collapse if struck by an aircraft. It may be necessary to restrict the height of such luminaires, and mounting no higher than 1.2 m may be appropriate.

20.4.2   At airfields handling international flights, the need for security lighting will be heightened by the need to defend against illegal entry and departure, hijacking, smuggling etc. The nature of the lighting at embarking/disembarking points will depend on the degree of sophistication of the airfield, and will support any system of providing 'locks' at checkpoints. Movements of passengers and cargo to and from aircraft are fraught with security risks, and the security lighting should prevent any areas of darkness that could conceal the movement of persons or contact between individuals across the customs line. The lighting should obviate any opportunity for parcels or luggage to be dumped for picking up later for illegal passage across the customs line. Glare-lighting may be used to enable airport officials and immigration officers to observe staff and passengers without themselves being readily seen.

20.4.3   Lighting at dispersal points and outdoor servicing bays should include security lighting to operate when work on aircraft is not in progress. A possible method is to provide area lighting at 5–30 lx according to the district brightness, and position the security guards' patrol path outside the lighted areas. Thus, an intruder would have to step into the lighted area to interfere with a parked aircraft, unless mounting an attack from a distance, for example, with firearms.

20.4.4 Because of the length of the perimeter of a typical airfield, perimeter defence can be difficult and costly. Closed circuit television may be used to aid the patrolling. A combination of infrared transmitters and cctv cameras sensitive to infrared, or an electronic intruder detection system, may be used, but use of such equipment does not remove the need for patrolling and the provision of suitable lighting for security purposes (chapter 17).

20.5 Dispersal points

20.5.1 Operational requirements vary greatly according to the country, the size of airfield and the types of aircraft concerned. However, on permanent airfields with paved dispersed aircraft parking areas, it is common for the lighting to provide several levels of illuminance. For example, the specification may call for 20–50 lx when work is being performed (to be augmented with task-lights), 10 lx for aircraft movement, and perhaps 5 lx for security only. Because of the distances involved, and the time it takes to get from one group of dispersal points to another, local switching of the lighting would be costly and time-consuming. What is needed is a system whereby lights can be switched remotely from the control tower or other control centre in advance of the aircraft's arrival at the dispersal point. This may be arranged by relay switching or mains-borne signalling switching – methods that avoid having to bring heavy cables and large switches into the control point.

A system with much to recommend it uses microprocessor control. All commands to the relay-operated switchgear adjacent to each dispersal are transmitted either over a single-pair telephone cable loop or by coded pulses transmitted via the mains cabling. In certain conditions, remote switching by radio control might be justified. By such means, a single luminaire, or the lighting for a single dispersal point or a group of dispersal points can be switched from a control point which may be up to 6 km away. The control panel will have means of selecting locations and the lighting levels to be provided according to need at the time, and a VDU display can indicate the state of the lighting at every point. Closed circuit television may be used for security supervision of unguarded dispersal points.

20.5.2 At less sophisticated airfields and at temporary airfields operating with the minimum of equipment, there may be no permanent lighting installation (section 20.6). Taxi-path lighting can be provided by battery-operated aviation ground lights (section 20.6), while lighting for servicing bays can be provided from generator sets powering portable lighting towers and task-lights, or by trailer-lights. At temporary airfields and helipad sites without facilities, it is possible for an aircraft to fly in with its own portable lighting, perhaps in the form of a jennylight (chapter 18).

20.6   Temporary flarepaths

20.6.1   At airfields that will be used for a limited duration, or where capital resources are limited, a safe and workable runway lighting system can be provided with portable equipment. The runway is marked out with portable aviation ground lights. Typically these contain a cold-cathode lamp, and are available in five interchangeable colours. The ground lights may be operated from their internal batteries or may be connected by cable to an electrical supply. Alternatively, the internal batteries of the ground lights can be kept in a good state of charge by a cable connection, and the charging cable can be the means of switching the lights on and off remotely. Other methods of control include photocell switching and radio control. In the latter case, control can be exercised from the air at unattended airfields.

20.6.2   In typical use, a runway is marked out by ground lights spaced at 60 m along its sides, with coloured ground lights arranged in transverse rows with a 6 m spacing to mark out the threshold, touchdown area and runway end. Runway dimensions vary according to the types of aircraft in use, the operating conditions and the altitude, being typically in the range 2–3 km in length, down to about 1 km for light aircraft. Coloured ground lights are used to mark out the taxi paths, spaced at 20–60 m along straight paths, and arranged on the outer side of bends at a spacing of 10–20 m.

20.6.3   The visual approach aid may be a portable battery-operated or mains-operated unit which gives a two-colour *angle-of-approach* indication. The airfield will also probably be fitted with an *identification light beacon* with a 30 km visibility, flashing a two-character Morse code signal. Additionally a radio beacon may be in use, and this, with the light beacon, will be operated from a mains supply if available, or from a generator set. Again, the beacons can be switched from the air at unattended airfields.

Airfield lighting equipment such as that mentioned in this section is provided by specialist suppliers such as Metalline Holdings Ltd (appendix G).

20.7   Helipad lighting

20.7.1   Conventional helipads for helicopters are marked in the form of an 'H' fitting into a square of side 60 m, the strokes of the 'H' being 20 m wide (Figure 20.1). Although the 'H' may be marked out using twelve runway marker ground lights (section 20.6), this may not be adequate in adverse conditions of mist or fog, smoke, snow, sandstorms, etc. when additional area lighting may need to be provided. One method of arranging this is to site two

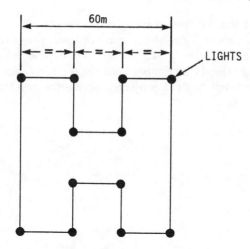

Figure 20.1 *Marking out a helicopter landing pad with portable aviation ground lights.*

8-m lighting towers at opposite corners, 15 m out on the diagonal line, each carrying two 1000-W SON-T asymmetrical floodlight luminaires (section 6.4) to produce a ground illuminance of around 30 lx over the 'H' area. An alternative method involves the use of small floodlights at a spacing of 20 m around the outside of the square enclosing the 'H', the floodlights being mounted on frangible masts of glass fibre or wood at a height of 1.2 m.

27.7.2   Under emergency conditions an improvized helipad with a smaller 'H' may have to suffice. The chosen position should be reasonably level, as far as possible from trees or buildings, free from large stones, brushwood etc., and be at least 20 m × 20 m and preferably up to 60 m × 60 m. A large clear 'H' should be marked out at the centre of the space chosen using, in the absence of anything better, sand, soot, small stones etc. The width of the strokes of the 'H' should be one-third of their length.

At one corner of the 'H', and spaced 50 m from the centre of the 'H' along a diagonal line, should be placed an improvized flag or wind-sock. This can be made from any available cloth etc., and should be around 0.5–1 m wide, and 3–4m long. If possible, this should be fixed to a pole about 5–6m high.

At night, if no other lighting is available, the 'H' can be illuminated with vehicle headlamps. The vehicles should be positioned well back from the 'H', say along an imaginary square of side at least 40 m whatever the size of 'H' marked. If they are available, four jennylights (chapter 18), each with two tungsten-halogen lamps of 500 W or 1000 W power, should be positioned, one at each corner of the 'H', and spaced 50 m from the centre of the 'H' along diagonal lines. If only two jennylights are available, they should be placed at opposite corners, with the windsock at one of the other corners. The objective is to light the 'H' so the pilot can see it clearly, but at the same time ensuring that no bright light is projected above the horizontal which could dazzle the pilot.

Part 6

Management of lighting

21 The cost benefits of good lighting

21.1 Lighting, productivity and quality of work

21.1.1 An important theme of this book is that the process of vision is aided by the provision of good lighting, and that this will beneficially affect worker performance. Generally, improvements in lighting tend to produce improvements in quantity or quality of output, or both. Within limits, this 'productivity effect' is related to the quantity of illumination up to certain optimum levels, as has been demonstrated by tests and measurements made in factories and offices. Better lighting also contributes to the wellbeing of the employees, and may result in even greater savings due to their better motivation, or to reductions in absenteeism and staff turnover (section 21.1.8).

21.1.2 Continual activity causes tiredness, and visual activity is no exception. Visual fatigue is of more rapid onset if the illuminance provided is insufficient, or if there is a significant degree of glare. A task which must be done for only a short time may be carried out with reasonable efficiency in a poor light, but, as the duration of the task is extended, visual performance declines, and the onset of fatigue is more rapid than would have occurred in 'sufficient and suitable' lighting. Thus, recommendations for the 'standard design illuminance' for a task may be weighted to arrive at the 'design service illuminance' to allow for the duration of the task and its dangers (appendix B).

21.1.3 Productivity is the beneficial result of applying resources (manpower, machines, materials, energy, capital, plant, buildings and land) to the achievement of an objective. The prime objective of management in organizations should be to maintain and improve productivity. An increase in volume of output may not be a reliable index, for such an increase might be obtained at higher operating costs or the goods or service produced might be of lower quality; nor would an increase in profitability be an acceptable index of productivity if the gain in productivity were obtained by imposing greater risks or discomforts upon the workers.

21.1.4 Good lighting can contribute to the achievement of true productivity. Studies have shown that if work is performed under poor lighting, the volume and/or quality of work performed are lower than might be achieved with the same labour force and resources but in optimum lighting conditions. It is well known that the disadvantage of adverse lighting conditions can be partially or temporarily overcome by dint of special effort by the workers, but that this cannot be maintained because of stress and the early onset of fatigue. In the past, the granting of short rest periods has been used as a substitute for improving the working conditions, and this led to bargains in which management agreed to pay the operatives 'relaxation allowances', i.e payment in lieu of rest periods, thus actually paying the operatives extra for working in a poor light!

21.1.5   It has been the consistent view of all observers that it is economically sound to provide lighting of suitable quantity and quality in any workplace, e.g. to the standards of the current edition of the *CIBSE Code*.[1] The cost of such lighting is justified by the expectation of greater output and better quality of work when the handicap of insufficient or unsuitable lighting is removed.

21.1.6   Where very poor lighting has been upgraded to modern standards, increases in performance of work, and decreases in defects in work, have been widely reported. It is true that the very marked improvement that occurs in the first few days after lighting improvement may not be maintained (the *Hawthorne Effect*), but generally the maintained improvement in perform- ance and quality following that initial period is at a level high enough to entirely justify the investment in better lighting.[5] In some cases the payback period has been a matter of a few months – quite apart from any savings in cost due to the employment of more efficacious lamps and luminaires.

21.1.7   In the design of new buildings, adopting a daylight factor that will enable effective use to be made of natural light incurs the cost of provision of windows, and may involve adopting an increased ceiling height in order that daylight may penetrate into large rooms from peripheral windows. Experience shows that a building that is lit by electric lighting alone with no provision of windows, is generally of significantly lower capital and running cost than a building in which it is attempted to make use of daylighting (section 9.6.3).

21.1.8   The benefits of good lighting in manufacturing industry are summarized in Figure 21.1. Quantifiable savings resulting from the adoption of better standards of lighting in a workplace include:

- savings in building energy;
- greater added value to manufactured goods due to improved quality;

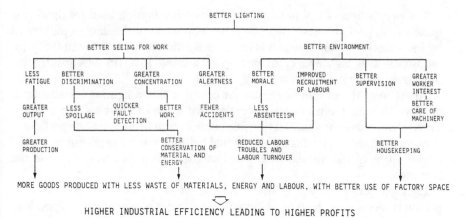

Figure 21.1 *How better lighting improves profitability in manufacturing industry.*

- net value of extra production due to improved productivity;
- savings by reduction in damage, scrapping and rectification of poor work;
- savings in running costs where lamps of higher efficacy are employed;
- savings in maintenance costs where lamps of longer life or better lumen maintenance are employed;
- savings stemming from better staff relations leading to reduced labour turnover, and thus to lowered costs of recruitment and training;
- savings due to reduced absenteeism, both *physical absenteeism* (including latecoming and early departure, extended breaks etc), and *invisible absenteeism* (where employee is present but not actually working);
- savings due to reduced accident rate (lost production time, lost management time, claims against the company) and the minor sickness rate (absences of three days or less).

21.2 Energy savings

21.2.1 Because of the 'greenhouse effect', the Earth is getting warmer, with disturbances of the natural weather patterns. The greenhouse gases include carbon dioxide (which is one of the prime causes of global warming), methane, chlorofluorocarbons and nitrous oxides. Nearly half of the carbon dioxide emission into the atmosphere in the UK is attributed to electricity generation. Lighting and electrical appliances account for almost half of the carbon dioxide emission attributed to building-related uses of electricity. More information on the greenhouse effect may be obtained from the publications of the Building Research Establishment (appendix G).

The proportion of UK national electricity consumption used for lighting is quite small, probably less than 6%. The proportion of the energy purchased by individual companies which is used for lighting may be between 0.5% (in industries with high energy usage) up to around 15% (in labour-intensive industries with relatively low energy usage).

Apart from the economic reason for ensuring that a lighting installation is energy-efficient, the user may have the satisfaction of knowing that efficient lighting will make the smallest possible contribution to atmospheric pollution.

For manufacturing industry, the annual lighting cost averages out at around 0.5 to 1.0% of the annual wages and salaries bill of the staff that use the lighting.

Lighting forms only a minute fraction of the total cost of producing products. Assuming the typical labour cost content in manufactured goods to be 40%, the lighting cost content is seen to be of the order of 0.2–0.4 pence per £1.00 of the cost of manufactured goods. There can be few factories with a costing system so tight that they could not afford to have the very best modern lighting – even ignoring the cost benefits that accrue in almost every case (Figure 21.1).

21.2.2   Energy-efficient lighting arises from:

- correct specification of illuminances and control of glare to ensure efficient work;
- the selection of lightsources of the highest possible efficacy for the duty;
- the use of luminaires of the highest possible utilance suitable for the duty;
- efficient and regular cleaning of lamps and luminaires, and the periodic cleaning or redecoration of building surfaces, to ensure that light is not needlessly absorbed by dirt;
- replacement of time-expired lamps at the proper time;
- avoiding wastage of lighting by instituting good switching routines, or installing automatic lighting controls.

21.2.3   Automatic lighting controls

It is possible to achieve a reduction in energy usage by employing automatic devices which switch off lighting in rooms which are unoccupied or which are receiving sufficient natural light (section 9.6). Devices such as the 'Liteminder' supplied by Setsquare Ltd (appendix G) combine passive-infrared (pir) occupancy detection with local light-sensing cells, to switch off all or some of the lamps in an area when the area is unoccupied or the daylighting is sufficient. Automatic dimming systems can also be used (section 22.4).

21.3   Calculating the payback period

21.3.1   A basis for comparison of the economics of two or more alternative lighting proposals is to compare their forecast costs over a period of say, ten years:

10-year cost $= C + 10(R + M)$

where $C =$ capital cost, i.e. cost of lighting equipment + cost of electrical installation + cost of initial set of lamps + cost of any permanent means of access.

$R =$ running cost per annum, i.e. cost of electrical units consumed + proportion of maximum demand charge on tariff.

$M =$ maintenance cost per annum, i.e. cost of replacement lamps and other consumable items + labour cost for cleaning and relamping + cost of preventive and corrective maintenance (materials and labour) + incidental costs (e.g. hire of access equipment).

A straight-line economic comparison between alternative proposals (between existing installation versus proposed, or between two alternative proposals) may be made by simple graphical means to discover the break-even period (Figure 21.2). Such comparisons do not take account of financial factors such as:

- increases in annual costs due to inflation;
- notional losses of interest on capital that is invested in the lighting system;
- depreciation of capital plant and any tax relief on investment.

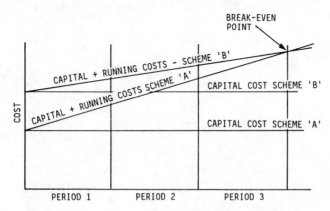

Figure 21.2 *Economic comparison of two lighting proposals without inflation adjustment.*

Because inflation rates, interest rates and tax reliefs (if any) vary over time and from place to place, no attempt is made here to deal with them. If large capital sums are to be invested in lighting, financial studies using the methods of discounted cash flow or future accounting might be employed. Such studies might show there was a net benefit from leasing the installation instead of buying it, or that it was beneficial to place the work of maintenance out to contract.

The net annual benefit from an investment in lighting will be the annual benefit from such savings and improvements that can be quantified (section 21.1) less the annual cost. For example, if the annual benefit from a lighting scheme was calculated at £5000, and the capital cost was £10000, then the payback period would be two years. Typical payback periods for refurbishments range from six months to three years. In calculating the annual benefit of a new installation, the capital cost should be discounted by the scrap value of the old system, if any. The annual benefit will continue throughout the life of the installation, say at least ten years.

In addition to the annual benefit from having installed more efficient lighting, there will in many cases be a productivity benefit. For reasons set out in section 21.1, and shown diagrammatically in Figure 21.1, it may be expected that improving the lighting will result in measurable productivity and profitability benefits from the factors listed in section 21.2.2. The increased profitability from such results should be taken into account when calculating the annual benefit from improving the lighting. It will be seen that an increase in productivity – even one of the order of a fraction of 1% – could bring an increase in profitability many times greater than any increase in lighting cost.

22 Electrical installations

22.1 Safety

22.1.1 Prevention of electrical accidents

It is generally appreciated that new electrical installations in the UK must comply with the latest edition (currently the 16th) of the *IEE Wiring Regulations*.[27] However, during the operations of installing lighting, it is not unknown for accidents involving electricity to occur. Throughout all the operations of stripping out old installations, modifications to wiring and installations, erection of lighting equipment, and the maintenance, testing and use of lighting, those responsible must show proper concern for the safety of personnel.

The most recent HSE statistics (1989/90) record that deaths at work from electric shock in the UK amounted to 23 for the year. Many of these tragedies could have been prevented by the use of a residual current device (RCD), the function of which is to cut the power if the currents in line and neutral become unequal because of leakage to earth, if connections are of incorrect polarity, or there is loss of the neutral or earth point connection.

Management responsibility in all these matters is clearly set out in the *Electricity at Work Regulations*, 1989 which require that persons shall be designated as 'competent' to perform duties of installation and testing of electrical installations.

22.1.2 Preventing other accidents

The practical work of installing lighting consists largely of placing luminaires at desired positions above the ground, and due attention should be paid to the risks of mishap or personal injury due to mechanical collapse of supports or means of access, damage by wind, or objects being dropped. If no emergency lighting is provided during the construction phases, personnel may be put in danger by unexpected outages of electrical supply. Attention must be given to the deleterious effects of rust and corrosion which can unexpectedly weaken supports and fixings.

There must be no short cuts in matters affecting safety. For example, while carrying out lighting installation or maintenance work in an existing flame hazard zone, the restraints and precautions specified for such zones must be respected by the installers (chapter 7). It may, for example, be necessary to close down, evacuate and thoroughly ventilate a flame hazard area in order that welding work can be carried out.

A common cause of accidents is the use of improvised temporary lighting. Some guidelines on safe practice are given in Section 13–2.

22.2   Interior installations

22.2.1   Pitch of luminaires

The choice of suspension and wiring method for interior luminaires should relate to the building structural details, the type of luminaires and the maintenance methods proposed, and in particular the method of access planned. Where there are exposed joists, the obvious step of suspending the luminaires from them is not necessarily the best or the cheapest method. It may happen that the pitch of the joists nearly coincides with the required spacing of the luminaire, but even so, mounting on joists usually means that the end luminaires are too far from, or possibly too close to, the end walls. (Luminaire suppliers will advise on the correct spacing of the end luminaires from the end walls; the spacing-off is usually half the pitch between the other luminaires in the line.)

It is usually possible to arrange the pattern of HID-lamp luminaires so that luminaires are not located over joists, which offers the possibility of mounting the luminaires so that their bottom rims are level with or slightly above joist level, thus giving some protection against damage in areas where fork-lift or clasp trucks are employed.

In interiors where there is a gantry crane installed, luminaires should not be disposed in rows across the space, for this will result in heavy shadowing when a whole row is hidden when the crane is beneath them. It is better to arrange the luminaires in diagonal rows across the space.

22.2.2   Lighting trunking

Because of its higher labour content for installing, traditional wiring in steel screwed conduit may be more costly than *lighting trunking* which provides physical support for the luminaires and an enclosure for the wiring. There are trunking systems suitable for carrying HID-lamp luminaires, and others in which the control gear for fluorescent lamps is housed within the trunking. Overhead combined trunking can contain channels for both lighting and power distribution, and can provide supply drops to machines and plant.

Trunking may facilitate access, in that, if suitably fixed and laterally braced, ladders may be leaned against it. The ladders will preferably be fitted with hooks to prevent slipping (section 23.3).

22.2.3 Lighting track

Lighting track provides physical support for luminaires, and also contains two or more electrical conductors to which luminaires are connected by a fitment. Track of suitable design can carry HID-lamp luminaires. Track is particularly suitable for overhead-mounted localized or local lighting (section 9.5), and patterns are available with multiple circuits to facilitate switching of single luminaires or groups of luminaires.

22.2.4 Facilities for maintenance

Some patterns of trunking have fitments which are the means of attaching the luminaires and connecting them to the supply. With this system it is possible to remove luminaires during maintenance, and replace them at once with clean luminaires. Similarly, some patterns of lighting track have fitments which provide a means of attaching the luminaires and connecting them to the supply for swift substitution for maintenance.

Some patterns of luminaire partially dismantle for cleaning, enabling the 'overlamp' reflector to be brought to the ground for cleaning.

Raising and lowering gear may be fitted so that luminaires can be lowered for cleaning and maintenance (section 23.3.3).

22.2.5 Suspensions

Rigid mounting of suspended luminaires is only suitable for short drops (say not more than 300 mm), longer drops being fitted with conduit swivel-plates, paired-hooks on conduit, or chain suspension. The cables should not carry the weight of the luminaire, and should be of a grade suitable for the duty. Preferably the cable will be protected in flexible metallic hose (which gives mechanical protection and prevents possible attack by rodents). If chain is used, ordinary bent-wire chain will not suit industrial use, and a heavy-gauge welded-link chain should be employed. Long suspension drops should not be used unless horizontal stay-wires are fitted to prevent the luminaires swinging in a draught.

22.2.6 Mineral-insulated metal-sheathed cable

Mineral-insulated metal-sheathed cable (MIC) has much to recommend it for industrial lighting installations, for it is cheaper per metre installed than PVC cables in screwed steel conduit, with low labour cost. However the following points should be noted:

- Copper- or aluminium-sheathed cable is not as mechanically robust as screwed steel conduit, and exposed runs must be given protection.
- Entry of moisture into this type of cable before installation will reduce its insulation value. Therefore cut ends should be sealed at once if they are not to be immediately connected. (Bending the end over and hammering it down flat is fairly effective for a short time in a dry location.)
- *Caution*: there are restrictions on the length of MIC that may be used for connecting between remote control gear and a HID lamp or luminaire. The lighting supplier should be consulted for further information.

22.2.7   Vibration

Vibration due to the oscillation of steel-framed lightly-clad buildings may occur due to wind or gantry movement. Note that vibration can affect lamp-life (section 7.9.1).

22.3   Exterior installations

22.3.1   Safety of electrical connections

Exterior lighting installations at industrial sites require careful planning to ensure a high standard of electrical safety. A common fault is for the connection from an underground cable at the base of a mast or tower to be vulnerable to damage. Cables should not be left exposed, but should be adequately mechanically protected, and any metallic enclosure should be earthed.

At locations where there is the possibility of flooding, the connection and isolation provisions at each mast should be located well above the possible level of floodwater.

Underground cables should be protected from damage by being laid in suitable conduit and/or covered with protective tiles, especially at points which may be traversed by vehicles. Temporary underground cable routes should be marked above ground with tapes etc and vehicle drivers instructed not to drive over them.

22.3.2   Switching and isolation

If remote switching of supplies to masts is achieved by radio or mains-borne signals, there should be local means of electrical isolation at each mast to ensure safety during maintenance.

23.3.3　Temporary mains supplies

Where exterior lighting is required for a relatively short period, mobile equipment may be used (chapter 18). For longer periods of use, mobile or portable lighting towers etc, may be mains-supplied, but such temporary installations should not be of a lower standard of electrical safety than is required for permanent installations.

22.4　Lighting controls and dimming

22.4.1　Dimming systems

Building energy management systems are of growing importance. We have seen that permanent supplementary artificial lighting (PSALI) (section 9.6.2) can effect significant savings in energy by making the best use of available daylight. Rather than switching-off some or all of the luminaires, an alternative is to dim them.

In energy-saving dimming systems, the concept is that when there is adequate daylight entering the building, the light output from the luminaires is reduced by automatic controls with the objective of reducing energy consumption. Dimming of all kinds of lamps used for general lighting is now a practical proposition, but it must be noted that a reduction in power consumption by dimming is usually accompanied by a rather greater reduction in lumen output.

In reviewing the validity of claims made by some manufacturers and installers regarding the benefit of using energy reduction methods, the following cautionary examples should be borne in mind. A few years ago, several companies were offering systems for reducing the voltage on lighting circuits to effect energy economies, some claiming that the reduction in power consumption resulted in 'negligible lumen loss' or even 'no lumen loss', both claims being false. Such systems were offered under descriptions such as 'energy limiters' or 'power misers' etc, and the advertising gave the impression that one could get something for nothing.

Another product which briefly showed its face was a 'power reducing button' to be placed within a lampholder and (so it was claimed) to 'reduce the power consumption without loss of light output'. This was another false claim for a technically unsound product which, incidentally, was highly dangerous and exposed the user to risk of electric shock.

The present author could not support the installation of power limiting devices for constant use, installed with the objective of saving on energy costs. It surely must always be better economy to design the lighting to produce the illuminance actually needed, and to select the most appropriate lamps and luminaires for the duty. No add-on device could ever improve the economy and performance of such an installation.

Having entered all these caveats, it must be stated that the dimming of fluorescent tubular lamps for reason of economy is now a well established practice, and, with the availability of solid-state voltage-reducing/dimming equipment of new designs, the dimming of HID lamps is now a practical possibility which may be economically justified in particular situations.

Brian McKiernan of Lutron EA Ltd (appendix H) reports that he has engineered a number of such dimming installations using HID lamps, and claims that considerable energy savings have been achieved by varying the lighting level according to the availability of daylight. Regarding the possibility that lamp manufacturers' guarantees would be affected by the introduction of dimming systems, McKiernan believes that the dimming controls produce no deleterious effects on lamps, and he considers that the lamp manufacturers' guarantees remain valid. He is also of the opinion that the generation of mains harmonics due to 'wave chopping' dimmers is not a major problem, suggesting that suitable filters can reduce waveform distortion in sensitive situations to an acceptable level.

22.4.2   Switching systems

The prudent switching-off of lighting to prevent wastage when rooms are unoccupied is strongly recommended, coupled with cautions to prevent danger being caused. The switching-off of lighting by automatic systems at set times could place people in danger by their being unexpectedly put in darkness. Even switching down to emergency lighting levels could place people in danger from moving machinery, hot plant etc. The automatic switching-on of lighting by automatic systems at set times could also place people in danger, for circuits could be unexpectedly energized while repairs were being carried out or while relamping or cleaning of luminaires was being performed.

Useful economies may be achieved by arranging for local lighting at workstations to be switched by the operators, but the general lighting which remains on when such local lighting is off must be sufficient for the safety and comfort of all occupants. In one method, all the local lights are switched off automatically at the commencement of lunchtimes etc, and must be reselected to 'on' by the operator when required.

The switching of general lighting systems is sometimes practised to save energy during lunchtimes etc when the premises are not occupied. The following points should be noted:

- The life of lamps will probably not be affected, for the nominal life of lamps assumes a three-hour switching cycle.

- It is untrue that the inrush current on starting a fluorescent lamp or HID lamp is of such a magnitude as to negate the energy saving to be made by switching-off for a short period.
- Most lamps do not come to full light output instantly, and also may undergo significant colour change as they come up to their correct operating temperature. Therefore time must be allowed for them to run up on switching-on, particularly in areas where colour-matching is carried out (section 10.5).

There are now available sophisticated systems for the energy management of lighting, for example the Thorn Lighting C-VAS system, which utilizes high-frequency fluorescent-lamp luminaires and a variable illuminance controller. Other systems combine sensing of occupancy, measurement of ambient light due to daylight, and an overall time system which can progress the lighting down to night-time pilot lighting (chapter 12) and is integrated with the emergency lighting of the premises (chapter 14).

22.5   Standby lighting and power

22.5.1   Security of power supplies

Mains electricity supplies may be made more secure by taking supplies from two feeders fed from different parts of the supply company's distribution network. It is normal good practice to divide distribution circuits so that lighting loads are separate from power circuits, and adequately protected against overload. Supplies to luminaires in each area of the premises may be fed from two or three phases to minimize stroboscopic effects (section 2.8.7). This method of installation also reduces the risk of total blackout.

22.5.2   Effects of long outages

Good engineering and maintenance will reduce the risk of breakdown of lighting, but failure may be due to causes beyond the occupier's control, for example, fire in a plant room, or a failure of the mains supply.

Quite apart from emergency conditions due to fire and other dangers which may accompany, or be the cause of, a failure of lighting, a blackout in any workplace can have serious effects. For example, loss of lighting for some hours would not only lose production, it could cause danger because activities such as chemical processes and the operation of high temperature plant simply cannot be stopped quickly. Failure of outside lighting at a steelworks could bring the whole plant to a standstill in a few hours. Failure of security lighting could put a company at risk. Such a failure at a key strategic target could have serious implications. On a civil engineering site, a long failure of lighting while pouring a massive monolithic concrete structure might halt pouring and result in a weakened structure. The cost of such interruptions can be high compared with that for the provision of stand-by generation to provide power for lighting and essential services during a long mains outage.

22.5.3   Emergency lighting

Emergency lighting (chapters 15 and 18) enables people to escape from premises or from situations of danger during a failure of the normal lighting. It may also enable essential tasks to be continued during a failure of normal lighting, e.g. bringing plant to shut-down, and continuing essential processes to avoid danger or major loss. If such activities are likely to extend beyond the battery duration of the emergency lighting (generally between one and three hours) then *standby lighting* may need to be employed.

22.5.4   Stand-by supplies

For some premises, the capital cost of installing generator plant for standby use is justified. In general, self-generation is more costly than buying energy from a power company, but continuous generation can be economic if it is possible to utilize the waste heat from the prime movers (e.g. for steam raising or environmental heating). However, the cost of having a mains supply to be used only to back-up self-generation may be surprisingly high.

The reliability of generating systems used as power supplies for emergency lighting is discussed in *BS 5266*.[47] In such applications, it can be arranged for there to be a bridging battery to provide power for a very short period while the generator is starting up. With suitable controls it is possible to provide an 'uninterruptible supply' so that there is no outage during change-over from the normal supply to the standby supply.

Under emergency conditions, mobile lighting units having their own generators and intended for outdoor use (chapter 18) may be used to provide temporary indoor lighting (section 13.2.4). Battery-operated handlamps (section 13.1.2) may be of great value under emergency conditions, as may chemiluminescent lightsticks (section 5.13.1).

22.5.5   No-break supplies

On large sites having special processes and dangerous operations, it may be desired to raise the reliability of the power supply for lighting and other essential loads. The economics may be favourable to what are somewhat misleadingly termed 'uninterruptible supplies', though correct design of equipment can give a very high degree of protection against interruption. Some ways of achieving this are briefly summarized in this section.

Rotating-machine no-break power system

This method (Figure 22.1) interposes an electrical rotating-machine between the external supply and the load. On failure of the mains a.c. supply, the d.c.

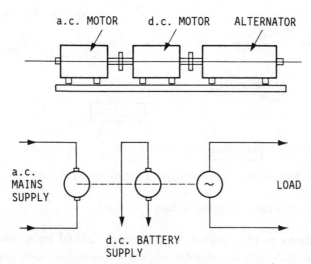

Figure 22.1 *Rotating-machine system with no-break power supply.*

motor takes an alternative supply from the battery, the loss of rotational speed (and hence the loss of frequency of the a.c. output) being minimized by provision of a flywheel to bridge the interval of a second or two while power is being automatically switched to the d.c. motor drive for the alternator. Variations on the theme include:

- a three-unit assembly (alternator, a.c. motor and d.c. motor, all on one shaft);
- a dual-wound motor (with a.c. motor and d.c. motor windings) driving an alternator;
- units as above in which the d.c. drive doubles as the charging unit for the storage batteries when the mains are healthy.

Any of the above variations can be backed-up with a prime-mover to drive the alternator during outages.

Virtual no-break standby system

An alternative to using a rotating-machine assembly is for power for the load to be taken as normal from the mains, with an alternative static power supply held in readiness and in frequency synchronism (Figure 22.2). Change-over by solid-state switching enables the supply gap to be limited to about 0.25 of a cycle, meaning that HID lamps will not be extinguished. On restoration of the supply, switching back is of about the same delay duration, and then the batteries return to charge.

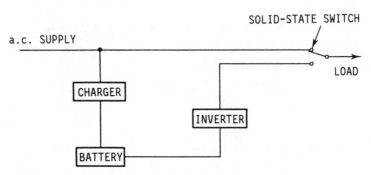

Figure 22.2 *Diagram of 'virtual no-break' stand-by power system.*

Solid-state no-break standby system

These systems employ solid-state switching initiated by a microprocessor which performs various monitoring and control functions (Figure 22.3). Typically the battery capacity might be 30 minutes. In normal mode the battery is 'floating', and the output from the static inverter is zero (or is provided with a small dummy load for operational reasons). The modes are:

- Normal mode. Switch 'A' connects main battery charger to mains; switch 'B' connects the maintained load to the mains.
- Failure mode. Switch 'A' operates; switch 'B' operates to disconnect the maintained load from the normal supply and connect it to the static inverter.
- Reversion mode. If normal supply is restored within, say, one minute, the microprocessor will bring the static inverter into synchronism with the mains supply before operating switch 'B' to reconnect the maintained load to the mains supply, and switch 'A' operates to reconnect the main battery charger to the mains.
- Emergency operation mode. If the normal supply is not restored within a short period, say greater than one minute, the microprocessor closes switch 'C' to start the prime mover, monitors the result, repeats if necessary, and opens switch 'C' again on engine start-up. It synchronizes the alternator

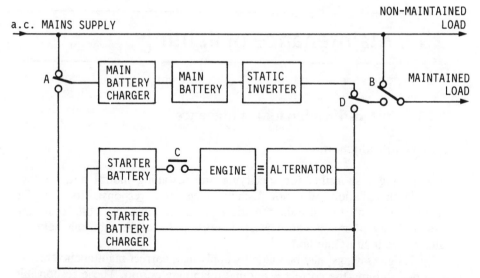

Figure 22.3 *Microprocessor-controlled virtual no-break system giving two-tier standard of reliability. Switches 'A', 'B', 'C' and 'D' are microprocessor controlled.*

and static inverter, and operates switch 'D' to transfer the maintained load to the alternator output.

A series of protective or recovering steps are then carried out according to circumstances. For example, if there is spare generator capacity, it will commence re-charging the main battery and starter battery. During the remainder of the outage, the static inverter has zero output but remains in synchronism with the alternator. If the alternator frequency should drop (possibly indicating failure), switch 'D' will operate, transferring the load back to the static inverter. When mains power is restored, the system reverts to normal mode.

The benefits of such a standby power system as described area:

- High reliability.
- At every stage the microprocessor controls the system to give the highest reliability to the maintained load.
- Automatic operation, but with the facility for manual intervention at every stage.
- Program takes account of malfunctions of plant, and will maintain essential load for as long as possible under all circumstances.
- Minimum size of standby battery required.
- Microprocessor can monitor parameters such as low fuel for prime mover, as well as operating annunciators and alarms, and switching lighting.
- Virtual no-break supply enables use of HID lamps.

23 Maintenance of lighting

23.1 Cost justification for maintenance

23.1.1 Introduction

The people responsible for specifying and installing indoor and outdoor lighting installations do not usually have the responsibility for the maintenance of the installation throughout its life, yet the cost and practicability of maintenance is an important factor in choosing between alternative lighting methods.

In design, savings may be made by applying a correct maintenance factor, and determining the correct economic relamping period. Those responsible for maintenance procedures should be properly briefed and the value of their work recognized so that an adequate budget is provided for it. Money spent on routine preventive maintenance is not wasted, for it will result in a safe, clean and efficient installation, operating at its planned cost-in-use, and probably with extended technical life.

23.1.2 Capital deterioration

The technical life of most permanent indoor and outdoor lighting installations is well in excess of ten years, but experience shows that after about ten years it may be economic to replace all or some of the luminaires to gain the advantage of improvements in lamps and luminaire performance that have occurred over that period. In some cases it will be economic to refurbish the luminaires, perhaps by replacing reflectors or adapting them to a new type of lamp.

As regards exterior lighting, the capital invested in lighting towers and electrical installation work must be regarded as a wasting asset, the life of which may be extended by good maintenance. The cost of good maintenance may be justified by improved safety or security, the additional hours to be worked, or the savings due to better or faster work. It is usual to assume that the value of an exterior industrial lighting installation depreciates to zero in about ten years even if well maintained, and that its scrap value is approximately equal to the cost of its removal.

As regards temporary, mobile and portable lighting used out of doors (chapter 18), this too has a theoretical technical life of ten years, but damage, deterioration and thefts may much reduce this. Nonetheless, the equipment may have a trade-in value. At the end of long construction or civil engineering contracts it may not be economic to transport mobile lighting equipment back to a distant depot, and it may be disposed of locally, the receipts from its sale offsetting part of the site lighting cost. Proper maintenance of the equipment will extend its life and enhance its disposal value.

23.1.3 The penalties for poor lighting maintenance

The failure of a lighting system may place personnel in danger, or bring about a cessation of work. Emergency lighting (chapters 14 and 19) will provide for escape from premises or descent from heights etc. and thus mitigate risks. Standby lighting and electrical systems (chapter 22) may be justified if failure of lighting could cause serious loss of customer service or production.

Neglecting to clean luminaires regularly and to replace lamps at the end of their economic life results in obtaining less light than is being paid for. The reduced illuminances may slow down production, or reduce the quality of work performed. Neglecting to replace sporadically failed lamps may lead to patches of poor illumination so that the stage is set for accidents to occur.

Fluorescent tubular lamps and HID lamps which have reached the end of life and exhibit 'cycling' (repeated attempts to start) may cause overheating of control gear which can lead to its failure or to fire in the luminaire. The use of high-frequency ballasts obviates this problem, and the use of electronic starter canisters in fluorescent luminaires will also prevent cycling (section 5.5.2).

23.2 Safety during maintenance

Lighting equipment of types suitable to the application and environment, and in a good state of repair electrically and mechanically, is very safe. Most accidents associated with lighting relate to lamp changing and the cleaning of luminaires, or when carrying out modifications to the system. Typical accidents include falls, injuries through items being dropped, electric shock, and fires.

Management has a duty to provide proper instructions and training to maintenance personnel. For example, staff entering a department to carry out maintenance of the lighting should be properly briefed as to any hygiene requirements, and any hazards. They may need to wear special clothing – or at least to put on clean overalls – to prevent contamination, or they may need to take precautions for their own safety, such as wearing face masks for protection against harmful dusts, or being careful not to transfer poisonous material to their mouths, and not to smoke.

23.2.1   Improvised means of access

It is clearly a management responsibility to provide suitable tools, equipment and means of access (section 23.3). A common cause of accidents during maintenance of lighting is improvised means of access to high-mounted lighting equipment. Appropriate access equipment should be purchased or hired, and the staff concerned instructed in its safe use.

23.2.2   Hosing non-hoseproof luminaires

In chemical works, pharmaceutical factories, abattoirs and food factories, it is common practice to hose-down the lighting equipment. It is important to note that luminaires which are certified to IP-1, IP-2 and IP-3 are protected from water descending on them, i.e. they are drip-proof or rainproof. Such luminaires may not also be hoseproof, in which case hosing would produce danger of electric shock.

In factories which have some areas fitted with IP-4 hoseproof luminaires and some fitted with ordinary non-hoseproof luminaires, it is a wise precaution to display prominent notices in the latter areas warning not to hose the luminaires.

23.2.3   Unexpected loss of power

When people are working at heights, a power failure due to any cause can place them in danger. Emergency lighting should be available, or portable lighting equipment used (section 13.1). A person stranded aloft in darkness should stay put, and shout for help. A whistle carried in the pocket can be of value in these situations, as may a torch or one or two chemical lightsticks (section 5.13).

23.2.4   Unexpected switching on

Electrical accidents due to equipment and circuits being live or becoming live unexpectedly are completely preventable. Before work on wiring and connections, a permit-to-work system should be implemented, and the relevant circuits properly locked off in accordance with the *IEE Wiring Regulations*[27] and the *Electricity at Work Regulations*.[18]

23.3   Access

23.3.1   Access method affects lighting cost

The greater part of the cost of periodic cleaning of luminaires is the labour cost of gaining access to them. Solving the access problem is part of the lighting design process, not a separate problem to be tackled later. In estimating the cost of a new lighting installation, the cost of access equipment should be included, and in calculating the annual running costs, the labour cost for maintenance should be estimated. It would be good accounting to allocate a notional value to the floor area which becomes unproductive because of the need to provide extra wide lanes between plant to permit access equipment to pass through. Thus, it would be justified to budget for some extra cost for access equipment that did not have this requirement, e.g. that folds to a narrow width for movement (section 23.3.2), or which is permanently installed at high level (section 23.3.3).

23.3.2   Mobile access equipment

In planning access to overhead luminaires, it may be difficult to bring mobile access equipment into position if there is fixed plant or other permanent obstruction at floor level. Overhead pipes, trunking or cranes may also obstruct access. Above hot processes, it may be too uncomfortable or too dangerous to go aloft when the plant is operating.

Some of the principal types of mobile access equipment are discussed in the following sections. Heights and details may vary between manufacturers.

Stepladders

These may be small enough to be carried by one person. They will give up to about 2 m to the top step, permitting reach to about 3.5 m or a little higher if a hand-steady is provided at the top. It is not possible to reach both ends of a 2-m fluorescent luminaire conveniently from one ladder position.

Fixed-length and extending ladders

For access up to about 9 m. Many industrial injuries arise from the use of such ladders which should be secured at the foot during use, and preferably at the head too. If such ladders are used to gain access to luminaires on trunking or suspended from roof joists, purpose-made hooks can secure the head, but the foot should still be secured or 'footed' by an assistant. A safety harness is available for use when working at 2.5 m or more above the ground.

Ladder towers

These consist of telescopic nesting frames constructed of wood or metal, which can be scaled up to around 7 m. They may be provided with extendable outriggers for stabilization when erected. If fitted with wheels, the manufacturer's instructions should be followed regarding locking before scaling. The equipment should be used only on firm ground. Some ladder towers are unstable if subjected to even moderate wind forces, and must be weighted down with sandbags at the base, or stayed.

Demountable scaffold-frame systems

Frames are placed one on the other to the required height, usually with built-in ladders, platforms at the head and at intermediate heights, and handrails. The manufacturer's instructions should be followed regarding the maximum height of assembly. Typical maximum un-stayed height is around 12 m. These systems must be used only on level firm ground, or be erected on solid planking. They may be moved while erected. The wheels should be locked, and any levelling feet or outriggers should be properly secured before climbing. Pairs of frame systems may be used to form a bridge to gain access to luminaires that are located over a floor obstruction. For example, two wheeled units can stand in adjacent aisles to give access to luminaires over a row of machines.

Hydraulic lift platforms

These are sometimes called 'cherry pickers' or 'beanstalks'. They extend up to about 20 m, and the cradle can accommodate two people. The system is extended by handpumping, or by an electric pump powered from a socket-outlet, or by rechargeable batteries housed in the base. Control of height may be from the cradle, which may have an offset permitting objects to be reached 2 m off-centre of the unit. The pivoted feet and wheels should be locked before ascending.

Trailer-towers and vehicle towers

There are various patterns giving access up to 25 m or higher. The units are usually for use out of doors. High access indoors is usually more conveniently arranged by permanently installed means of access.

23.3.3   Permanently installed means of access

A decision to build-in means of access to lighting equipment should preferably be made before the steelwork drawings for a new building are completed; then high-level walkways, roof walkways and hatches etc can be incorporated at the lowest possible cost. Individual means of fixed access to high-mounted luminaires are discussed in the following sections.

High-level walkways

These may be constructed within the roof framing to provide a safe means of access even at great heights. Walkways should be wide enough for two people to pass each other (say at least 750 mm wide), and provided with waist-high rails each side (say 1250 mm high). At the edges of the walkway a kickboard should be provided, say 150 mm high to prevent people and objects slipping off the walkway and falling. It should be possible to reach the luminaires from the walkway without placing the engineer in a precarious position. Preferably the luminaires will be pivoted so they can be brought to a convenient position for attention (Figure 23.1).

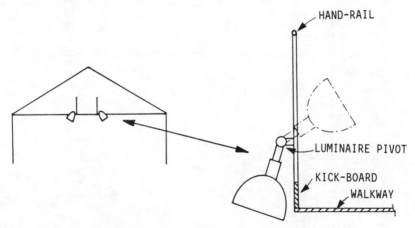

Figure 23.1 *Internal high-level walkway to give access to luminaires.*

Roof walkways

In hazardous zones, luminaires may be mounted above the roof glazing, with access by roof walkways (section 7.10.4). In buildings having normal atmospheres (using ordinary luminaires), and in those having hazardous atmospheres (using protected luminaires), luminaires may be mounted within the building on hinged hatches which can be opened from above, again with access by roof walkways.

Access from gantry cranes

Where there is an internal gantry crane, it might be used to gain access to the luminaires. It is important that there is proper provision for safety of personnel, with a safe platform complete with handrails and kickboards provided as specified for walkways. Particular attention should be paid to guarding of the crane mechanisms, and to prevent personnel coming into contact with electrical conductors supplying the crane drives. A permanent ladderway for access to such a gantry crane should be provided at the 'parking position', and be safety-hooped from 2.5 m above the ground in accordance with *BS 4211*.[52]

Raising and lowering gear

Raising and lowering gear is a practical and widely-used form of built-in access to high-mounted luminaires. Each luminaire has a small hand-operated winch by which it can be lowered for maintenance (Figure 23.2). The weight

Figure 23.2 *Raising-and-lowering gear for high-bay luminaires.*

of the luminaire when 'parked' is not carried by the cable, but by the connecting unit above the luminaire; this also connects and disconnects the power automatically. To lower a luminaire, the handle is inserted in the winch, and is turned as though to raise the luminaire. This disengages the luminaire from its electrical connections and releases the physical support. To return the luminaire to service it is wound up to its 'parked' position; and further pressure on the winding handle re-engages the mechanical lock and electrical contacts. Ensure that any gantry crane is not operating before lowering the luminaire. Such gear is generally reliable, but there must be some means of gaining access to the connecting units and the pulleys etc, if they should require attention.

Pole lamp-changers

These are poles having a device at the head which will grasp an HID lamp or GLS lamp and enable its removal and replacement by an operator without climbing. Patterns are available from suppliers such as No Climb Products Ltd (appendix G), and it is claimed that they may be used to replace lamps at heights up to 9 m. Such devices have serious limitations; they can only be used for replacing single-ended lamps in open-construction luminaires which do not have an enclosure, louvre or protective grille, and in which the lamp is mounted vertically. Although they may be useful for dealing with sporadic lamp failures, the availability of a pole lamp-changer in a works tends to encourage the practice of lamp replacement without cleaning the luminaires.

23.4   Maintenance, cleaning and relamping

23.4.1   Objectives

How a lighting installation will be maintained affects its cost-in-use. It will be cheaper to use luminaires of designs that tend to keep clean and are easy to clean than luminaires which are not so resistant to soiling and which require a greater labour cost to clean them. If luminaires are easy to maintain, then possibly a higher maintenance factor could be adopted, with savings in the capital cost of the installation. It can never be economic to 'over light' to minimize the need for routine maintenance, and allow the lighting level to fall by neglect to the actual required illuminance. If a lesser illuminance will suffice, it will be cheaper to design for it and carry out a suitable programme of maintenance. Using a smaller lighting load will save energy as well cost.

The essence of good maintenance is preparation and anticipation. Improvisations and hasty interventions by the maintenance department in response to complaints about lamp failures, dirty luminaires and inadequate lighting are evidence of a failure to manage the lighting well.

23.4.2   Ensuring correct replacement lamps are used

There is considerable risk in inserting a lamp of the wrong type into a luminaire. An elementary preparation for future good maintenance is to compile a schedule of lamps which will be required for replacement purposes, and to place a copy of this where it is sure to be found, for example in a sealed plastic envelope attached to the switchgear. Modern lamps last a long time, and it can be five or six years between relampings, by which time memory cannot be relied upon, records may be lost, and the original personnel departed.

23.4.3   Safety

A prime requirement for safe lighting maintenance is the provision of proper means of access (section 23.3).

23.4.4   Avoiding damage to equipment

The use of torque-limiting spanners and screwdrivers will prevent distortion of luminaires when tightening closures.

23.4.5   Use of lubricants

Use of unauthorized lubricants may promote corrosion, electrical tracking or failure of seals. In practice, most luminaire lubrication requirements can be met by applying a very slight smear of petroleum jelly to fixing threads and hinges, though a silicone lubricant spray may be preferred if there is risk of contact with electrical insulation materials or flexible seals.

The threads of ES and GES lampcaps may be *very lightly* smeared with a trace of conductive graphite grease (wipe on/wipe off). Do not allow the grease to get onto the body of the lampholder where it could cause short circuiting.

23.4.6   Disposal of lamps

The disposal of lamps is controlled by a variety of regulations and laws which are enforced in the UK by Her Majesty's Inspectorate of Pollution (HMIP). Under UK Local Authority regulations, it is illegal to put lamps in ordinary garbage. Dangers arise from broken glass, but a greater concern is the effect of the chemical substances which may be released into the environment.

It is convenient to crush lamps in order to dispose of them, but the act of crushing can release minute particles of chemical substances (aerosols) into the atmosphere, and these can be very damaging to people and to the environment.

Fluorescent lamps and HID lamps contain cadmium and mercury, and minute traces of a variety of other chemical substances. If the debris from lamp crushing were washed away, these poisonous and harmful substances would find their way into the subterranean water systems and could eventually appear in our drinking water. The effluent generated by spraying water over the crushed debris of such lamps must not be flushed into drains, but must be disposed of as described in the UK *Trade Effluents (Prescribed Process and Substances) Regulations, 1989.*[65]

Users of small quantities of lamps should enquire of their wholesaler or lighting provider to see if they will take expired lamps for disposal. It will generally be found that the Local Authority will provide information, or may provide disposal facilities. Users having large quantities of expired lamps to dispose of should consider purchasing one of the modern types of lamp crushers which are available. These enable the user to crush lamps safely into very small fragments for disposal, but without releasing aerosols, and retaining wash water for disposal without releasing chemicals into the environment. Equipment of this type is available from Balcan Engineering Ltd (appendix G).

23.4.7 Bulk replacement versus spot replacement of failed lamps

Spot replacement

This is the method in which lamps in a lighting installation are replaced individually as they fail, a method which has the following disadvantages:

- Interruption of work while the lamp is defective (i.e. out, or worse still, 'cycling').
- Disturbance because of the need to bring access equipment to the luminaire.
- Lamps in the installation, being of various ages, will not all give the same output, thus some will seem brighter than others.
- Lamps tend to have some colour shift as they age, thus there will be differences in lamp colour appearance (and possibly of colour performance) from lamp to lamp.
- If they have not failed, lamps tend to go on lighting up long after they have passed the end of their economic life. Therefore, either less than the designed illuminance will be received, or the user will be buying the light dearly by employing lamps that are time-expired and therefore of low efficacy.

Bulk replacement

This is the method in which all the lamps in an installation are replaced at one time at the end of their economic life, a method that has the following advantages:

- Minimum disturbance to persons working in the area, for failure of any lamp will be a very rare event.
- The bulk replacement can take place outside the normal working hours or during a shut-down.
- The access equipment has to be brought only once to the area.

23.4.8 Combined relamping and cleaning

If it were possible for the routine cleaning of the luminaires to take place at the same time as the bulk replacement of the lamps, the labour expended would be less than would be required for two separate operations. A periodic cleaning of luminaires combined with replacement of all lamps is a task that is convenient to put out to an electrical installation contractor or specialist cleaning company, freeing staff for other work.

Unfortunately, it is not always possible to bring these two events to the same time. The life of lamps depends to some extent on the frequency of switching. If the switching cycle is less than three hours, the lamp life for fluorescent and HID lamps will be reduced. Switching on for one-hour periods would result in a reduction from the theoretical 100% to about 65% of life. Switching on for 45-minute periods would bring this down to 45%, while five-minute switching cycles would reduce the life expectancy to a mere 20% of theoretical life. This variability in life expectancy is coupled with variability in the frequency with which cleaning of luminaires is required. This depends on the dirt-resistant properties of the luminaires and the propensity of their environment to soil them.

In an installation using lamps with a 7000-hour projected life, for premises working 8 hours per day for 245 days per annum, the period before relamping is required is about 3.5 years, yet cleaning might be required annually or more frequently. If the hours of lighting use were continuous, bulk relamping

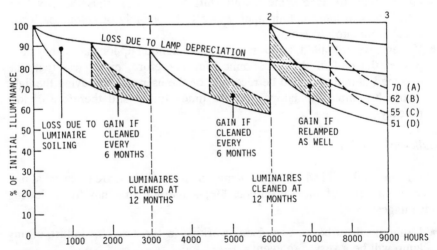

Figure 23.3 *Depreciation of illuminance. Typical diagram for 3000 h per annum. (A) cleaned twice a year and lamp renewed; (B) cleaned once a year and lamp renewed; (C) cleaned twice a year and original lamp; (D) cleaned once a year and original lamp.*

would be required at intervals of less than a year. The best that can be done in such cases is to make the time for a relamping or cleaning a little earlier or later, so that cleaning and relamping operations can be done together when ever practicable.

23.4.9 Monitoring

The maintenance operation will be facilitated by careful monitoring. When a new interior lighting installation is complete and operating, allow the lamps to operate for 100 hours before taking check lightmeter readings during the acceptance procedure (section 24.4.8). Take these readings with care, noting the exact positions of measurements and the height above floor etc. Thereafter take periodic readings in exactly the same manner, and plot the results of a few key readings (Figure 23.3). This will enable the depreciation of light output due to aging lamps and luminaires becoming soiled to be monitored, and for judgements regarding the time for relamping and the time for luminaire cleaning to be made. Remember that soiling of the room surfaces will also depreciate the illuminance (section 9.2.8).

24 Procurement of a lighting system

Industrial users of lighting do not usually get enough practice in buying lighting installations to have the procedures at their fingertips, so study of this chapter may give some useful guidelines. For the professional specifier, the checklists may be helpful to ensure that all important matters are considered. The first step is the preparation of a *design brief* in which the objectives and constraints are identified (section 24.1). Next, an *outline lighting specification* is needed to form the basis for schemes and quotations from potential suppliers (section 24.2). The order should be placed on the basis of a *final specification* agreed between the buyer and seller (section 24.3). After the work is completed, there should be approval and *acceptance* of the installation by a responsible person before the work is paid for (section 24.4).

24.1 Setting objectives and recognizing constraints

24.1.1 On a new construction project, it will be part of the duties of the architect and the building services engineering consultant to provide the specification for a suitable lighting installation. For speculative developments, a basic lighting system may be provided, with sufficient capacity in the main wiring and switchgear for the system to be developed or extended to suit the needs of the first occupier. Some new industrial and warehouse buildings are not provided with lighting in main areas until they are let or sold, though a basic system of lighting may be provided in the offices and corridors. The nature of the occupancy determines the kind of lighting system needed. Generally the recommendations of the *CIBSE Code*[1] will give sufficient guidelines for initial proposals to be formulated.

24.1.2 In existing buildings, a different situation may obtain, for lighting systems tend to last longer than the term of the typical occupier. Thus, new management will find a lighting installation in existence. In most cases, the installation will simply be accepted without analysis or criticism – at least for a while. Indeed, considerable energy and money may be expended trying to cope with problems of low productivity, poor quality work, and high accident rate, absenteeism rate and labour turnover rate, before realizing that at least some of the problems may be attributable to insufficient or unsuitable lighting (chapter 21).

Every factory and office block should have a lightmeter (appendix E) and a copy of the *CIBSE Code*.[1] It cannot be claimed that touring the premises with these two indispensable aids will reveal every lighting defect. They cannot, for example, be used to tell anything about glare conditions or the need for improved colour-rendering lightsources, but they will certainly deal with the question of sufficiency.

Helpful and constructive advice may be available from the sales engineer of the electricity supply company, from the representative of a reputable company of electrical installation contractors, or from the representative or lighting expert of a reputable lighting manufacturer. (For trade associations see appendix G, and for lighting manufacturers see appendix H.)

If an employer does not have a suitably qualified person on the staff to deal with the specification, design and purchase of a lighting installation, a consultant may be employed. If approaching a general consulting engineering practice, the client should insist on the work being handled by a person specially qualified in lighting.

Unbiased advice can only be obtained from a qualified person who will not benefit from any business decisions made by the client. In the UK, an appropriately qualified person would be a Member or Fellow of the Chartered Institution of Building Services Engineers (CIBSE) (Lighting Section), or of the Institution of Lighting Engineers (ILE) (appendix G). The consultant should certify that he will act in an unbiased manner in the best interests of the client, that he is remunerated only by professional fees from his clients, and has no other business connections. The help of such an advisor would be of value at practically every stage of dealing with a lighting project for any kind of premises.

24.1.3   In setting about a project of creating a lighting installation for a new or existing building, the person responsible should define or be given clear objectives as to what is to be achieved, and the criteria by which the success of the project is to be judged. The following factors should be considered:

- Is the highest priority to be accorded to getting the installation completed in a restricted period of time (for example, during a factory summer shut-down) even at the expense of not being able to use better equipment which is on long delivery?
- Is it more important to complete the task within a very tight capital budget, even though this will result in an installation which will be greedy in energy use and maintenance cost?
- Will the achievement of desirable lighting parameters (e.g. illuminance, glare control, uniformity etc.) be regarded as paramount, even if the installation will turn out to be rather unsightly?
- Is a future high cost of maintenance going to be preferred to carrying out overhead works (e.g. installing catwalks or lowering-gear)?

Compromise may be possible, the first step towards making good decisions being the ranking of features in priority order, and the setting of a realistic budget. An early meeting of the affected parties within the organization is advisable. Those consulted should include the budget controller, premises manager, production manager, and those responsible for internal installation work and maintenance. The meeting should define what is desirable as to resources, time and practicalities. Investigation will reveal that some desirable courses of action are barred by constraints (section 24.1.4), and these should be evaluated, quantified and recognized as forming the framework within which decisions about the lighting project must be made. Constraints are of two kinds, removable and rigid.

24.1.4   Removable constraints are usually those imposed by higher authority (e.g. budget constraint, or perhaps planning consent cannot be obtained for certain proposals). Such constraints might be removed by logical argument, by appeals to Local Authorities, and above all by the cogent marshalling and presentation of facts upon which a revised decision might be based. The strongest argument for the removal of financial limitation is the presentation of a soundly-reasoned cost justification (chapter 21).

24.1.5   Rigid constraints may be imposed by building dimensions, the strength of roof construction, and limitations imposed by the laws of physics. Some rigid constraints are of a legal nature. For example, it is unlawful to 'shine a light to seaward, confusable with a navigation light or likely to be a danger to mariners'. Other legal constraints relate to fire regulations and the instructions of insurers and inspectors from the Health and Safety Executive (appendix A).

24.1.6   When the project team has determined the objectives and set them in priority order, and when the constraints have been recognized and identified as rigid or possibly removable, an *outline lighting specification* (section 24.2) can be written. This will be the framework for the lighting design.

24.2   Lighting specification and scheme preparation

Because every lighting installation is unique, the processes of design and bases for the selection of equipment cannot be standardized, though the procedures can. Having decided the objectives and identified the constraints (section 24.1), the following factors should be reviewed.

24.2.1 Factors affecting interior lighting

The constraints of the building:

- Interior dimensions: length breadth and height of each space to be lit.
- Mounting heights of luminaires in each area: maximum, minimum and optimum.
- Reflectances of interior surfaces (ceiling, walls, floor) and furnishings (e.g. reflectances of bench-tops).
- Obstructions in the spaces; ceiling or roof features; competition for overhead space from heating and ventilating plant, fire sprinklers, electrical and other services distribution runs etc.
- Possible weight of lighting equipment; load-bearing capacity of roof structure or ceiling.
- Availability/desirability of daylight in the interior; combination of electric lighting and daylighting.
- Layout of plant or furniture; effect of this on access to lighting equipment overhead (section 23.3)

The lighting objectives

- Purpose of the interior; tasks to be lit.
- Tasks with very small details. Could there be visual difficulties because of low reflectances or low contrasts? Is task lighting required? (section 9.4).
- Special inspection tasks. Need to view luminous tasks (chapter 10).
- Need for directional lighting to produce modelling shadows and highlights (section 9.3).
- General illuminance required in each area (refer to *CIBSE Code*[1] and appendix B.)
- Colour-rendering and colour-appearance requirements (chapter 4).

Environmental considerations

- 'Normal clean dry environment', or hostile to the lighting equipment. Need for 'proof' equipment (chapter 7).
- Unusually high or low ambient temperatures (chapter 7).
- Need for lighting to be integrated physically or thermally with the heating, ventilating or air-conditioning equipment (chapter 22).
- Need to co-ordinate the lighting with other building services

Lighting design parameters

- Luminaire spacing/mounting-height ratio (chapter 9).
- Need for enhanced vertical or directional lighting (section 9.3).

- Lamp types that meet the colour requirements (chapter 5, section 10.5).
- Lamp outputs related to powers and efficacies, suitable for selected types of luminaires etc. (manufacturers' data).
- Need to reduce flicker or the danger of stroboscopic effect (section 2.8).
- Distribution characteristics of possible luminaires; utilization factor (chapter 6).
- Need for non-standard equipment to suit special requirements.

Installation and maintenance

- Availability (present and future) of replacement lamps and components.
- Method of wiring; trunking; lighting track (chapter 22).
- Access during installation, and for future maintenance (section 23.3)
- Maintenance factor to be adopted in design calculations; frequency of cleaning (chapter 23).
- Method of switching and location of switches (chapter 22).

Safety matters

- Statutory requirements, official recommendations (appendix A); insurer's requirements.
- Requirements for emergency lighting, standby supplies (chapter 14, section 22.5).
- Sufficiency of colour-rendering of chosen lamps to enable identification of dangers, and to read coloured information and mandatory and prohibitive signs (chapter 4).
- Run-up time of lamps. Need for bridging lighting (chapter 14). Availability of instant-output lamps or hot-restrike lamps.
- Safe means of access during installation and for future maintenance work. Need for permanent built-in means of access (section 23.3).
- Need for zones of intermediate illuminance at entrances (section 9.6).
- Any special dangers during the installation period. Safety for welding to be done in the installation zone (chapter 7).

Contract operation considerations

- Need to harmonize methods, use of labour, equipment or means of access with those employed in earlier, concurrent or following contract.
- Timing of contract; earliest and latest starting and finishing dates; possible hours of work.
- Availability of secure dry storage for lamps and luminaires before installation on site.

24.2.2 Factors affecting exterior lighting

Many of the factors listed in section 24.2.1 will apply equally to exterior lighting installations, but the following additional factors should be considered:

- Strength of buildings to which luminaires may be fixed, giving consideration both to weight and windage.
- Constraints on siting of poles and towers carrying luminaires.
- Constraints on locating and aiming luminaires which could cause danger or nuisance to others.
- Protection of power lines, luminaires etc from vandalism or accidental damage.

24.2.3 Outline lighting specification

When all the foregoing matters in this chapter have been given consideration, the decisions should be recorded to enable the compilation of an 'outline lighting specification' as agreed between the client and the designer. This document will form the basis of enquiries to lighting providers and installers and thus ensure that competing offers are made on an equal basis. The minimum headings are listed below.

Part A – Basic lighting parameters

The client or his consultant shall specify the following requirements:

- the general illuminance levels (lux) for each area;
- the plane of measurement for general illuminance (usually taken at 0.85 m above floor level);
- the limiting glare index for each area;
- the uniformity of illuminance for each area;
- the minimum measured illuminance requirements;
- the requirements for colour appearance and colour rendering.

Part B – Data to be provided by tenderers

The candidate lighting provider shall be required to state:

- the total electrical load including gear losses;
- the power factor of the whole lighting installation and its parts;
- the utilization factor and light loss factor employed in the calculations for each area, and the proposed frequencies of cleaning and relamping;

- that the equipment will be suitable for the environmental conditions in the places of use;
- the ten-year cost (section 21.3) of the installation at agreed present-day value rates for energy and labour;
- agreement to abide by an arbitration clause (section 24.3.3) in the event of a contract to supply being offered.

24.3   Management of the tender

24.3.1   The general handling of a tender for interior or exterior lighting is not basically different from that for other goods and services except for the twin problems of 'two-tier quoting' and 'skinning'.

Two-tier quoting refers to the practice of calling in a lighting provider early in the progress of the scheme, and getting the benefit of his advice, e.g. preliminary estimates of cost, help with drafting the outline lighting specification (section 24.2), and generally treating the supplier as a 'free consultant'. This would be unremarkable if the supplier was then automatically made the nominated supplier and could be certain that the cost of any pre-order efforts for the client would be compensated for by receiving the order. Sadly, it often happens that, having got all the problems solved by the helpful supplier, the client goes out to tender and awards the order to another lighting provider. Sometimes the client may not even include the first supplier in the list of those invited to tender. There is a high probability that another supplier (who has not had to bear the cost of advising and preparing a preliminary scheme) will be able to offer a marginally lower price. The system is thus unfair to the obliging lighting provider who was contacted first, and may in time rebound to the disadvantage of the client, since it is not unknown for suppliers to quote high to such customers. It would be illegal for suppliers to collaborate in this, and in time another supplier will come along who will underbid them.

To avoid this unethical and unbusinesslike procedure, it is suggested that in those cases where no consultant is employed, the client should attempt to devise his own outline lighting specification on the basis of the guidelines given in this book, and should then call for schemes and quotations against that specification. Without such a specification, the client will have difficulty in trying to compare the value of unlike schemes, and will lay himself open to the second malpractice of 'skinning' (section 24.3.2).

24.3.2   *Skinning* refers to the malpractice of lighting suppliers quoting the prospective client for a lighting scheme, but being deliberately vague as to the parameters of their design. Thus, the scheme may appear to be attractive because of its low capital cost, but the client is not made properly aware that the overall cost, e.g. the ten-year cost (section 21.4) may be far higher than for another scheme which yielded lower energy consumption or cost of maintenance, etc. If selection of schemes was always made solely on capital cost, factories would be lighted again with filament lamps – just as they were 50 years ago! The data proposed for the potential lighting supplier to provide (section 24.2.3, Part B) should be matched against the minimum basic data provided by the client (section 24.2.3 Part A), plus the client's other specific requirements.

24.3.3   While it is expected that the client will undertake purchasing according to his customary practices and the laws of the country, it will be sound practice to insert an arbitration clause into the contract, worded generally as follows or as advised by a lawyer:

> In the event of a dispute between the purchaser and the vendor, either may require the other to enter into arbitration to settle the matter in contention. The procedure shall be that each party shall appoint a single arbitrator to act on his behalf, and the two arbitrators so appointed shall appoint a third arbitrator agreeable to them both who shall act as chairman of the arbitration, and who shall have an equal vote in any matter voted on by the three arbitrators. The decisions of the three arbitrators as decided by a majority vote shall be binding on both the purchaser and the vendor.

24.3.4   The client may decide to appoint a lighting consultant (section 24.1), who may arrange for the work to be carried out in phases according to the needs of the project:

> *Phase 1*: Assessment and preliminary report. This may include first estimates of cost, and preparation of an outline lighting specification.
> *Phase 2*: Scheme preparation. (Unless deputed to one or more lighting manufacturers to submit their proposals.) In smaller projects, Phases 1 and 2 are sometimes combined.
> *Phase 3*: Tender management on behalf of the client.
> *Phase 4*: Assisting with the selection of the successful tender, and placing of orders.
> *Phase 5*: Acceptance and commissioning the installation.

In the above work, the consultant will collaborate and consult with those concerned in the project, the architect and other consultants, Health and Safety Executive officers, Fire Officers and insurers. The consultant's duty is to protect the client's interests in all matters relating to the project, and to try to achieve the objectives agreed with the client at the outset at the lowest cost.

Remuneration of the lighting consultant is usually on a daily basis rather than the fee being a percentage of the project cost, but practices vary from place to place across the world. One system of remuneration that is sometimes used is for the consultant to work on a daily basis until the design brief is settled, and then to quote a fixed fee for the subsequent stages of the project.

24.3.5   Although there is an obvious advantage in the person who advises on the lighting having no financial connections with any supplier, there are reputable lighting providers who will undertake design work on a fee basis, often with the proviso that, if they are the successful tenderers, the fee for design services will be rebated against the cost of the installation. With companies of good repute, this will work excellently, and provides a method of avoiding two-tier quoting and skinning (sections 24.3.1, 24.3.2).

24.3.6   Quotations may contain alternative offers, or contain matters to be decided by the client. It may therefore be necessary to provide a final specification which may modify some of the data in the outline lighting specification after consideration of the offers received, and this will be the basis for the order on the chosen lighting provider. If the installation of the lighting equipment and electrical installation wiring is not to be carried out by the lighting provider, the client or his consultant can at this stage invite tenders for the installation work from suitably qualified electrical installation contractors. The appointed electrical installation contractor will be made responsible for ensuring compliance with the details of the final specification, and for ensuring that the installation meets all requirements for electrical safety (chapter 22).

24.4   Supervision of the contract and acceptance tests

24.4.1   Because the person responsible for designing a lighting installation is not usually also responsible for supervising the installation of the equipment, the need for such essentials as providing means of safe and easy access to overhead luminaires and wiring may be overlooked until necessity forces attention to the problem (section 23.3). Commonly, there are unnecessary delays in the progress of installation work because the activity clashes with other work going on in the same space. For example, electricians may be unable to work overhead because the floor is being screeded, or a convenient system of scaffold platforms erected for the plasterers (and which would be ideal for working on the high-level wiring and installing luminaires) is removed at the end of the plastering contract – perhaps just a day or two before the electricians arrive. Such wasteful and irritating occurrences might be avoided by obtaining greater collaboration between all parties on site and co-ordinating their activities. One method of scheduling activities in their best sequence is to prepare a critical path analysis.

24.4.2 For the duration of the installation contract, there should be provision made for the secure storage of all the materials and components to be used, not only for their protection against pilfering, but also against damp and the weather.

24.4.3 If the building has not dried out and is unheated, installation of any luminaires that are not adequately rustproofed should be delayed until just before the area is to be occupied (section 7.1).

24.4.4 If a newly lighted area is not to be immediately occupied, or cannot be made secure against unauthorized entry, it would be advisable to remove HID lamps from the luminaires, and store them elsewhere securely until the installation is about to be brought into use. These lamps are prone to being stolen, particularly before they show obvious signs of use. Some lamp manufacturers will permanently mark lamps with the user's name or identifying number or symbol to prove ownership in the event of theft.

24.4.5 When refurbishing existing premises, it may be necessary to leave an old lighting installation in operation until the new one is complete. Precautions must be taken to avoid inadvertently energizing incomplete circuits (thus exposing personnel to risk of electric shock), by instituting a permit to work scheme and locking off supplies to unfinished sections in accordance with the *Electricity at Work Regulations*.[18]

24.4.6 No circuit should be energized for the first time until it has been checked with a low-voltage continuity tester, and the insulation between lines and from lines to earth has been checked with an insulation test instrument in accordance with the *IEE Wiring Regulations*[27], and the appropriate certificate completed by a competent person. Luminaires which contain ignitors or other solid-state circuitry (control-gear components or components of the emergency-lighting facilities) will be damaged by over-voltage or reversed voltage applied during insulation testing. Such circuits should be shorted-out before any insulation test is performed.

24.4.7 At the completion of the electrical installation work, the name, address, and telephone number of the electrical installation contractor (and of the local electricity supply company) should be permanently recorded and displayed prominently near the main switch of the installation. It may be a requirement of the inspecting Fire Officer that a circuit diagram and a location diagram of switches etc also be displayed, and that a copy be readily available to the Fire Brigade in any emergency.

24.4.8   If the lighting provider supplies the required data, it will be possible to carry out a lightmeter check to see if the installation produces the specified illuminance. This is a contentious matter, over which purchasers are often short-changed by unscrupulous providers. If the initial lumens, the lighting design lumens and the end-of-life lumens (terminal lumens) per lamp are known, as well as the utilization factor and maintenance factor used in the calculations, the light loss factor can be determined to enable a forecast to be made of future actual minimum illuminance or service illuminance by the method given in the *CIBSE Code*.[1] Lightmeter tests must be delayed by 100 operating hours to allow for the decay of the initial 'bonus lumens' from new lamps. It may be a contract condition for there to be a 10% or 15% 'hold-back' of the provider's bill, payable only after satisfactory acceptance testing.

Appendix A
UK legislation on industrial lighting

Before British manufacturers came under the EEC Low Voltage Directive 73/23/EEC in 1973, they had no legal compulsion to comply with any safety standard other than the requirements of the Common Law. As the UK is a member of the EC, there is a statutory requirement for all electrical equipment for low voltage (under 1000 volts) sold and used in the UK to be safe by common safety standards. This requirement was originally imposed by the Electrical Equipment (Safety) Regulations, 1975, as amended 1976, and the Consumer Protection Act, 1961, now replaced by the Low Voltage Equipment (Safety) Regulations under the new Consumer Protection Act, 1987.

Thus, as David Bertenshaw has pointed out,[44] after 94 years of public use of electric lighting, it became for the first time a legal duty to provide luminaires to a specific safety standard. The requirements are now embodied in EuroNorm EN60958 (1989), which contains hundreds of requirements which cannot be reviewed here beyond noting a typical snippet from Part 2.5 (Floodlights). That section, after imposing extra safety requirements regarding construction and marking, requires floodlights intended for mounting over 3 m high to have two independent fastenings.

The general situation in the UK is that the law requires occupiers to make certain provisions for the health and safety of all occupants. Both the *Health and Safety at Work Act*, 1974[42] and the *Offices, Shops and Railway Premises Act*, 1963[63] stipulate that the occupier shall provide 'sufficient and suitable lighting'. If prosecuted for failure to provide this, a good defence would be to prove that the lighting in fact complied with the provisions of the current edition of the *CIBSE Code for Interior Lighting*.[1] The *Fire Precautions Act*, 1971[51] requires the occupier to provide means of escape which are 'capable of use at all relevant times', and this is construed as meaning that emergency lighting shall be provided in compliance with *BS 5266:Part 1:Emergency Lighting*.[47]

The safety of persons on industrial premises is dependant on the provision of electrical installations and lighting equipment certified as suitable for the environmental conditions in which they are situated. For example, there may be hazardous atmospheres requiring the use only of flameproof electrical enclosures. Parts of the premises may be designated as Zone 0 (may contain no electrical apparatus other than equipment that is certified as 'intrinsically safe'), or as Zone 1 or Zone 2 (in which only equipment with appropriate certification may be employed). It is the responsibility of the occupier (i.e. the management) to ensure that these important regulations are properly observed. If as is usual, the personnel of the factory are not qualified to declare the nature of the zoning and the extent of the zones, expert help must be obtained. Some advice may be given by the Health and Safety Executive Inspector, but it is normal good practice to employ expert professional consultants who will issue certificates designating the zoning. It is important to note that in addition to the hazards due to flammable atmospheres, some atmospheres are also corrosive, so that the lighting equipment and installation components may need to be corrosion resistant as well as certified for the appropriate Zone of use.

As regards the *Electricity at Work Regulations*, 1989,[18] it may not be generally realized that it is only necessary for the HSE inspector to prove that an electrical injury has occurred to establish an offence under this legislation. An excellent first step towards ensuring compliance will be to ensure that all luminaires comply with the relevant provisions of *BS 4533: Electric luminaires*[30] as well as being enclosed in electrical casings which comply with the appropriate protection grade under the IP System as regards resistance to impact by solid bodies, and penetration by foreign bodies, dusts and liquids.

Lighting used by any occupier of premises must not cause hazard or nuisance to others. In particular it should be noted that at coastal situations, it is an offence to project light to seaward which is 'confusable with a navigation light or a hazard to mariners'.

Persons who suffer pollution by someone else's light can obtain relief and damages at law. Many successful cases have been fought in UK courts relating to:

- *Nuisance*. For example, light shining into a bedroom.
- *Hazard*. The glare from lighting outside a property prevented the owner from walking safely on his own land; light projected towards a railway handicapped the drivers of trains; and similarly drivers of road vehicles suffered embarrassment of vision.
- *Interference with crops*. Polluting light caused chrysanthemums to bloom too early for the Christmas trade and the flower grower suffered loss.
- *Interference with livestock*. Chickens exposed to stray light from adjacent premises could not maintain their daily laying rhythm, with resultant loss to their owner because of small and premature eggs.

In the rapidly changing situation as the EC Single Market comes into existence, the reader is advised to obtain frequent updates on the new EC rules. These are available from the Department of Trade and Industry (Hotline tel: 081-200-1992) which has booklets available on many new rules including:

- *Machinery Safety Directive*. Relates to 'functioning machines', that is, assemblies of mechanically linked parts, at least one of which moves.
- *Mobile Machinery and Lifting Equipment Directive*. This widens the coverage of the Machinery Safety Directive to include most mobile machinery and lifting equipment.
- *Personal Protective Equipment Directive*. Relates to any appliance designed to be worn or held for protection against any safety or health hazard.
- *Electromagnetic Compatibility Directive*. Applies to almost all electrical and electronic appliances, equipment and apparatus.
- *Construction Products Directive*. Relates to construction products produced for incorporation in a permanent manner in construction works including buildings and civil engineering works.

The efficient use of energy in providing lighting for industry will be of increasing importance in future years. An extensive framework of EC legislation is in preparation in relation to carbon dioxide emmissions and energy economy. Directives have been proposed or are being prepared in the following areas:

- energy taxation;
- energy auditing and energy management;
- energy certification and insulation of buildings;
- environmental auditing;
- energy conversion.

It is not claimed that this appendix gives a definitive review of the subject of current and future regulations and standards relating to interior and exterior industrial lighting; nor would such a review remain up to date for long. However, a most useful current reference is a paper by Baker[45] which lists over fifty UK Acts of Parliament, Statutory Orders and Statutory Rules and Orders, as well as a similar number of British Standards and Codes of Practice, all of which relate to lighting.

Appendix B
CIBSE Code for Interior Lighting

B.1 Introduction

Throughout this book frequent reference is made to the *CIBSE Code for Interior Lighting*.[1] The document referred to is the 14th edition which was published in 1984 with reprints in 1985 and 1989. For over half a century, the former Illuminating Engineering Society (IES) and its successor the Chartered Institution of Building Services Engineers (CIBSE) have brought out successive editions of this invaluable guide to good interior lighting practice. At the time of writing, a task group of the CIBSE is drafting the 15th edition which is expected to be published in 1992.

B.2 Development of the *Code*

The recommendations of successive editions have reflected changes in lighting practice arising out of developments in lamps and luminaires, changes in architectural practice, and changes in the lighting needs of industry and commerce. In editions prior to 1968, illuminances were specified in foot-candles ($1\,\text{fc} = 1\,\text{lm/f}^2 = 10.76\,\text{lx}$). The lighting industry changed to metric units and adopted the use of the lux in 1968. That edition included recommendations for a limiting glare index, and contained guidance on the selection of lightsources for their colour properties. It was also the last edition to specify illuminance recommendations as 'minimum service illuminance' values; later editions have quoted instead a *standard service illuminance* for each type of location or task, this being a basic recommended illuminance level which may be adjusted upward or downward according to circumstances.

B.3 Validity of the *Code*

There have been times in the past when the *Code* has been attacked by critics who accused the compilers of advocating wastefully high illuminances. It has even been suggested that the *Code* was merely a means of promoting sales of lamps and lighting equipment. It was argued that, if the basis of the illuminance recommendations in the original *Code* were sound, it would not have been necessary to increase the recommendations through successive editions. It must be stated that the former IES and the CIBSE have always been strong advocates of the wise and economical use of energy, and have consistently indicated to their readers the most suitable and economical way of achieving lighting objectives. The suggestion that there has ever been any commercial bias to promote excessive lighting levels is utterly refuted.

The very minimal illuminances specified in earlier editions have been gradually upgraded, reflecting the increasing need for better lighting standards with the development of industry. Users of lighting today have higher expectations for both the quantity and quality of lighting. The development of lamps of higher efficacy and longer technical life has resulted in a steady fall in the true cost of electric lighting, making possible the adoption of enhanced illuminances. Improvements in lamp efficacy have been accompanied by the availability of lamps having better colour properties, and of luminaires which have better light control and higher utilance of light – these factors again contributing to the decreasing true cost of lighting.

The relationship between task illuminance and the visual performance of workers is clearly established, but the relationship is not absolute. Illuminances which are a little lower than the *Code's* recommendations will not noticeably change worker performance – at least not in the short term. However, for older workers, and for those with a visual handicap, and for all subjects performing visually critical tasks, any reduction of illuminance below the recommended levels could bring the resolution of the subject's eyes closer to the point where efficient vision was not possible, increasing the stress and fatigue experienced by the subject. Conversely, illuminances higher than the *Code's* recommendations are unlikely to bring about significant enhancement of task performance in the short term, but well-engineered lighting provided at levels above the recommended illuminances can only be beneficial.

These days the user gets some four to five times as much light per unit of cost (viz, lumen-hours per pound) than 30 years ago, and the curve is still rising. Thus, the illuminances currently employed would have been totally impracticable in former times, when only less efficacious sources were available.

The true cost of lighting today is minute compared with other business overheads. Even the very best industrial lighting only costs typically between 0.2 and 0.4 pence per pound's-worth of goods or services sold (section 21.2.1).

B.4   The next edition of the *Code*

International liaison and consultation on lighting standards is proceeding. In 1989, CEN Technical Committee TC169 commenced its work by comparing the existing illuminance recommendations of the professional lighting associations of eight EC and EFTA countries and its work will continue far beyond 1992. There are indications of a good level of general agreement.

As regards indices of discomfort glare, some countries favour the luminance curve system which results in a letter representing a degree of glare, while other countries prefer the glare index system as used in the UK. It seems that the outcome will be harmonization of the two methods, with the former method being used for luminaire specification purposes and the latter method being used for complete installations.

The Committee is considering other factors of lighting design, including uniformity of task illuminance. Colour appearance and colour rendering have been defined by the CIE.

B.5   Illuminances; examples from the *Code*

While some information on the present *Code's* recommendations is given here, no summary or condensation can equal the value to the reader of studying the full text which – apart from the tables of recommended illuminances – contains a great deal of sound guidance on the technology of interior lighting.

The *Code* recommends a standard service illuminance for each application or location, and this value must be adjusted for special conditions to arrive at the design service illuminance as explained in the following procedure. First, refer to Table 18 to find the standard service illuminance appropriate to the activity or interior. Second, make the cumulative adjustments to the value as indicated below in accordance with the answers to the following questions. The steps of illuminance used are 50, 100, 150, 200, 300, 500, 750, 1000, 1500 and 2000 lx.

1. Are the task details unusually difficult to see? (If 'NO', take no action; if 'YES', *increase* the illuminance one step.)
2. Are the task details unusually easy to see? (If 'NO', take no action; If 'YES', *reduce* the illuminance one step.)
3. Has the task to be done for an unusually long time? (If 'NO', take no action; If 'YES', *increase* the illuminance one step.)
4. Has the task to be done for only an unusually short time? (If 'NO', take no action. If 'YES', *reduce* the illuminance one step.)
5. Is visual impairment widespread among those doing the work? (If 'NO', take no action; If 'YES', *increase* the illuminance one step.)
6. Do errors have unusually serious consequences to people, plant or product? (If 'NO', take no action. If 'YES', *increase* the illuminance one step.)

It is the author's suggestion that, after completing the foregoing adjustment procedure, the illuminance should be raised one further step if eye protection (section 8.4) must be worn continuously.

If the design service illuminance is more than two steps on the illuminance scale above the standard service illuminance, consideration should be given to whether changes in the task details, the organization of the work or the people doing the work would be more appropriate than adopting a higher illuminance.

If the design service illuminance arrived at is 1000 lx or more, note that such illuminances are usually best provided by a combination of a reasonable level of general lighting (say, 500 or 750 lx) plus suitable arrangements of local or localized lighting (section 9.5) to provide the design service illuminance which has been determined by the foregoing procedure.

Table 18 *Illuminances for interior lighting. Examples of standard service illuminance values appropriate to particular activities and interiors, extracted from CIBSE Code for Interior Lighting. (Numbers in brackets are Limiting Glare Indices.)*

Standard service illuminance (lux)	*Characteristics of the activity or interior*
50	Interiors visited rarely; visual tasks confined to movement; casual seeing without perception of detail. *Examples:* cable tunnels; indoor storage tanks; walkways.
100	Interiors visited occasionally; visual tasks confined to movement; casual seeing calling for only limited perception of detail. *Examples:* corridors; changing rooms; bulk stores.
150	Interiors visited occasionally; visual tasks requiring some perception of detail or involving some risk to people, plant or product. *Examples:* loading bays – large materials (25); medical stores; switchrooms; boiler houses (25); pump houses; stairs; gangways in steelworks; kitchens – food stores; rest rooms (19).
200	Continuously occupied interiors; work without difficult visual tasks; movement and casual seeing. *Examples:* assembly shops, casual work (25); generator station turbine halls (25); general lighting for automatic processes (25); warehouses – small materials, racks (25); chemical raw material stores (25).
300	Continuously occupied interiors; moderately easy visual tasks, i.e. rough work; rough machining and assembly; simple tasks with large details, e.g. >10' arc, and/or high contrast. *Examples:* glassworks – mixing rooms (25); steelworks – mould preparation (28); laundries – receiving, sorting, washing (25); leather works – general lighting (25); office print rooms (19).

Table 18 *Continued*

Standard service illuminance (lux)	*Characteristics of the activity or interior*
500	Routine work with moderately difficult visual tasks, i.e. details to be seen are of moderate size, e.g. 5–10′ arc, and may be of low contrast; colour judgements may be required. *Examples*: engine and vehicle body assembly (22); aircraft fabrication and inspection (22); kitchens – working areas (22); general lighting of drawing offices (16); general offices – clerical (19).
750	Difficult visual tasks, i.e. details to be seen are small, e.g. 3–5′ arc, and of low contrast; good colour judgements may be required. *Examples:* drawing offices – on boards (16); ceramic decoration; meat inspection; printing machine room – presses (22); stores issue counters (22); woodworking – medium bench and machine work (22); deep-plan general offices (19).
1000	Very difficult visual tasks, i.e. details to be seen are very small, e.g. 2–3′ arc, and may be of low contrast; accurate colour judgements may be required. *Examples*: inspection of telecommunication equipment (19); paintworks – colour matching (19); fine bench and machine work (22); printed sheet inspection (19); jewellery – fine processes (16).
1500	Extremely difficult visual tasks, i.e. details to be seen are extremely small, e.g. 1–2′ arc, and of low contrast; visual aids may be an advantage; fine colour discrimination may be needed. *Examples*: fine, intricate gauging and inspection (16); gem cutting, polishing and setting (19); upholstery – cloth inspection (16); hand tailoring (19); fine die sinking (19); assembly of minute mechanisms; fine soldering, work on scientific instruments (19).

Note: Reference to the *CIBSE Code*[1] itself is recommended to obtain detailed lighting recommendations for industrial locations and processes, together with guidance as to position for measurement of illuminance, and the colour properties of the lamps to be used, etc.

Appendix C
Illuminances for exterior lighting

C.1   Illuminances for outdoor applications

Occupiers must satisfy the requirements of local legislation and the lighting codes of the country in which a lighting installation is located, as well as contractual requirements. In general, the lighting requirements for minimum standards of safety and health are greatly exceeded by the levels of lighting commonly employed for ordinary efficient operation of plant and premises. There is no general agreement as to the levels of illuminance required for outdoor security lighting, for outdoor emergency lighting, or for many types of outdoor work, and anomalies exist between recommendations for the same function when performed indoors or outdoors.

The minimum measured illuminance (MMI) values given in Tables 19 and 20 are used for checking the installation, and are the values (measured on the

Table 19   *Illuminances for outdoor emergency lighting*

Activity	*Standard design illuminance (lux)*	*MMI (lux)*
Escape along safe and known routes: emergency exit lanes, or walkways and paths where the users are familiar with them or the route is level and not dangerous to traverse	1, or 1% of the normal illuminance (whichever is the greater)	0.2
Escape along dangerous or unknown routes: emergency exit lanes, walkways and paths where the users are not familiar with them or the route is uneven or possibly dangerous to traverse, and involves risk of falls or contact with hot or sharp objects etc.	5, or 5% of the normal illuminance (whichever is the greater)	1

ground or on the working plane as appropriate) below which the illuminance at any point should not be allowed to fall. They are values which should be sought for and checked regularly with a lightmeter, and the appropriate actions taken if the MMI level is not found at any point in the lighted space.

C.2 Outdoor emergency lighting

The determination of illuminances for emergency lighting for outdoor workplaces (chapter 15) is not a subject upon which lighting experts are yet generally agreed. The illuminance values given in Table 19 are the author's proposals.

C.3 Lighting for outdoor work

The recommendations for illuminances to be employed for outdoor work (Table 20) are based on recommendations given in the publications of the Chartered Institution of Building Services Engineers, the publications of the former Electricity Council, the International Labour Organization, and the British Standards Institution, together with the author's interpretations of their intentions and his own recommendations based on modern practice.

To understand why the steps between the levels of recommended illuminances in Table 20 are so arranged, it is necessary to appreciate that the response of the eye is substantially logarithmic over its normal range, i.e. the eye must be presented with an illuminance that is ten times its predecessor before we judge it to be twice as bright.

Making order out of a number of conflicting attempts at standardization (and not using the commonly employed 150-lux step), one can select eight levels of illuminance that will satisfy practically every requirement for outdoor industrial work, the levels being 5, 20, 30, 50, 100, 300 and 500 lx.

The *standard service illuminances* in Table 20 must be adjusted for special conditions to arrive at the design service illuminances. Refer to Table 20 to find the standard service illuminance appropriate to the activity etc. Then adjust the value as indicated in accordance with the answers to the following questions to arrive at the *final design illuminance*.

1. Will the task be performed for a short duration only? (If 'YES', *reduce* the illuminance by one step.)
2. Are contrasts and reflectances unusually low? (If 'YES', *increase* the illuminance by one step.)

3. Will errors have serious consequences? (If 'YES', *increase* the illuminance one step.)
4. Must eye protection be worn continuously? (If 'YES', *increase* the illuminance by one step.)

Table 20   *Illuminances for outdoor work*

Activity	Standard service illuminance (lux)	MMI (lux)
Non-working areas: for supervision of site; vehicle movements; pedestrian movement on safe paths	5 general lighting	1
Normal movement: movement of people, powered machines and vehicles; loading and unloading of bulk materials; walkways and ladderways, staircases and access routes; car-parks and lorry parks	20 general lighting	5
Stores and stockyards: general lighting of stores and stockyards, container parks, yards where goods and stock are stored	30 general lighting	5
General work areas: site clearance, excavations and soil work; stacking and supervision of outdoor stores; washing-down vehicles; loading and unloading large items; checkpoints, weighbridges, site entrances	50 general lighting	20
Simple visual tasks: setting reinforcing rods; concreting, erecting shuttering; bricklaying (not facings), blockwork; erecting and dismantling scaffolding; ordinary work with tools; stuffing and stripping containers; operating hoists and cranes; intermittent reading; loading conveyors; stowing cargo	100 on the task plus 50 general lighting	50

Table 20.   *Continued*

Activity	Standard service illuminance (lux)	MMI (lux)
More difficult visual tasks: on workbenches; all work with power tools or circular saws, grinders, sanders etc; plastering, screeding, terrazzo; ordinary painting; brickwork facings; joinery; pipe-fitting; masonry work	300 on the task plus 100 general lighting	200 at the task 50 general lighting
Demanding visual tasks: paintwork, rubbing-down, spraying; machining metals and wood; reading micrometers and instruments; continuous reading, writing, typing; inspection of craft work	500 on the task plus 200 general lighting	300 at the task 100 general lighting

Note: The illuminance recommendations in this table relate to typical tasks and their visual demands; if more difficult visual tasks are to be served, then the illuminance recommendations of the *CIBSE Code*[1] should be followed. Whether a task is to be performed outdoors or indoors, the illuminance recommendation is the same.

The final design illuminance is the value which should be used in planning the scheme, and is the illuminance which will be received as an average over the area when the light loss due to dirt on lamps and luminaires is at its average for the chosen cleaning cycle, and when the lamps are emitting their 'service lumens' (their average lumens through life). It will be seen that the questions above allow for a reduced illuminance to be employed if tasks are performed for short durations only.

Providing general lighting at illuminances higher than 200 lx out of doors is not practicable if applied over large areas. For outdoor workshop areas, 500 lx might be provided, but generally it is better to light the general working area to 200 lx, and to provide task lighting to bring the illuminance at the point of work to the desired level.

C.4   Illuminances for security lighting

Refer to Chapter 17. The recommendations for illuminances to be used for outdoor security lighting (Table 21) are considerably lower than were advocated when the first recommendations were made in 1969. The present recommendations are based on experimental field work carried out under the auspices of the former Electricity Council.[60]

Table 21   *Illuminances for security lighting*

| Degree of risk | District brightness[a] | | |
	High (lx)	Medium (lx)	Low[b] (lx)
Class A, extreme risk	20	10	5
Class B, high risk	10	5	2
Class C, moderate risk	5	2	1

[a] District brightness: *high* with adjacent main-road lighting, lighted land nearby, nearby floodlighting etc; *medium* with adjacent secondary road lighting or footpath lighting nearby; *low* with no lighting on roads or land nearby.
[b] Somewhat lower illuminances may be employed in conditions of low district brightness if the flow of light is substantially horizontal (as explained in section 17.5).

Appendix D
Conversion factors

Length

	Metre	Foot	Inch
One metre =	1	3.281	39.37
One foot =	0.305	1	12
One inch =	0.0254	0.0833	1

Inches to millimetres

Inches	0	0.25 (¼)	0.50 (½)	0.75 (¾)
0	–	6.35	12.7	19.05
1	25.4	31.75	38.1	44.45
2	50.8	57.15	63.5	69.85
3	76.2	82.55	88.9	95.25
4	101.6	107.95	114.3	120.65
5	127.0	133.35	139.7	146.05
6	152.4	158.75	165.1	171.45
7	177.8	184.15	190.5	196.85
8	203.2	209.55	215.9	222.25
9	228.6	234.95	241.3	247.65
10	254.0	260.35	266.7	273.05
11	279.4	285.75	292.1	298.45
12	304.8	311.15	317.5	323.85

Illuminance

	Lux	1 m/ft^2 or footcandle
One lux =	1	0.093
One lumen per square foot* =	10.76	1

* formerly termed a footcandle

Conversions of luminance units

	cd/m^2	stilb	cd/in^2	apostilb	lambert	footlambert
Candela per square metre (cd/m^2)	1	0.0001	0.000645	π	$\pi \times 10^{-4}$	0.292
Stilb (cd/cm^2)	10000	1	6.452	$\pi \times 10^4$	π	2919
Candela per square inch (cd/in^2)	1550	0.155	1	4869	0.487	452
Apostilb (lm/m^2)	$1/\pi$	$1/(\pi \times 10^4)$	0.000205	1	0.0001	0.0929
Lambert (lm/cm^2)	$10^4/\pi$	$1/\pi$	2.054	10000	1	929
Footlambert (lm/ft^2)	3.426	0.0003426	0.00221	10.76	0.001076	1

Multiply a unit in the left hand column by the factor shown to convert to a unit in the top line.

Appendix E
Lightmeters

Illuminance may be measured with a portable instrument called a lightmeter (also called a luxmeter or portable photometer). The light cell of the instrument is held in the plane of measurement, and the illuminance in lux read from the scale. Such instruments commonly have ranges of 10 lx to 5000 lx full scale deflection, and are made to an accuracy of ±15%. Accuracy of ±5% may be achieved in use if the lightmeter is regularly calibrated and the readings are weighted with correction factors supplied by the test house. Higher accuracy instruments are available, but are not needed for ordinary lighting work. Lightmeters reading to much lower levels (e.g. 5 lx full scale deflection) are available for measuring emergency lighting and exterior lighting. Lightmeters are covered by *BS 667*:1968.[24]

It is recommended that a suitable lightmeter be held available in every factory, for it is as easy to use as a thermometer, and just as essential. Its use enables existing lighting to be measured and compared with the recommendations of the *CIBSE Code*,[1] and enables the user to ascertain if cleaning, relamping or upgrading of existing installations is necessary.

E.1 Measuring average illuminance

A lightmeter indicates the illuminance at the point of measurement, not the average in the space. To find the average over an area, it is necessary to divide the area into a number of equal areas which should be as square as possible. The illuminance at the centre of each square is then measured, and the results averaged. The minimum number of equal areas required for accuracy is determined by first calculating the room index (k), thus:

$$k = \frac{\text{Length} \times \text{Width}}{H_m \times (\text{length} + \text{width})}$$

where H_m is the height of the luminaires above the plane of measurement. The working plane is taken at 0.85 m (common bench height) unless specified, or at floor level (if work is performed down to floor level).

The required number of measurement points relates to k thus: if k is less than 1, 4 points; between 1 and 2, 9 points; between 2 and 3, 16 points; and, if 3 or more, 25 points. If the proposed points coincide with luminaire positions or are in constant relationship with them, increase the number of measurement points.

When the average illuminance has been calculated, compare the result with the recommendations of the Code[1] for the type of activity in the area.

E.2   Measuring illuminance in daylight-lit rooms

To take a lightmeter reading in a room into which daylight penetrates, take the reading at a selected point, then switch off the lighting and take the reading again. Do not delay in this, for daylight is very variable. Subtract the second reading from the first.

E.3   Measuring reflection factors

A lightmeter can be used to determine the approximate reflection factor of a non-glossy surface of reasonably uniform colour.

Method 1

1. Measure the illuminance at the surface (E_1).
2. Take a lightmeter reading at the same point, but with the lightmeter cell facing the surface and held 300 mm from it (E_2).
   Then

$$\mathrm{RF} = \frac{E_2}{E_1}$$

Method 2 (slightly more accurate)

1. Take a lightmeter reading at a point on the surface but with the lightmeter cell facing the surface and held 300 mm from it (E_2).
2. Take a sheet of clean white non-glossy paper of A4 size or a little larger – say 300 mm × 300 mm, of assumed reflection factor 0.9, and place it on the surface being tested with its centre approximately over the previous point of measurement.
3. Take a lightmeter reading at the centre of the paper in this position with the lightmeter cell facing the paper and held 300 mm from it (E_3).

Then

$$RF = \frac{E_2}{E_2 \times 0.9}$$

For best results the light should flow reasonably normal to the surface and be from a source of large area. Accuracy can be improved if the cell 'sees' the surfaces in method 2 through a matt-black-lined tube (say, of cardboard) which excludes ambient light.

The principle of this method can be used at a higher level of accuracy if, instead of a sheet of white paper, one uses a photometric test tile of known reflection factor (as may be obtained from the National Physical Laboratory), and if the measurements are made not with a lightmeter but with a *luminance meter* as supplied, for example, by Minolta (UK) Ltd.

E.4   Care and use of lightmeters

Avoid subjecting a lightmeter to excessive vibration or shock. Do not allow it to become hot by, for example, leaving it on a radiator or in direct sunshine. Take care not to expose the cell to over-bright sources, e.g. the sun or any source that drives the needle beyond the scale limit or drives a digital indicator to overload; such use can permanently damage the cell. Keep the lightmeter in its case when not in use.

Do not shade the cell with your body when taking readings. For accuracy it may be necessary to repeat readings at a point, but with the operator standing in a different position relative to the cell. Lightmeters fitted with a pointer-lock may be used to take readings in difficult situations where the operator would shadow the cell if attempting to read it. Place the cell at the point of measurement, stand as clear as possible and apply the lock, then take the instrument from that position and read the indication. Follow the manufacturer's instructions as to whether the pointer-lock should be applied when the instrument is stored or carried about.

Shading of the cell by the operator's body can be avoided by use of a 'wand' (Figure E.1). This is also a convenient way of taking a succession of readings at ground level without fatigue.

Before using a lightmeter that has not been in use for weeks, expose the cell for ten minutes or so to an illuminance that moves the pointer or digital indicator to about the centre of the scale; then subject it to several swings of up to full scale deflection by facing the cell towards a suitable lightsource, taking care not to expose it to a greater illuminance than that catered for by that scale.

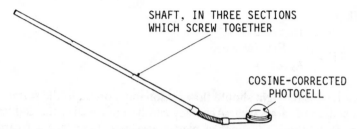

SHAFT, IN THREE SECTIONS
WHICH SCREW TOGETHER

COSINE-CORRECTED
PHOTOCELL

Figure E.1 *Photocell wand to hold cell in position without shadowing it with the operator's body. Convenient also for taking floor-level readings.*

If a scale-change switch is fitted, note that high-resistance in its contacts can cause inaccuracy of readings. To check, cover the cell (for zero reading), and move the scale-change switch between all its settings several times. Then uncover the cell and take readings in a position where the measured illuminance can be read on either of two scales, and compare the readings. If scale change is obtained by placing a mask over the cell, the latter test can similarly be performed by making measurements with and without the scale-change mask.

If the cell is remote from the instrument, take care not to reverse the polarity when connecting.

E.5   UK lightmeter suppliers

Some suppliers of lightmeters located in the UK:

Belvoir Lighting Consultancy*
Applegarth, Burton Lane, Whatton-in-the-Vale,
Nottinghamshire NG13 9EQ
Tel/Fax: 0949 50660

GEC Alsthom Measurements Ltd
St. Leonards Works, Staffordshire ST17 4LX
Tel: 0785 223251   Fax: 0785 212232

Megatron Ltd**
165 Marlborough Road, London N19 4NE
Tel: 071-272 3739   Fax: 071-272 5975

*Suppliers of narrow-angle-of-acceptance cells used with the Lyons goniophotometer (appendix F), and of light-cell 'wands' (section E.4).
**Undertake repair and calibration of lightmeters.

Minolta (UK) Ltd
Tanners Drive, Blakelands North, Milton Keynes MK14 5BU
Tel: 0908 211211   Fax: 0908 613497

Permic Emergency Lighting Ltd
PO Box 3, Chesterfield, Derbyshire S40 1EX
Tel: 0246 270914   Fax: 0246 275879

Appendix F
Goniophotometry

Photometry is the measurement of light, and the term is especially related to measuring the light output of a luminaire or lightsource in various directions and compiling polar distribution curves. Photometry to Certified standards requires the use of costly equipment in space dedicated to the function.

Users may need to perform photometry in-house: (a) to compare their own products with those of competitors; (b) in the development of new lighting products – especially in developing new reflector profiles and louvre designs; and (c) to check cut-off angles or to solve specific lighting problems. It is not usually necessary to perform these photometric processes to the high standards of Certified Photometric Laboratories, and experience shows that simple equipment will satisfy many requirements. For this application the Lyons goniophotometer has been developed.

The Lyons goniophotometer is a low cost instrument that enables the user to measure the light output from any luminaire or lightsource and compile a polar distribution curve. It can be operated in a small floor area (say 4 m × 3 m) without the need for full blackout. It is demountable, i.e. it can be dismantled for storage in a small space.

The equipment enables designs for specular and matt reflectors and louvres to be simulated by analogues and experimentally modified and developed. With care, fluorescent luminaires up to 2 m can be measured; shorter luminaires, and HID luminaires can generally be readily measured. Directional luminous intensities through 360° in any plane are measured in arbitrary units by use of a special variable angle-of-acceptance photocell.

It is not claimed that the Lyons goniophotometer is of the accuracy achieved by Certified Photometric Laboratories. However, it is a practical tool and, used skilfully, there is a good correlation between its results and those produced by standard photometric equipment.

A full specification is available from Belvoir Lighting Consultancy (appendix G).

Appendix G
Useful names and addresses

Absolute Action Ltd
Mantle House, Broomhill Road, Wandsworth, London SW18 4JQ
Tel: 081 871 5005   Fax: 081 877 9498
(Fibre optic systems)

Balcan Engineering Ltd
Woodhall Spa, Lincolnshire LN10 6RW
Tel: 0526 53075   Fax: 0526 52256
(Lamp crushers)

Belvoir Lighting Consultancy
Applegarth, Burton Lane, Whatton-in-the-Vale, Nottingham NG13 9EQ
Tel/Fax: 0949 50660
(Consultancy; suppliers of Lyons goniophotometer)

Blakley Electrics Ltd
Connington Road, Lewisham, London SE13 7LJ
Tel: 081 852 4383   Fax: 081 318 5284
(Temporary lighting systems)

British Electrotechnical and Allied Manufacturers' Association (BEAMA)
Leicester House, 8 Leicester Street, London EC2H 7BN
Tel: 071 437 0678   Fax: 071 734 2406

British Standards Institution (BSI)
Enquiry Department, Linford Wood, Milton Keynes MK14 6LE
Tel: 0908 221166   Fax: 0908 320856

Building Research Establishment (BRE)
Garston, Watford WD2 7JR
Tel: 0923 894040   Fax: 0923 664010
(Publications relating to wise use of energy and the greenhouse effect)

Chartered Institution of Building Services Engineers (CIBSE)
Delta House, 222 Balham High Road, London SW12 9BS
Tel: 081 675 5211   Fax: 081 675 5449
(Publications giving recommendations on practice relating to many aspects of
lighting – a list is available on application)

Commission Internationale de l'Eclairage (CIE)
(The UK representative may be contacted through CIBSE)

Cyanamid of Great Britain Ltd
3 The Potteries, Wickham Road, Fareham, Hants, PO16 7HZ
Tel: 0329 221 664
(Chemiluminescent light-sticks)

Department of Trade and Industry (DTI)
Radiocommunications Agency, Waterloo Bridge House, Waterloo Road,
London SE1 8UA
Tel: 071 215 5000
(Information on radiofrequency interference; agency to which complaints
should be directed)

Electrical Contractors' Association (ECA)
32/34 Palace Court, London W2 4HY
Tel: 071 229 1266   Fax: 071 221 7344

Electricity Association
30 Millbank, London SW1P 4RD
Tel: 071 834 2333   Fax: 071 931 0356

ERA Technology Ltd
Cleeve Road, Leatherhead, Surrey KT22 7SA
Tel: 0372 374151   Fax: 0372 374496

Health and Safety Executive (HSE)
Baynards House, 1 Chepstow Place, London W2 4TF
Tel: 071 243 6000 and 071 221 0870   Fax: 071 727 2254.
(Information and Publications Department open 10 am–3 pm).

Other HSE offices:

Sheffield – Tel: 0742 768141 and 0742 752539

Bootle – Tel: 051 951 4381

For a free short guide on EC Health and Safety Directives,
call 071 221 0870 or 0742 752539 between 10 am and 3 pm.

Institute of Environmental Engineering
The South Bank Polytechnic, 103 Borough Road, London SE1 0AA
Tel: 071 928 8989   Fax: 071 261 9115
(Continuing professional education)

Institution of Lighting Engineers
Lennox House, 9 Lombard Road, Rugby CV21 2OZ
Tel: 0788 76492

Lighting Industry Federation Ltd (LIF)
Swan House, 207 Balham High Road, London SW17 7BQ
Tel: 081 675 5432   Fax: 081 673 5880
(Publication list on application)

Metalline Holdings Ltd
Winster Grove, Birmingham B44 9EJ
Tel: 021 360 2222   Fax: 021 366 6003
(Portable airfield lighting equipment)

National Inspection Council for Electrical Installation Contracting (NICEIC)
Vintage House, 37 Albert Embankment, London SE1 7UJ
Tel: 071 582 7746   Fax: 071 820 0831

No Climb Products Ltd
15a Alston Works, Alston Road, Barnet, Hertfordshire EN5 4EL
Tel: 081 440 4331   Fax: 081 449 4029
(Pole lamp-changers)

Pilkington Security Systems
Colomendy Industrial Estate, Rhyl Road, Denbeigh, Clwyd,
North Wales LL16 5TA
Tel: 0745 814771   Fax: 0745 815933
('Sabretape' fibre-optic barbed aggressive tape)

Prima Security and Fencing Products Ltd
15 Aubrey Avenue, London Colney, St. Albans, Hertfordshire AL2 1NE
Tel: 0727 822222   Fax: 0727 826307
(Mesh and palisade fencing)

Setsquare Ltd
5a Valley Industries, Hadlow Road, Tonbridge, Kent TN11 0AH
Tel: 0732 851888   Fax: 0732 851853
(Automatic lighting controls)

Warwick Evans Optical Co Ltd
22 Palace Road, Bounds Green Road, London N11 2PS
Tel: 081 888 0051   Fax: 081 888 9055
(Vision screening equipment)

Appendix H
Buyer's guide to UK lighting suppliers

The letter codes indicate the following products and services supplied:

A = Tubular fluorescent lamps
B = Compact fluorescent lamps
C = Inductive fluorescent ballasts
D = High-frequency ballasts
E = High intensity discharge lamps
F = HID lamp control-gear
G = Industrial low-bay luminaires
H = Industrial high-bay luminaires
I = Hoseproof luminaires
J = Luminaires for Zone 1
K = Luminaires for Zone 2
L = Adjustable task lights
M = Emergency lighting luminaires
N = Exterior lighting equipment
O = Roadlighting equipment
P = Security lighting equipment
Q = Sells directly to users
R = Offers a design service
S = Undertakes installation
T = Undertakes maintenance

Abacus Municipal Ltd
Oddicroft Lane, Sutton-in-Ashfield, Nottingham NG17 5FT
Tel: 0623 511111   Fax: 0623 552133
(G H I N O P Q R S T)

ABB Control Ltd
Grovelands House, Longford Road, Exhall, Coventry CV7 9ND
Tel: 0203 368524   Fax: 0203 364499
(J M P Q)

Absolute Action Ltd
Mantle House, Broomhill Road, Wandsworth, London SW18 4JQ
Tel: 081 871 5005   Fax: 081 877 9498
(Fibre-optic lighting; Q R S T)

Andrew Chalmers and Mitchell Ltd
338 Hillington Road, Glasgow, G52 4BL
Tel: 041 882 5553   Fax: 041 883 3704
(E F H I J K N O P Q R)

Anglepoise Lighting Ltd
51 Enfield Industrial Area, Redditch, Worcestershire B97 6DR
Tel: 0527 63771   Fax: 0527 61232
(L M R)

Beta Lighting Ltd
383/387 Leeds Road, Bradford BD3 9LZ
Tel: 0274 721129   Fax: 0274 305007
(A B C D E F G H L M N Q R)

CCT Lighting Ltd
Windsor House, 26 Willow Lane, Mitcham, Surrey CR4 4NA
Tel: 081 640 3366   Fax: 081 648 5263
(E F Q R T)

CEAG Ltd
CEAG Centre, Pontefract Road, Barnsley, South Yorkshire S71 1AX
Tel: 0226 206842   Fax: 0226 731645
(A B E G H J M N Q)

Crompton Lighting
Woodlands House, The Avenue, Cliftonville, Northamptonshire NN1 5BS
Tel: 0604 28882   Fax: 0604 26982
(A B C D E F G H I J K L M N P Q R T)

Cryselco Ltd
Cryselco House, 274 Ampthill Road, Bedfordshire MK42 9QJ
Tel: 0234 273355   Fax: 0234 210867
(A B C D E F G H I M N P Q R)

Davis Cash and Co Ltd
Alexandra Road, Ponders End, Enfield, Middlesex EN3 7EN
Tel: 081 804 4028   Fax: 081 805 2896
(L Q R)

Designplan Lighting
Wealdstone Road, Kimpton Industrial Estate, Sutton, Surrey SM3 9RW
Tel: 081 641 7070   Fax: 081 644 422253
(B E I M N P Q R)

Fitzgerald Lighting Ltd
Normandy Way, Bodmin, Cornwall PL31 1HH
Tel: 0208 75611   Fax: 0208 74893
(A B C D E F G H I M N P R)

GE Thorn Lamps
Miles Road, Mitcham, Surrey CR4 3YX
Tel: 081 640 1221   Fax: 081 640 2842
(A B E N O P)

GE-Tungsram Lighting Ltd
Nene House, Drayton Fields Industrial Estate
Daventry, Northamptonshire NN11 5EA
Tel: 0327 77683   Fax: 0327 76386
(B E)

G.T. Lighting Ltd
Bunkers Hill, Modbury, Devon PL21 0RH
Tel: 0548 830189   Fax: 0548 830078
(M P R; luminous arrow signs)

GTE Sylvania Ltd
Otley Road, Charlestown, Shipley, West Yorkshire BD17 7SN
Tel: 0274 595921   Fax: 0274 597683
(A B E G H I N P R)

Harvey Hubbell Ltd
Ronald Close, Woburn Road Industrial Estate, Kempston,
Bedfordshire MK42 7SH
Tel: 0234 855444   Fax: 0234 854008
(B E G H I M N O P Q R)

Holophane Europe Ltd
Bond Avenue, Bletchley, Milton Keynes MK1 1JG
Tel: 0908 649292   Fax: 0908 270006
(G H I M N O P R)

Illuma Lighting Ltd
24–32 Riverside Way, Uxbridge, Middlesex UB8 2YF
Tel: 0895 72275   Fax: 0895 70024
(B L R)

Industrolite Ltd
Unit 4, Brazil Close, Beddington Farm Road, Croydon, Surrey CRO 4XB
Tel: 081 665 7070   Fax: 081 684 1407
(G H I M N Q R; illuminated guidance handrails, inspection pit lighting)

JSB Electrical plc
Manor Lane, Holmes Chapel, Cheshire CW4 8AB
Tel: 0477 37773   Fax: 0477 35722
(M R T)

Kestron Lighting Ltd
6 Merse Road, North Moons Moat Estate, Redditch,
Worcestershire B98 9HL
Tel: 0527 584123   Fax: 0527 68332
(A C D F G I Q R)

Labcraft Ltd
Bilton Road, Chelmsford CM1 2UP
Tel: 0245 359888   Fax: 0245 490724
(M R T)

LDMS
191 High Road, South Benfleet, Essex SS7 5HY
Tel: 0268 755511   Fax: 0268 755445
(N)

Linolite GTE Ltd
Buettell Way, Tetbury Hill, Malmesbury, Wiltshire SN16 9JX
Tel: 0666 822001   Fax: 0666 824954
(I M N)

Lumitron Ltd
Chandos Road, London NW10 6PA
Tel: 081 965 0211   Fax: 081 965 8629
(I J M Q R)

Lutron
6 Sovereign Close, Wapping, London E1 9HW
Tel: 071 702 0657   Fax: 071 480 6899
(Lighting controls; D R T)

Marlin Lighting Ltd
Hanworth Trading Estate, Feltham, Middlesex TW1 36R
Tel: 081 894 5522   Fax: 081 755 1215
(A B E I M N O P Q R)

Moorlite Electrical Ltd
Burlington Street, Ashton-under-Lyne, Lancashire OL7 0AX
Tel: 061 330 6811　Fax: 061 330 2815
(C D F G H I J K M N O Q R)

N.E.I./Victor
P.O. Box, Wallsend, Tyne and Wear NE28 6PP
Tel: 091 262 8331　Fax: 091 234 4550
(I J K M Q R T)

Orbik Electronics Ltd
Orbik House, Aldridge, Walsall, West Midlands WS9 8TH
Tel: 0922 743515　Fax: 0922 743173
(Ballasts and control-gear for emergency lighting)

Osram Ltd
PO Box 17, East Lane, Wembley, Middlesex HA9 7BR
Tel: 081 904 4321　Fax: 081 904 1081
(A B D E F)

W J Parry and Co (Nottingham) Ltd
Victoria Mills, Draycott, Derbyshire DE7 3PW
Tel: 03317 2321　Fax: 03317 4035
(C F)

Petrel Ltd
Thimblemill Lane, Birmingham B7 5HT
Tel: 021 328 5055　Fax: 021 326 6290
(J K Q R)

Philips Lighting
City House, 420/430 London Road, Croydon, Surrey CR9 3QR
Tel: 081 689 2166　Fax: 081 684 0136
(A B C D E F G H I J K M N O P R)

Poselco Ltd
Walmgate Road, Perivale, Middlesex UB6 7LX
Tel: 081 998 1431　Fax: 081 997 3350
(C D F G H I M N P Q R)

Rada Lighting Ltd
Hollies Way, High Street, Potters Bar, Hertfordshire EN6 1UL
Tel: 0707 43401　Fax: 0707 45548
(G H I J M N P Q R)

Reggiani Ltd Lighting
12 Chester Road, Borehamwood, Hertfordshire WD6 1LT
Tel: 081 953 0855   Fax: 081 207 3923
(B E F L M N R T)

Ring Electronics Ltd
Gelderd Road, Leeds LS12 6NB
Tel: 0532 798887   Fax: 0532 792591
(M R; luminous arrow signs)

Siemens Lighting Ltd
Lea Green Road, St. Helens, Merseyside WA9 4QQ
Tel: 0744 850850   Fax: 0744 814320
(G H I J M N O R)

Strand Lighting Ltd
Grant Way, Isleworth, Middlesex TW7 5QQ
Tel: 081 560 3171   Fax: 081 568 2103
(E F Q R T)

Surelux Lighting Ltd
Surelux Works, New Hey Road, Marsh, Huddersfield,
West Yorkshire HD3 4BR
Tel: 0484 540425   Fax: 0484 512466
(A B G M N P Q R)

Taison Lighting
Taison Industrial Park, Great Horton Road, Bradford,
West Yorkshire BD7 4EN
Tel: 0274 521550   Fax: 0274 521481
(A B C D E F G H I M N O R)

Thorn Lighting Limited
Elstree Way, Borehamwood, Hertfordshire WD6 1HZ
Tel: 081 905 1313   Fax: 081 905 1278
(C D F G H I J K M N O P R)

F W Thorpe plc
Merse Road, North Moons Moat, Redditch, Worcestershire B98 9HH
Tel: 0527 584058   Fax: 0527 584177
(C D E F G H I K M N P Q R)

Tridonic Limited
Thomas House, Hampshire International Business Park,
Crockford Lane, Chineham, Basingstoke, Hampshire RG24 0NA
Tel: 0256 707000   Fax: 0256 707002
(C D R)

JEL Building Management Ltd
Worton Drive, Reading, Berkshire RG22 0TG
Tel: 0734 752000   Fax: 0734 861064
(Lighting/security control systems; Q R S T)

Transtar
Victoria Industrial Estate, Hebburn, Tyne and Wear NE31 1UB
Tel: 091 4832797   Fax: 091 4280262
(C D F Q R)

D W Windsor Ltd
Pindar Road, Hoddesdon, Hertfordshire EN11 0EZ
Tel: 0992 445666   Fax: 0992 440493
(N O Q R)

References

1. Chartered Institution of Building Services Engineers, *CIBSE Code for Interior Lighting*, CIBSE, London (1978) (new edition in press); defines and explains standards for interior lighting and gives specific recommendations for various kinds of interiors).
2. Lord Rayleigh, Night myopia, in *Collected papers*, 6 vols., Cambridge University Press, Cambridge (1899–1920).
3. Pritchard, D. C., *Lighting*, Longmans, Green & Co, London and Harlow (1969); reviews basic physics of vision and light.
4. Civil Aviation Authority, *Licencing of Aerodromes*. Publication no. CAP-168, CAA, Cheltenham, 4th edition (1990); sets out the physical requirements for aerodrome safety, including lighting.
5. Lyons, S., *Management guide to modern industrial lighting*, Butterworths, London, 2nd edition (1983); informs non-engineering managers about lighting for industrial premises, their duties, the economics of lighting.
6. Health and Safety Executive, *Effective policies for health and safety*, HMSO, London (1980); report of a five-year study by the Accident Advisory Unit.
7. Lighting Industry Federation Ltd, *Lamp Guide*, LIF, London (1990); describes lamp types and gives their designations, powers etc.
8. Lighting Industry Federation Ltd, *Interior Lighting Design*, LIF, London, 6th edition (1986); gives design procedures and calculation methods for the design of interior lighting including the lumen method.
9. Lighting Industry Federation Ltd, *Factfinder: Lighting and Energy*, LIF, London; gives valuable guidance on the wise use of energy by choice of lighting equipment and methods.
10. Lighting Industry Federation Ltd, *Factfinder: Benefits of Certification* LIF, London; explains the BSI safety mark certification.
11. Lighting Industry Federation Ltd, *Factfinder: Hazardous Area Lighting*, LIF, London; gives guidance on selection of lighting equipment.
12. Lighting Industry Federation Ltd, *Application Guide: High Frequency Ballasts for Tubular Fluorescent Lamps*, LIF, London.
13. Lighting Industry Federation Ltd, *Application Guide: Lighting Controls and Energy Management Systems*, LIF, London.
14. Lighting Industry Federation Ltd, *Application Guide: Emergency Lighting (ICEL:1003)*, LIF, London.
15. Lyons, S., *The influence of lighting on industrial and domestic accidents*, Journal of the Junior Institution of Engineers, **59**(9), pp. 257–264 (1949).
16. Chartered Institution of Building Services Engineers, *Daytime lighting in buildings* in *Applications Manual: Window Design*, CIBSE, London (1987).
17. Chartered Institution of Building Services Engineers, *Calculation of Glare Indices, Technical Memorandum No. 10*, CIBSE, London.

18. *Electricity at Work Regulations*, HMSO, London, 1989.
19. *BS 5252:1976 Framework for colour co-ordination for building purposes*, British Standards Institution, London.
20. *BS 1710:1984 Identification of pipelines*, British Standards Institution, London; gives colour specifications to *BS 4800*.
21. *BS 1853:- Tubular fluorescent lamps for general lighting purposes*, British Standards Institution, London.
22. Bellchambers, H. E. and Phillipson, S. M., *Lighting for Inspection*, IES Trans., **27**(2) (1962).
23. Health and Safety Executive, *Annual Reports of HM Chief Inspector of Factories*, HMSO, London.
24. *BS 667:1968 Specification for portable photoelectric photometers*, British Standards Institution, London (1965).
25. *Health and Safety at Work Act*, HMSO, London.
26. *Protection of Eyes Regulations*, 1974. No. 1681, as amended 1975 No. 303, HMSO, London.
27. *IEE Wiring Regulations* (Regulations for the wiring of buildings), Institution of Electrical Engineers, London, 15th edition and 16th edition both in force until 31 December 1992; 16th edition alone in force from 1 January 1993.
28. Lightoller, B., *Upwardly mobile* in *Electrical Design*, November (1990).
29. Editorial in association with Chris Pritchard, Ups and downs for airport services, in *Electrical Design*, November (1990).
30. *BS 4533:- Luminaires*, British Standards Institution, London.
31. *BS 5394: Part 1: 1988: Specification for limits and methods of measurement of radio interference characteristics of fluorescent lamps and luminaires*, British Standards Institution, London.
32. *CISPR-15: Limits and methods of measurement of radiointerference characteristics of fluorescent lamps and luminaires*, International Special Committee on Radio Interference (CISPR). (Available from Lighting Industry Federation, London and British Standards Institution, Milton Keynes).
33. *Harmonics and waveform distortion due to lighting* Publication No. G5/2. Distribution Engineering Section, of the former Electricity Council (1985). (Refer to The Electricity Association).
34. *The Food Hygiene (General) Regulations*, HMSO, London (1970).
35. *BS 5750 Quality Systems*, British Standards Institution, London.
36. *Wireless Telegraphy (Control of Interference from Fluorescent Lighting Apparatus) Regulations 1979*, Amd No. 807 (1985), Amd No. 561 (1989), HMSO, London.
37. *List of Approved Equipment*, British Approvals Services for Electrical Equipment in Fiery Atmospheres (BASEEFA), London. (Available from Health and Safety Executive.)
38. *BS 5345:- Code of practice for selection, installation and maintenance of electrical apparatus for use in potentially explosive atmospheres (other than mining applications or explosive processing and manufacture)*, British Standards Institution, London.
39. Chartered Institution of Building Services Engineers, *Calculation and use of utilization factors*, Technical Memorandum No. 5, CIBSE, London (1980).
40. *BS 8206: Part 1:1985 Code of Practice for Artificial Lighting*, British Standards Institution, London.

41. Royal Society for the Prevention of Accidents, *Guidance on danger of fatigue in using display screen equipment*, RoSPA, Birmingham (1985).
42. *Health and Safety at Work Act*, HMSO, London (1974).
43. Lyons, S., *Writing Non-fiction.*, published by the author (1991); a manual outlining the functions of technical authors and giving guidelines on technical writing.
44. Bertonshaw, D. R., *Recent Developments in Luminaire Safety*, ERA Technology Ltd, Leatherhead, *Conference Proceedings*, 91–0001, (1991).
45. Baker, J. E., *Overview of Regulations and Standards – Lighting*, ERA Technology Ltd, Leatherhead, *Conference Proceedings*, 91–0001 (1991).
46. Dawson-Tarr, E., *Fibre Optic Lighting*, ERA Technology Ltd, Leatherhead, *Conference Proceedings*, 91–0001 (1991).
47. *BS 5266:Part 1:1988 Code of Practice for the emergency lighting of premises other than cinemas and certain other specified premises used for entertainment*, British Standards Institution, London.
48. Lyons, S., *Emergency Lighting*, Butterworth-Heinneman, (in preparation).
49. *BS 4363:1968 Specification for distribution units for electricity supplies on construction and building sites*, British Standards Institution, London.
50. *BS4343:1968 Specification for industrial plugs, socket-outlets and couplers for a.c. and d.c. supplies*, British Standards Institution, London.
51. *Fire Precautions Act*, HMSO, London (1971).
52. *BS 4211:1987 Specification for ladders for permanent access to chimneys, other high structures, silos and bins*, British Standards Institution, London.
53. Chartered Institution of Building Services Engineers, *Emergency Lighting, Technical Memorandum TM-12*, CIBSE, London.
54. *BS 5499:- Fire safety signs, notices and graphic symbols*, British Standards Institution, London.
55. *BS 5490:- Specification for classification of degrees of protection provided by enclosures*, British Standards Institution, London.
56. *BS 6467:- Electrical apparatus with protection by enclosure for use in the presence of combustible dusts*, British Standards Institution, London.
57. *Guide to Fire Precautions Act 1971 applied to existing places of work*, HMSO, London.
58. Chartered Institution of Building Services Engineers, *Lighting Guide for Building and Civil Engineering Sites*, CIBSE, London (in press).
59. Lyons, S., *Security of Premises – a Manual for Managers*, Butterworth, London, (1988).
60. Baker, J. E. and Lyons, S. L., Lighting for the security of premises, *Lighting Research Technology*, **10**(1): 10 (1978).
61. *Essentials of Security Lighting*, a publication of the former Electricity Council. (may be available from The Electricity Association).
62. *BS 1722:- Fences*, British Standards Institution, London.
63. *Offices, Shops and Railway Premises Act*, HMSO, London (1963).
64. *Electromagnetic Compatibility (EMC) Regulations*, HMSO, London, (1990); effective from 1 January, 1992.
65. *Trade Effluents (Prescribed Process and Substances) Regulations*, HMSO, London (1989).

Index